The Common Agricultural Policy

Second Edition

Edited by

Christopher Ritson

and

David Harvey

Department of Agricultural Economy and Food Marketing
University of Newcastle upon Tyne
UK

CAB INTERNATIONAL

CAB INTERNATIONAL
Wallingford
Oxon OX10 8DE
UK

Tel: +44 (0)1491 832111
Fax: +44 (0)1491 833508
E-mail: cabi@cabi.org

CAB INTERNATIONAL
198 Madison Avenue
New York, NY 10016-4314
USA

Tel: +1 212 726 6490
Fax: +1 212 686 7993
E-mail: cabi-nao@cabi.org

A catalogue record for this book is available from the British Library, London, UK.

Library of Congress Cataloging-in-Publication Data

The common agricultural policy / edited by Christopher Ritson and David Harvey. -- 2nd ed.

p. cm.

Rev. ed. of : The common agricultural policy and the world economy. c1991.

Includes bibliographical references and index.

ISBN 0-85198-988-8 (alk. paper)

1. Agriculture and state--European Economic Community countries. 2. Produce trade--Government policy--European Economic Community countries. 3. Food industry and trade--Government policy--European Economic Community countries. I. Ritson, Christopher. II. Harvey, David R. III. Common agricultural policy and the world economy.

HD1920.5.Z8C62 1997

338.1'84--dc21 97-14633

CIP

ISBN 0 85198 988 8

Printed and bound in the UK by Biddles Limited, Guildford and Kings' Lynn.

Contents

PART IV: THE CAP AND THE WORLD

PART V: THE CAP AND THE FUTURE

Contributors

Allan Buckwell
Department of Agricultural Economics, Wye College (University of London), Wye, Nr Ashford, Kent TN25 5AH, UK

Andrew Fearne
Department of Agricultural Economics, Wye College (University of London), Wye, Nr Ashford, Kent TN25 5AH, UK

Simon Harris
British Sugar Plc, PO Box 26, Oundle Road, Peterborough PE2 9QU, UK

David Harvey
Department of Agricultural Economics and Food Marketing, University of Newcastle, Newcastle Upon Tyne NE1 7RU, UK

Lionel Hubbard
Department of Agricultural Economics and Food Marketing, University of Newcastle, Newcastle Upon Tyne NE1 7RU, UK

Tim Josling
Food Research Institute, Stanford University, Stanford, California 94305 USA

Michael Keane
Department of Food Economics, University College, Cork, Ireland

John Lingard
Department of Agricultural Economics and Food Marketing, University of Newcastle, Newcastle Upon Tyne NE1 7RU, UK

Philip Lowe
Department of Agricultural Economics and Food Marketing, University of Newcastle, Newcastle Upon Tyne NE1 7RU, UK

Denis Lucey
Department of Food Economics, University College, Cork, Ireland

Arie Oskam
Department of Agricultural Economics, Wageningen Agricultural University, P O Box 8130, 6700 EW Wageningen, The Netherlands

Christopher Ritson
Department of Agricultural Economics and Food Marketing, University of Newcastle, Newcastle Upon Tyne NE1 7RU, UK

Spiro Stefanou
Department of Agricultural Economics and Rural Sociology, Penn State University, University Park, PA 16802, USA

Alan Swinbank
Department of Agricultural and Food Economics, University of Reading Whiteknights Road, P O Box 237, Reading RG6 2AR, UK

Stefan Tangermann
Institut für Agrarökonomie der Universität Göttingen, Platz der Göttinger Sieben 5, 3400 Göttingen, Germany

Martin Whitby
Department of Agricultural Economics and Food Marketing, University of Newcastle, Newcastle Upon Tyne NE1 7RU, UK

Preface

The first edition of this book (Ritson and Harvey, 1991) arose as a consequence of a series of public lectures on the Common Agricultural Policy, held at the University of Newcastle Upon Tyne in honour of the memory of John Ashton, who had been Professor and Head of Agricultural Economics for 24 years. Subsequently we invited a number of former friends and colleagues of John to join the lecture-givers in contributing chapters to a book dedicated to his memory. The book was very well received by reviewers and has been widely adopted as a student text. Almost by accident we had created a book which, in addition to being a collection of essays, was approaching a textbook on the Common Agricultural Policy.

This volume is rather more than a second edition. We have deliberately attempted to produce a textbook, drawing on the expertise of a range of specialists. In this new edition, only seven chapters (1, 2, 4, 11, 12, 15, 16) are a revision and updating of chapters in the 1991 book. Three (3, 9, 13) cover broadly similar topics, tackled by different authors; and eight (5, 6, 7, 8, 10, 14, 17 and 18) are new, to fill gaps in the first edition and reflecting the major developments in the CAP during the 1990s.

The editors would like to thank all who have contributed, and in particular are indebted to the three who during the course of 1996 have dedicated much time and skill to the production of the final text - Kirsti Karlsen, Nicola Forster, and Wilma Lister.

Christopher Ritson
David Harvey

December 1996

Reference

Ritson, C. and Harvey, D.R. (1991) *The Common Agricultural Policy and the World Economy.* Essays in Honour of John Ashton: CAB International, Wallingford.

Chapter 1

Introduction

Christopher Ritson

This short introductory chapter attempts only a very brief description of some central features of the operation of the Common Agricultural Policy (CAP). Some of the authors have assumed that readers will be familiar with the basic principles and mechanics of the Policy, but these are not dealt with succinctly in any of the other chapters. The chapter concludes with an outline of the way the book has been structured to provide what we believe to be a comprehensive and informed account of contemporary aspects of the European Union's Common Agricultural Policy.

The Treaty of Rome

The Treaty of Rome, which established the European Economic Community (EC) in 1958, required that "The Common Market shall extend to agriculture and trade in agricultural products" (Article 38) and stated that "The Community shall be based upon a Customs Union..." (Article 9). A customs union is a form of economic integration in which all barriers on trade between member states are removed, and a common barrier is established on trade with third countries.

It was clear from the outset that this would require a *common* policy for agriculture. All the original six member states had adopted complicated mechanisms for controlling agricultural product markets. The simple procedure of removing barriers on intra-Community trade, and establishing a common customs barrier on imports of agricultural products from outside the EC, would have involved developments in market prices for farm products which would have been unacceptable to most of the six member states.

Article 39 of the Treaty of Rome specifies a set of objectives for the CAP which are similar to those adopted by most of the developed countries. The policy seeks:

a) to increase agricultural productivity by promoting technical progress and by ensuring the rational development of agricultural production and the optimum utilisation of the factors of production in particular labour;

b) thus to ensure a fair standard of living for the agricultural community, in particular by increasing the individual earnings of persons engaged in agriculture;

c) to stabilise markets;

d) to assure the availability of supplies;

e) to ensure that supplies reach consumers at reasonable prices.

Arguably, this is the most cogent statement of objectives for agricultural policy ever made, and several of our authors have found it helpful to refer to them when reviewing a particular aspect of the CAP.

What is perhaps unusual about the CAP, however, is the severity with which one objective - the desire to protect rural living standards - came to dominate the way the policy was implemented. In particular, support prices for the major farm commodities were set at levels well in excess of those at which supplies could normally be bought or sold on world markets. Partly as a consequence of this, European farmers have been supplying a growing proportion of domestic requirements.

Table 1.1 gives some estimates by the European Commission of the extent to which CAP support prices exceeded those applying on world markets. The figures in Table 1.1 were calculated by deducting annual average import taxes or export subsidies from EC support prices. Concern that this was exaggerating the degree to which EC farmers were supported was probably responsible for the Commission ceasing publication of this information. However Table 1.2 takes up the story from 1980 using data published by OECD.[1] Table 1.3 indicates the way self-sufficiency in farm products has tended to increase. The Policy now covers all the major agricultural products produced within the EC, with the exception of potatoes.

An alternative way of characterising the CAP is by its 'three principles' rather than its five objectives. These were formulated somewhat obscurely during the early 1960s, but it quickly became *de rigueur*, as far as some member states were concerned, that any reform of the CAP must not call into question the three principles upon which it was founded, as follows:

a) free intra-Community trade: no barriers to trade in farm products between EC member states;

b) Community preference: supplies from within the Community to be given preference in the market over those from outside the EC;

c) common financing: funding for the CAP would be through a European budget responsible for all revenues and expenditure generated by the Policy.

Table 1.1: Prices of certain agricultural products in the EC as a percentage of prices on world markets (1968–1980).

Product	1968/ 69	1970/ 71	1972/ 73	1974/ 75	1976/ 77	1978/ 79	1979/ 80
Common wheat	195	189	183	107	204	193	163
Rice	138	210	115	81	166	157	131
Maize	178	141	143	106	163	201	190
White sugar	355	203	127	41	176	276	131
Beef and veal	169	140	112	162	192	199	204
Pigmeat	134	134	147	109	125	155	152
Butter	504	481	249	316	401	403	411
Skim-milk powder	365	na	145	139	571	458	379
Olive oil	173	155	125	113	192	200	193

Source: Eurostat.

Table 1.2: Prices of certain agricultural products in the EC as a percentage of prices on world markets (1979–1994).

Product	1979	1982	1985	1988	1991	1994
Common wheat	140	139	117	184	231	155
Maize	165	147	119	170	211	140
Barley	186	159	135	184	219	214
Rice	129	146	188	198	194	209
White sugar	133	108	161	144	137	106
Milk	231	156	253	215	259	241
Beef and veal	161	174	194	203	186	208
Pigmeat	123	118	111	131	134	130
Poultry	123	118	126	161	122	118
Sheepmeat	243	197	193	315	228	156

Source: OECD.

Table 1.3: Degrees of self-supply in certain agricultural products (EC production as a percentage of EC consumption).

Product	1956/60 Ave.	1972		1981/82		1985/86		1992/93
	EC-6	EC-6	EC-9	EC-9	EC-10	EC-10	EC-12	EC-12
Wheat	90	111	99	125	127	132	124	133
Maize	64	68	58	72	73	94	77	94
Rice	83	112	92	131	130	82	75	75
Sugar	104	122	100	144	na	135	123	128
Fresh veg	104	100	94	97	100	100	107	106
Fresh fruit	90	87	76	81	85	83	87	85
Citrus	47	52	34	36	45	47	75	70
Cheese	100	102	102	108	107	107	106	107
Butter	101	124	106	123	122	133	110	121
Eggs	92	81	84	104	103	108	107	108
Beef & veal	92	81	84	104	103	108	107	108
Pigmeat	100	99	100	101	101	102	102	104
Poultry	93	100	102	110	110	107	104	105
Veg oils	19	31	na	na	na	40	56	65

na = not available

Source: Commission of the European Communities, the Agricultural Situation in the Community, European Union (annual reports).

It was originally envisaged that the CAP would have two arms of approximately equal weight — a market arm and a structural arm. In the event, the market policy has dominated, taking the major part of expenditure on the CAP. The policy is financed by a special section of the Community budget, (usually known by its French acronym FEOGA) — the European Agricultural Guidance and Guarantee Fund. Guarantee expenditure is for the market policies; the Guidance section finances structural reform. Details of the way the CAP has developed and is financed are given in Chapter 2.

Policy mechanisms

The core of the CAP is a set of mechanisms which attempt to control the markets for agricultural products within the European Union. These vary significantly from product to product, and have been subject to repeated modification. It is nevertheless helpful to an understanding of the operation of the CAP market policy to construct a simplified 'ideal' model of a CAP market support system, as originally envisaged, to which all the specific commodity market regimes have approximated to a greater or lesser extent.

Fig.1.1 therefore describes the essential features of a typical CAP support system for a farm product. Each year the Council of Ministers sets a target price for the commodity in question. This price is intended as a guide to producers and a reference point for the operation of the Policy. The main mechanism which ensured that internal market prices were kept near the target level was a levy on imports which varied in such a way that imported products could not undercut the target price. This levy was calculated by reference to a minimum import price — sometimes known as the threshold price — set a little below the target price to reflect the cost of transport from port to internal market centre.

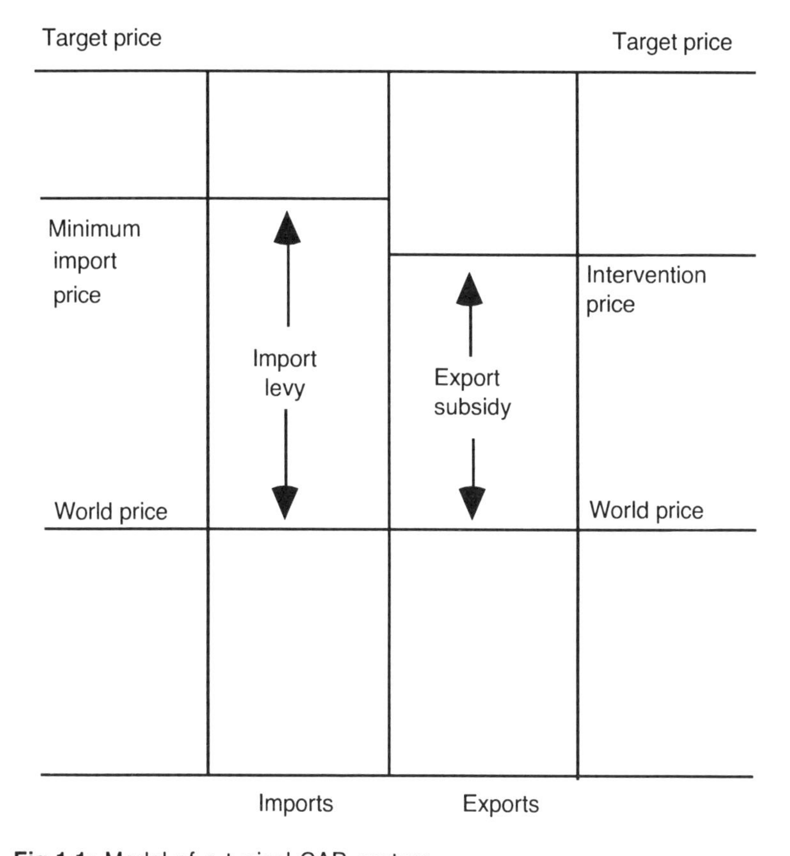

Fig.1.1: Model of a typical CAP system.

A tax (the levy), equal to the difference between the threshold price and the world market price, was then charged on imports. 'World price' usually means, in this context, the lowest price at which a consignment of produce is being offered at a particular port during some specified time period.

An import levy will, on its own, keep internal market prices near the target level as long as the Community is less than self-sufficient in the commodity. If, however, EC farmers supply more of the commodity than can be sold on domestic markets at the target price, internal prices will begin to drop below target levels. For this eventuality, a second line of defence is required to prevent excess supplies (known as 'surpluses') from depressing producer prices. An intervention price is set, somewhat below the target price. If the internal market price should now fall to the intervention level, official intervention agencies will buy produce offered to them at the intervention price. The agency will then store the produce and subsequently export it at a loss. Alternatively, private traders receive an export subsidy (known as a refund or restitution) equal to the difference between the intervention price and the world price, and in fact the bulk of surplus produce has been disposed of in this way. For some commodities, other methods are used to dispose of surpluses. For example, both wheat and skimmed milk powder have been subsidised for use as animal feeding-stuffs.

It is evident from Tables 1.1 and 1.2 that for most agricultural commodities, for most of the time, CAP support prices have exceeded 'world' prices. However, during the 1973/74 commodity boom, some prices for agricultural products on world markets did move above CAP support levels, and this happened again with sugar during the winter of 1980/81, and grain during 1995/96. For commodities in surplus, the EC has been able to restrain domestic price levels by imposing export taxes, as applied to cereals during 1996. As a general rule import subsidies have not been introduced into the CAP system.

The reforms of the 1990s

Two extremely important and interrelated developments have affected the CAP during the 1990s. These are the so-called Mac Sharry reform of the Policy (implemented between 1992 and 1995) and the General Agreement on Tariffs and Trade (GATT) Agreement on Agriculture (being implemented between 1995 and 2001). Two chapters in this book are directed specifically at these developments (5 and 17), but many others make extensive reference to them. In order for individual chapters to be coherent, this makes an element of repetition in describing the CAP reform and the GATT Agreement inevitable and this has been retained quite explicitly as a matter of editorial policy. In a very over-simplified way these developments have involved revision of Fig.1.1 in three ways.

i) Some internal market support prices have been reduced and direct payments to producers introduced, as arable area or livestock headage payments.
ii) The (variable) import levy has been replaced by a 'tariff equivalent'. In the case of cereals, the new arrangements continue to operate so that a variable import tax supports a minimum import price (see Chapter 5), but for most other products fixed import taxes now apply, often at very high levels, similar to those indicated for 1988 in Table 1.2.
iii) There are quantitative and financial restrictions on the degree to which exports can be subsidised.

Nomenclature

The CAP and the EU are littered with often well-known and widely used acronyms. We have adopted an editorial policy of, usually, spelling out a term when first used in any chapter, with the exception of a few which are very well known. A current source of confusion for students is the various descriptions of the European Community/European Union. The Common Agricultural Policy was part of the European Economic Community (EEC) established by the Treaty of Rome. However, two other communities were established by the original six countries (for coal and steel, and atomic energy) and it was a Treaty of Accession to the European Communities which applied to Denmark, Ireland and the United Kingdom in 1973. This led to the widespread use of the term EC (or ECs) often used interchangeably with EEC (and 'Common Market'). More details of the various communities and how they have been described will be found in Chapter 3.

The act creating the Single Market has led to the introduction of the term European Union in the 1990s. In this book we have tried to use the term EU when referring to contemporary events, but to continue to describe the countries applying the Common Agricultural Policy as 'the Community', or 'EC', when referring to earlier events.

Structure of the book

We have divided the various aspects of the CAP considered in this book into four sections. Part I — entitled 'Understanding the CAP'— begins with a (deliberately) long chapter which traces the history and development of the CAP from its origins in the 1950s through to the enlargement to include twelve member states in the mid-1980s, and the growing financial cost of price support. This is followed by three further chapters intended to provide a basic understanding of the Policy. Chapter 3 discusses the process by which CAP policy decisions are taken; Chapter 4 looks specifically at the pressures for CAP reform and the way the Policy has reacted; and Chapter 5 describes the 'new CAP' of the 1990s.

Part II consists of four chapters which analyse the CAP from different perspectives. Students will find this part of the book the most demanding in terms of economic background and analytical skills though, with the exception of a section of Chapter 9, diagrammatic rather than algebraic analysis is deployed. Part III considers the relationship between the CAP and four specific interests, namely, the consumer, the environment, farmers, and the food industries.

In Part IV we turn to the external face of the Policy. Thus the CAP looks east (Europe), west (North America) and south (less developed countries). Throughout this section, and in earlier chapters, frequent references are made to the conflict between the trade impact of the CAP, and the attempt to liberalise world trade in agricultural products under the General Agreement on Tariffs and Trade (GATT).

Chapter 17 then provides a more detailed treatment of the GATT Agreement on Agriculture and the implications for world trade policy. The concluding chapter speculates about the future of the Common Agricultural Policy — the CAP in the 21st century.

---

[1]The difference between the final column of Table 1.1, and the first column in Table 1.2 gives some idea of the extent to which the way Table 1.1 was constructed may have overestimated the real gap between EC and world prices.

PART I

Understanding the CAP

Chapter 2

The History and Development of the CAP 1945–1990

Andrew Fearne

Introduction

This chapter is concerned with the stages leading to European integration and a common policy for agriculture. It discusses the influences emanating from the political parties and interest groups within the six original members of the Community, and the development of the CAP during a period in which the European Community (EC) grew from six to twelve member states.

The origins of the CAP

1945–1954: Post World War II – The preamble to European union

When examining the factors which shaped the CAP from the outset, it is important to remember that a common policy for western Europe's agricultural sector was the result of (rather than the reason for) a common desire amongst the nucleus of west European countries to establish a political and economic union.

Economic union was to be a means of political unification and a guarantee for peace following the resolution of world war. However, the impetus for economic and political co-operation came, not from within the Continent, but from the United States (US), which considered west European integration as the only way for Europe to recover from the aftermath of the War, and to construct an effective barrier to communism and the expansion of the Soviet Block.

Greater co-operation was achieved in western Europe through the implementation of the Marshall Plan through which, between 1948 and 1952, some 25 billion dollars of aid from the US and Canada was channelled. Even more important than this financial assistance (as far as European integration is concerned) was the emergence of the Organisation for European Economic Co-operation (OEEC), which formed the European Recovery Programme (ERP) and established the priority of genuine European co-operation through the reduction of trade controls.

In the same year as the formation of the OEEC (1948), the elimination of customs duties and the institution of a common tariff for imports took effect in Belgium, Netherlands and Luxembourg, under the agreement which formed the Benelux Customs Union. The agreement, signed on September 5, 1944, called for co-operation in the field of tariff policies, which would lead to an economic union providing for "...the free movement of persons, goods, services and capital between the three countries" (De Vries, 1975). While the Benelux partners could not have envisaged in 1944 the degree of economic integration which was to emerge in western Europe, Benelux was clearly a forerunner to the wider European union which was later established.

The impetus created by Benelux was carried still further, later in 1948, with the signing of the Treaty of Brussels, designed to encourage economic, social, cultural and defence co-operation. Significantly, Britain took part in this, along with France and the Benelux countries, but West Germany and Italy were not invited to sign the Treaty because those involved believed it was too soon after the War for active reconciliation between these two recent enemies.

A year later, following the Congress of Europe, the Council of Europe was created, involving a committee of ministers of national governments and an assembly of members of national parliaments. The Congress had adopted a resolution requiring the surrender of some national sovereignty prior to the establishment of economic and political union in Europe. The creation of a European parliamentary assembly, in which resolutions would be carried by majority vote was proposed by the French, with the support of the Belgians. Britain was fully opposed to this form of supranationalism and managed to water down considerably the original proposals, with the Assembly and Council being granted no legislative powers but merely the role of a forum — a debating society designed only to provide recommendations to national governments (Swann, 1981).

Britain's opposition to any form of supranationalism was reflected in her contribution to the first discussions on the common organisation of agricultural markets. A special committee was set up by the Council, in 1950, to examine the prospects for the integration of European agriculture. France was particularly keen to open up Europe's agricultural markets and proposed the creation of a high authority for agriculture, with extensive supranational powers; production was to be controlled, prices were to be fixed and all barriers to agricultural trade were to be removed. However, while Britain accepted the concept of an authority, it insisted on it being intergovernmental, with the modified role of reconciling differences in national agricultural policies.

As Tracy (1989) points out, these 'Green Pool' proposals were largely doomed from the start. The fifteen western European countries which took part in the negotiations between 1952 and 1954 failed to reach any agreement, and Britain's opposition to the general idea of a supranational organisation was instrumental in this. However, as a further step towards European integration, the proposals and subsequent discussions served to identify the differences which existed between the countries involved, at least as far as agriculture was concerned, with France in particular committed to the European Movement and Britain anxious to maintain her links with the Commonwealth and her sovereignty over policy formulation.

A much higher level of co-operation was established through the setting up of the European Coal and Steel Community (ECSC), in 1951. This time Germany and Italy were involved, along with France and the Benelux countries, but Britain did not participate. Again, the plan was essentially a French one, designed by Jean Monnet (then the head of the French State Planning Board) and based on the desire to integrate a rapidly reviving German economy into the rest of western Europe while simultaneously ensuring that war between France and Germany would become not merely inconceivable but physically impossible (Kitzinger, 1967).

The result, through the removal of customs duties, quotas, subsidies and price agreements, was the creation of a common market in coal, steel and iron, with a high authority endowed with substantial direct powers. The high authority was, however, responsible to a European assembly, consisting of members elected by national parliaments, and on most matters the Council of Ministers representing member governments would have to be consulted. This was effectively the essence of the Community method of gradualist integration — progress towards political unity by the integration of one sector at a time, with explicit political objectives being sought via economic co-operation. The Six were in agreement, but the UK, although invited to join the Community, declined. With Monnet as the president of the first High Authority, the Six were firmly set on a course to which the British were fully opposed.

The ECSC was an undoubted success with significant increases in the output and productivity of coal, steel and iron achieved from 1951 onwards. However, while the visible success of the Community lies in the impact on production, productivity and prices in the three industries, its real significance lies in the fact that it marked reconciliation between recent enemies and was the first grouping of the Six. It also signified a stage in the recent history of the European Movement when personalities came to play a major role in the composition and nature of an agreement designed primarily to further the federalist cause. Monnet and Schumann of France, Konrad Adenauer (the first Chancellor of West Germany), Alcide de Gasperi (Prime Minister of Italy) and Paul-Henri Spaak (Foreign Minister and former Prime Minister of Belgium), along with their political parties, were all agreed that nationalism should be contained while legitimate national and minority interests were safeguarded (Broad and Jarrett, 1972). The creation and subsequent success of the ECSC served to prove that the most effective solutions to the economic and political problems in western Europe could be sought collectively, through the development of the European Movement.

1955–1957: The Treaty of Rome and objectives for agriculture

It was the Benelux countries who, aware of the difficulty of establishing a political union, eventually outlined a series of economic proposals for the creation of a fully integrated European market, covering a wide range of commodities. The memorandum which they presented recommended the convening of an intergovernmental conference to draw up treaties covering a general common market. Such a conference, involving the foreign ministers of the Six, but excluding a disinterested Britain, was set up in Messina, Sicily, in June, 1955, where it was agreed to carry out the necessary preparatory work. The result of this was a report drawn up in 1956, under the chairmanship of Spaak, which formed the basis on which the Treaty establishing the European Economic Community (EEC) was built.

The chapter devoted to agriculture made it clear that the establishment of a common market in Europe which did not include agriculture was inconceivable (Comite Intergouvernemental, 1956). However, the final resolution of the Six to proceed with the construction of Europe included no specific reference to agriculture. As Neville-Rolfe (1984) points out, this omission was probably made to facilitate British participation in the steering committee, set up by Spaak. Britain did take part for a while, but within the year the British representative on the committee was withdrawn, with the Government publicly rejecting the proposals for economic unification and warning the six foreign ministers of the ECSC not to divide Europe by setting up an economic organisation separate from the OEEC. Nevertheless, the UK withdrawal left the way open for the committee to draw up its proposals, published in April 1956, for a common market which included agriculture.

The Spaak report outlined the special circumstances of European agriculture — the social structure of the family farm, the need for stable supplies and the problems resulting from climatic conditions and the inelastic demand for food. Moreover, it was recognised that the removal of tariffs and quotas would not be sufficient to allow the free movement of commodities between member countries. Those problems which demanded market intervention at a national level would not simply disappear with the creation of a common market. Thus, a common solution to the problems of European agriculture was sought (Fennell, 1987).

To that end, the Spaak Report laid down a number of objectives for future agricultural policy, four of which were to be reflected in the Treaty of Rome a year later: (a) the stabilisation of markets; (b) security of supply; (c) the maintenance of an adequate income level for normally productive enterprises; and (d) a gradual adjustment of the structure of the industry.

However, specific mechanisms for tackling these objectives were not detailed in the Report. The long term goal of a united Europe superseded those of a more temporal nature and few among those in favour of European integration wished to delay the creation of a common market while particular policy measures for agriculture were discussed.

National intervention in the agricultural sector varied among the Six and, to avoid distortion, national policies should have been dismantled entirely. Had this been agreed, then competitive forces would have redistributed resources and changed the structure of agricultural production in western Europe — an outcome totally unacceptable to the governments of the Six (Marsh and Swanney, 1980). Thus, when the heads of delegations met under the chairmanship of Spaak to negotiate the European Community Treaty, at Val Duchesse in 1957, the manner in which the decisions relating to agriculture were reached established that "...the common policy would be more a matter of accommodating national interests than of requiring radical adjustments" (Pearce, 1983).

It is evident that during negotiations on the Treaty, agriculture was not considered a major priority. When the delegations of the Six got together in Brussels to negotiate the Treaty, the working parties formed did not include one for agriculture. In the Treaty itself, agriculture is only one of ten sectors in which the range of measures towards co-operation and integration were designed to apply. This lack of (agricultural) interest among the heads of government was largely due to the overwhelming desire for the creation of the Community not to be held up by wrangles over specific sectoral issues. For this reason, while the objectives for agricultural policy were detailed in the Treaty, the specific mechanisms by which they were to be achieved were not.

Not surprisingly, the Treaty itself deals more explicitly with the general goals of the Community, aimed at "...establishing a common market and gradually approximating the economic policies of the member states, to promote throughout the Community a harmonious development of economic activities, a continuous and balanced expansion, an increased stability, an accelerated raising of the standard of living and closer relations between the member states." (European Communities, 1987).

Article 3 details how these objectives should be achieved: via the elimination of customs duties, quotas and so on; the establishment of a common customs tariff (CCT) and a common commercial policy (CCP) towards third countries; the removal of obstacles to the free movement of persons, services and capital, and the co-ordination of economic policies. Article 3 also refers directly to the creation of common policies for transport and agriculture. However, the member states were not obliged to complete a common policy for agriculture until the end of the twelve year transition period, laid down for the achievement of the Common Market itself. Thus, it is not surprising that the explicit objectives described for agriculture (articles 38–47) are fairly broad and allow for a range of interpretations.

Article 38 defines the scope of the Common Market as it applies to agriculture (covering products of the soil, of stock farming, of fisheries and of first stage processing directly related to these products) and states that the Common Market for agriculture should be accompanied by a common agricultural policy. The objectives of the policy as detailed in Chapter 1 are set out in article 39.1. These objectives were to provide the yardstick by which all measures relating to agricultural policy would be judged acceptable, and indeed legal.

The vague reference to policy instruments is contained in article 40. It is here that the establishment of a market organisation is stipulated. The form of the organisation, however, was not detailed, but guidelines were provided. These included the regulation of prices, production and marketing aids, storage and carry-over facilities and the stabilisation of imports. Discrimination between producers or consumers within the Community was to be strictly avoided and any price policies adopted were to be based on common criterion and a uniform method of calculation. Article 40 also stipulated that a fund (or funds) should be created to finance the common organisation of agricultural markets.

The process by which the CAP was to be established was outlined in article 43. The Commission was to submit proposals on the CAP to the Council of

Ministers within three years of the signing of the Treaty and, following consultation with the European Parliament, the Council was required to make regulations, issue directives or take decisions thereon. Significantly, these regulations, directives or decisions were, for a short period of time, to be made unanimously, with (qualified) majority voting to be established thereafter. This was an important federalist input, aimed at removing the inevitable tendency for member states to think of nation before Community.

Recognition of the need to devote more time to the consideration of the machinery behind the CAP was also made explicit in article 43, which recommended the convening of a conference between member states, to discuss their existing agricultural policies and formulate a statement of individual resources and requirements. If, as it has been suggested, agriculture started out near the bottom of the signatories' list of priorities, this provision certainly confirmed that the consideration of agricultural problems and the construction of a common policy with which to tackle them, was to become a major factor influencing the structure of the Community during its infancy.

1958: The Stresa Conference — A definitive policy framework

In accordance with article 43 of the Treaty, delegations from each member state, including (for the first time) representatives from the main farming organisations and the food industry, assembled at Stresa, Italy, in July 1958, to outline formally the problems to be tackled and the means by which they were to be resolved. Although the agreement reached at Stresa was not legally binding, the final resolution offered a more coherent view of the CAP than was presented in the Treaty of Rome.

Amongst the points agreed at the conference, the following were of particular importance. As mentioned in the Treaty, agriculture was to form an integral part of the overall economic strategy; trade was to be developed within the Community without threatening established political and economic ties with third countries; policies designed to manage the market were to be supported by structural measures, aimed at evening out production costs and ensuring a rational resource allocation, thereby stimulating productivity; equilibrium was to be sought between production and market outlets and it was hoped that increased productivity would allow the application of a price policy without the encouragement of over-production; aid to disadvantaged farmers was seen as a way of easing the necessary adaptations and a high priority was attached to increasing the efficiency of the family farm unit, which was to be safeguarded at all costs. Finally, it was hoped that the resultant improvement in the structure of the industry would enable capital and labour in the agricultural sector to receive remuneration comparable to that obtained in other sectors of the economy (Commission, 1958a).

It was an impressive declaration of objectives and incentives for agricultural policy, but still lacking the precision which was a fundamental prerequisite to the eventual implementation of the CAP. This was largely due to the inevitable conflict of opinion which surfaced at the conference.

The French and Italian delegations stressed the importance of mutual preference and favoured a high degree of market organisation, whilst the Dutch and the German delegations were preoccupied with the continued development of trade links with third countries. Indeed, the Dutch argued that preference was a consequence and not an aim of the Treaty of Rome (Tracy, 1994).

However the slick management of the conference by the secretariat of the Commission ensured that from conflict came compromise, and from ambiguity came cohesion. This in turn was largely the result of the determined effort from the commissioner responsible for agriculture, Sicco Mansholt. He was to play a major role not only in the agreement at Stresa, but more significantly in the following years, during which time the CAP would officially come into being.

Like most of the representatives present at the conference, Mansholt was still in doubt over many of the key issues. He expressed his scepticism over the usefulness of price policies and his concern over the potential creation of surpluses and the effect this might have on the EC's trading partners. But as far as the farming representatives were concerned, by far the most important issue which was covered was the principle that the family farm should remain the foundation of agriculture in the Community — a view to which Mansholt strongly subscribed and reiterated in the closing words of his final address "...it is particularly encouraging that the conference has provided the opportunity for a frank discussion on doctrine and on the goals of our agricultural policy, that is to say, on the need to guide agriculture in the direction of sound family farms...In my view this must be so because...there can be no structural policy, or market policy, if we lose sight of this starting point, which in the long run is our final destination as well." (Commission, 1958a).

Following the conference at Stresa, the Commission, in its first report on the activities of the Community, was to outline its views on the problems facing agricultural policy. Following the lines of the Stresa resolution, it considered the central problem to be the disparity existing between the level of income in agriculture and that in other sectors of the economy. The economic and political necessity of maintaining trade relations outside the EC meant that the Community could not become a self-sufficing entity and the Commission, like the delegates at Stresa, warned of the potential dangers of price support, stating that "It would serve no useful purpose to ask for improvements in the structure of agriculture if prices were at the same time fixed at a level which enabled even those enterprises to cover their expenses, which owing to their inferior structure were producing at high costs." (Commission, 1958).

The Commission was clearly still not prepared to produce a blueprint for the CAP, particularly as it was still undecided over the specific policy instruments to adopt. Nevertheless, it was by this time evident that the EC was seeking to formulate a policy which would safeguard the family farm and support farm incomes, while simultaneously avoiding surpluses and maintaining trade links with third countries. The conflicting elements were still to be reconciled and the difficulties inherent in such a task meant that it was another four years before a policy sufficiently flexible to accommodate such diverse constraints could be established.

1959–1962: Conflict and compromise — The birth of the CAP

Following the Stresa conference, the Commission gave itself a limit of two years from the Treaty's signature within which to submit proposals to the Council of Ministers on the working and implementation of the CAP. In November 1959 a draft of its general proposals was submitted to the Economic and Social Committee (ESC)[1] and from March 1958 to December 1959 the European Parliament debated the issues raised. A final set of revised proposals were submitted to the Council in June 1960.

The proposals which the Commission presented to the Council, on the shape of the CAP, included the free circulation of agricultural products within the EC; progressive development over the transition period, in *harmony* with general economic and social activities; the close inter-dependence of structural, market and trade policies in agriculture; the eventual adoption of a system of common prices; and the encouragement and co-ordination of national structural policies. The detailed proposals for market organisations for most of the main products rested on the principle of variable levies, both on third country imports and (during the transition period) intra-Community trade, with target and threshold prices as a means of harmonising existing policies.

In July 1960 the Council met to consider the proposals and created the Special Committee on Agriculture (SCA), giving it a continuing mandate to prepare future Council decisions on agricultural issues. The proposals were further debated by the European Parliament in October and twelve months later, following negotiations between the SCA and the Commission, the Council eventually accepted the substance of the proposals for a system of levies, to be applied to intra-Community and third country trade. The Commission was then called upon to submit draft regulations applying the levy system to a series of products over the following year.

Throughout the year (1961) a number of draft regulations incorporating the mechanisms proposed by the Commission were circulated. These outlined the system of support prices, import levies and export refunds, intervention buying and so on, all of which were subsequently ratified. However, the Commission did not have it all its own way. For example, the idea of a central organisation responsible for calculating daily import levies, restitutions and so on — the mechanisms illustrated in Fig.1.1 — was unacceptable, largely because member states wanted to keep intervention boards firmly under state control (Neville-Rolfe, 1984). Equally unacceptable was the proposal for support policies to be financed by separate product stabilisation funds, with levy revenue providing the main source of finance, backed up (if necessary) by producer participation. The Commission realised, even at this early stage, that for certain products (notably milk) financial problems were likely and that making sectors accountable for their own financing would help to avoid this. However, most member states (particularly France) were in little doubt that such an idea would be impossible to sell to their farmers and, as a result, no more was heard of producer participation in the financing of the CAP for a further fifteen years.

Following the draft regulations for cereals, pork, eggs, poultry, fruit, vegetables and wine, submitted in July 1961, the Council finally agreed, after more than two hundred hours of intensive negotiations, on January 14, 1962, to adopt a series of regulations giving legal effect to the levy system and instituting a common market organisation for each product. The levy system took effect from July 1, 1962 and from that date "...agriculture formally ceased to be a subject of purely national administration and control." (Lindberg, 1963).

It is interesting to note that at the outset, export refunds were not considered a major element in the system. Originally applied to processed products and designed to compensate for the increased raw material costs resulting from import levies, the Commission expected levy revenue to exceed the cost of export refunds. It was not until later, when the Community began to expand its exports of cereals and sugar, that the cost of export refunds became a major policy issue (Tracy, 1994).

Overall, the elements of conflict, not only between member states but across policy objectives meant that the Council had rejected the strict organisation of trade and markets based on quantitative restrictions, but they did agree on the three fundamental principles, referred to in the previous chapter, upon which European agricultural policy was to be organised: (a) market unity — a single agricultural market, a common marketing system and common pricing; (b) Community preference — the competitiveness of Community producers should not be threatened by third country imports; and (c) financial solidarity — expenses incurred to be financed by the Community, and income generated to form part of the Community's own resources.

These three principles have (to varying degrees and with fluctuating emphasis) been adhered to throughout the CAP's existence and have been resolutely defended by the Commission (Ritson and Fearne, 1984). Thus, while a number of issues were still to be resolved, notably the initial level of support prices, the agreement reached in 1962 undoubtedly signified the official birth of the CAP.

1963–1967: Common prices and common financing — the final hurdle

To the extent that the objectives of the policy and the regulations outlining the market organisations for the various products were agreed in 1962, it can be said that the CAP was created in that year. However, before the policy could begin to operate effectively, let alone have an impact on the Community's agricultural sector, a number of crucial issues had still to be resolved: the level and seasonal scale of common support prices, the location of intervention centres, and quality standards.

Because of the inherent difficulties associated with establishing a common price level, progress on this issue was slow, with Germany in particular reluctant to adopt any form of common prices until the end of the transition period. However, pressure was brought to bear from outside the EC, in February 1963, when the Council asked the Commission to negotiate on the Community's behalf in the forthcoming GATT negotiations. Without a common internal price level it would have been difficult to adopt a common negotiating position on the level of prices. The most important product under consideration was grain, and with the US and other grain exporting nations eager to maintain trade flows with the EC, internal price levels could not be set too high. It was important that a common price level should be agreed as soon as possible and the Commission began to apply pressure internally to this end, with a view to achieving the alignment of cereal prices to a common level by the beginning of the 1964/65 marketing year.

In the event, the deadline was not met and in October 1964 the French Government (under pressure from the French farm lobby) delivered an ultimatum to the Germans, threatening to withdraw from the EC if an agreement on cereals (crucial to France and the completion of the CAP) was not reached by the end of the year. Some concessions were forthcoming, notably the basic acceptance of a reduction in German cereal prices, to come into force in July 1967; but these were accompanied by demands for compensation and a revision of the price level in 1967, to account for the increase in production costs. This was unacceptable to both the French and the Italians and a deadlock seemed inevitable. However, on December 13 the German minister for economic affairs, Schmucker, declared that his Government would accept the price reductions of between 11 and 13%, as proposed by the Commission, provided that they were not applied until July 1, 1967 (as opposed to 1966) and that they were accompanied by increased compensation and a revision clause. Three days later, after another marathon Council session, a final compromise was reached, largely due to French agricultural minister Pisani's decision to accept the proposed increase in compensation for German farmers. The Community had recovered from the brink of collapse and, following the agreement on cereals, the Commission set a similar deadline of July 1, 1967 for the alignment of national support prices for the other products. However, before celebrating the removal of the ultimate hurdle, the Commission still had to settle a number of issues involving the financing of the CAP.

There were two main problems with regard to the financial arrangements for the Community: the first concerned the allocation of national contributions to the Community budget during the interim period, before 'Own Resources' fully financed the Community budget; and the second referred to the deadline by which the Community was to be dependent upon those Own Resources.

The Commission's proposals on common financing were submitted to the Council in March 1965. They concluded that the share of the budget attributed to the Guarantee and Guidance Fund (FEOGA) of the CAP should be increased, in order that total FEOGA expenditure could be met by the budget, from July 1, 1967. This meant that the single market should have been achieved in advance of the period laid down in the Treaty. In line with the regulations provided under the Treaty, the Commission proposed that the distribution of national contributions should be based on each country's share of agricultural imports from third countries. They also proposed that, under article 201 of the Treaty, member governments should be asked to surrender their control over levy revenue and duties formerly accruing to national treasuries, and that this power should be granted to the European Parliament, thus increasing their control over the Community budget.

To this fourth proposal, the French Government was totally opposed, claiming that it was (at that time) unnecessary, and that it was in any case not the role designed for the European Parliament. At the same time Germany was against the proposed method of calculating national contributions on the basis that it discriminated against the net importing countries, of which it was one. Once again the positions of France and Germany were to provide the key to a solution. The French Government wanted an agreement over the method of financing the CAP, but refused to accept the strengthening of the Community's institutional powers. The conflict finally came to a head during the French Presidency, when at the Council of Ministers meeting on June 28, the French Foreign Minister, Couve de Murville, left the Chair, at two o'clock in the morning of July 1, without an agreement over the finance issue. The French Government then announced it would take no further part in Council meetings.

It was six months before the 'empty chair' was re-occupied, when an extraordinary session of the Council took place at Luxembourg on January 17 and 18, 1966. With careful manipulation and interpretation of the constitution and regulations of the Treaty, a compromise was eventually found ten days later, when the meeting resumed. The financial arrangements were agreed and the allocation of national contributions was established on the basis of a fixed key, with ceilings for each country, applicable until 1970. This satisfied the Germans who negotiated a ceiling of 31.4%, compared to 32% for France and 20.3% for Italy. On the parliamentary issue, the debate was broadened to include the consideration of majority voting in the Council and the relations between the Council and the Commission. Thus, not only was it agreed to postpone ratification of the transfer of budgetary power until the end of the transition period, but the Council declared that in future where very important interests were at stake, a unanimous agreement would be necessary.

This 'Luxembourg Compromise' was a triumph for President De Gaulle, who foresaw the problems which could have arisen from the imposition of a majority decision contrary to the vital interests of a member state and came to assume central significance in the CAP decision-making process (Chapter 3). However, in establishing the means of protecting national interests, the pre-eminence of the Commission's proposals, envisaged by the Treaty of Rome, was substantially impaired. Nevertheless, as Tracy (1984) points out, the agreement did pave the way for work to resume on the remaining aspects of the CAP. Indeed, the agreement on common prices and the method of common financing was hailed as a considerable success, not only because the resolution (albeit temporary) of these problems allowed the effective implementation of the CAP, but also because it represented a considerable acceleration of the timetable laid down in the Treaty. As Raup (1970) succinctly puts it "...the Common Agricultural Policy, which had been regarded as a bottleneck, became the motor that drove forwards the total integration process and significantly expanded the area covered by Community law."

National influences on policy formation

During the post-war period, the nucleus of west European countries (governments and interest groups alike) were united in the belief that the political and economic problems which had emerged would be best resolved through co-operation and integration. Throughout the preliminary work, negotiations over the Treaty and subsequent agreements over the various policies adopted, one feature stood out clearly — the recognition of the need for a Community solution.

The agreements reached (inevitably) reflected the extent to which national interests among the Six conflicted. As far as the CAP is concerned, the nature and objectives of the policy established were not only a function of the contrasting agricultural policies and the relative importance of agriculture in the member states, but were equally dependent upon the distribution of political and economic power among the Six. Thus, in order to gain further insight into the factors shaping the CAP and a better understanding of its importance within the EC, it is necessary to examine the agricultural situation and the type of policies operating in the member states prior to and during the policy-forming period.

France

The attitudes of the French Government and the French farm lobby were of crucial significance in the establishment of the CAP. Just as Germany played the leading role in directing the Community's industrial policies, so France was instrumental in ensuring the agreement on an agricultural policy which would enable her to offset (to a greater or lesser extent) the underlying weakness and uncompetitiveness of her industrial base.

Table 2.1 illustrates the dominant position of French agriculture in the EC during the policy-forming years: France accounted for over 45% of the six countries' total agricultural area and around 40% of total food production in the Community. It also had the lowest population density of the Six and was by far the biggest exporter of agricultural products, particularly grain. Moreover, France was (and largely remains) particularly bound by agricultural tradition and as a country at the heart of the European Movement, it is not surprising that the CAP was seen by many as a 'French victory' (Clerc, 1979).

There are a number of factors (geographic, demographic and political) contributing towards the important position of agriculture in the French economy. The strong concentration of industry in a few regional centres has meant that in large regions (notably Brittany and the Midi) agriculture provides the economic and thus political base. However, perhaps the most significant factor emanating from the structure of the industry is the fact that French farmers appear less committed than in other member states to a particular political party or doctrine (Tangermann, 1980). Thus they constitute an effective body of floating voters who are consequently accorded political attention far in excess of their relative importance in the French economy, in terms of both population and economic activity.

As far as the Commission's proposals on the formation of the CAP were concerned, French farm lobby groups in general reacted favourably. Eager to exploit their export capabilities, they insisted on a system based on Community Preference, with both levies and quotas on imports from third countries. Price reductions were not ruled out, providing safeguards were included to protect farm incomes while structural changes were made. Overall the attitude of French farmers was an opportunist one, best illustrated by a statement made by the vice-president of the Federation Nationale des Syndicats d'Exploitants Agricôles (FNSEA) — the main farm union organisation: "Let us not forget the importance of the potential of French agricultural production, which must have markets. At our door there is a market of 170 million consumers which will become 180 million in ten years' time...It would be a grave error not to profit from the occasion and to let ourselves be put off by the obstacles that the Common Market will meet." (Lindberg, 1963).

Table 2.1: The relative importance of agriculture in the Six.[1]

	France	Italy	Germany	Neth.	Bel-Lux
Total arable area (sq.km)	346,330	209,650	143,320	23,100	18,720
Proportion of EC arable area (%)	46.7	28.3	19.4	3.1	2.5
No. employed in agriculture (m)	3.7	5.0	3.0	0.4	0.2
Proportion of total workforce (%)	19.0	25.0	11.0	10.0	20.0
Proportion of EC food prod. (%)	39.4	26.3	23.0	6.4	4.9

1) Rows 1 and 2 = 1965, rows 3 and 4 = 1964 and row 5 = 1955/56.

Source: Adapted from De la Mahotiere (1970, p.140) and Lindberg (1963, p.272).

The securing of outlets for their agricultural surpluses was also a priority for the French Government. The four year plan for 1962 to 1965 aimed at increasing agricultural output by 28% and domestic consumption was not expected to keep pace. As a result, the French Government's stance reflected the position of the farm lobby and, as Lindberg (1963) points out, subsequently became the closest of the Six to that of the Commission. They insisted on the rapid implementation of the CAP and made this a condition for action on other areas (Neville-Rolfe, 1984). The CAP was to be clearly preferential and, while accepting the system of levies on intra-Community trade, the French argued for minimum prices, as a safeguard in exceptional cases.

As it turned out, French concern over price levels was shown to be misplaced. Table 2.2 clearly illustrates the divergence of national farm support prices for the Six and the position of France as the country with the lowest average support prices meant that French farmers would benefit from a common price level nearer the upper than the lower range of existing support prices.

Table 2.2: Average wheat, barley and milk prices in the Six in 1958/9, 1966/67 and the common target price established in 1967/8 (UA per 100 kg).

	Belgium		France		Germany		Italy		Neth.		EC (Target)
	58/9	66/7	58/9	66/7	58/9	66/7	58/9	66/7	58/9	66/7	67/68
Wheat	10.0	9.3	6.7	8.5	10.8	10.5	10.5	11.7	7.2	10.2	10.6
Barley	7.9	8.3	5.7	7.7	10.0	10.6	7.0	8.6	6.8	9.0	9.1
Milk	5.9	9.8	6.2	8.4	7.9	10.1	7.7	9.6	7.5	9.6	10.1

Source: Marsh & Ritson (1971).

Thus as Kindleberger (1965) points out, the CAP was to be the main force behind the impetus which improvements in the French agricultural sector gave to the impressive overall economic performance of the French economy during the 1960s. Despite a considerable decline in the agricultural population and a reduction in the total cultivated area, output increased by over 50%, the number of farms declined at an annual rate of 2.8% per annum and the average farm increased in size from 20.4 hectares in 1963 to 27.6 hectares in 1970.

The effects of the CAP were there for all to see and, significantly, the gains made by the French from increased agricultural exports exceeded those made by Germany from industrial sales: between 1960 and 1966 French exports of foodstuffs to Germany increased in value from 142 million ua (unit of account) to 417 million ua, with no reciprocal advantage established in industrial goods (De la Mahotiere, 1970). However, while this confirmed French hopes of the impact which the CAP would have on intra-Community trade, it caused Germany to take a much closer look at the financing of the CAP, in particular the level of German contributions. The 1960s were clearly a period of accelerated growth in the French agricultural sector (and in the French economy overall), but the financial imbalances and growth of surpluses which this expansion induced meant that the economic emphasis of the CAP would have to change over the following decade.

Italy

While the importance of agriculture was as great in Italy as in France during the post-war years (see Table 2.1), the influence of the Italian Government on the structure of the CAP was much less evident. This was largely due to the contrasting structure of Italian agriculture — self-sufficient in wine, olive oil and fruit and vegetables, but requiring substantial imports of meat, cereals and dairy produce. The CAP offered an extended market for Mediterranean products, from which the Italians could benefit, but high common prices for the northern products threatened to exacerbate the balance of payments problems which had long confronted the Italian Government.

Prior to the establishment of the Community, around 70% of the country's total area was classed as agricultural land. Holdings were extremely small in size compared to the other five members and labour productivity was particularly low. However, as in France (though to a lesser degree), the introduction of the Common Market and the CAP shocked Italian agriculture into accepting drastic structural changes, with much marginal land going out of agricultural use and the agricultural population falling by around 40% during the 1960s, thus considerably improving the productivity of Italian agriculture, per man and per hectare (Parker, 1979).

Despite the importance of agriculture in the Italian economy, there was a much weaker tradition of support for agriculture than elsewhere in the Community (Neville-Rolfe, 1984). Apart from limited support for rice production and a high guarantee price for wheat, there was no system of direct aids to farmers and there were no duties or quotas on agricultural imports. Immediately after the Second World War, the Christian Democrat Government instituted a programme of extensive land reform. However, while this achieved little social or economic success, the political value of capturing widespread agricultural support for the Christian Democrats meant that further considerations of agricultural policy were effectively curtailed. The farm unions in Italy are more closely linked to the party political system than elsewhere in the Community and the land reform programme won the support of the Associazone Italiana dei Coltivatori Diretti (the largest farm union), representing some two million small scale owner occupiers who exercised a strong influence in parliament.

The Italian farming organisations in general were enthusiastic about the CAP and were the only allies of the French over the four main issues — common prices, the balance between structural and price policies, third country trade and the length of the accession period. Not surprisingly, they were especially concerned over exploiting their comparative advantage in the production of fruit and vegetables and pushed hard for the elimination of internal restrictions on trade in these products within the Community (particularly in France and Germany). However, as far as the other products were concerned, the Italian farm lobby was much more reticent about trade liberalisation and looked for extensive structural programmes as well as financial aid from the Community.

When the Italian Government did make its voice heard during negotiations on the CAP, its position also reflected its concern that Italian agriculture would find it difficult to respond to rigid price policies without structural measures to support the necessary adjustments. Thus, it was opposed to an accelerated harmonisation of prices on the grounds that Italian agriculture would need a longer period to adjust and become competitive. It also feared that price policies would merely lead to over-production of the northern products, the subsidisation of which would be part-financed by the Italian treasury, and thus demanded compensation in the form of definitive structural programmes.

One feature which rendered the Italian Government's bargaining position even less persuasive during negotiations was their apparent contradictory attitude towards common prices. As Tangermann (1980) stresses, the high proportion of food imports and the economic significance of low food prices meant that the Italians were in favour of low support prices for some products (particularly the northern ones), while the desire to maintain agricultural employment and farm incomes, as well as to exploit its exporting capacity in fruit and vegetables, meant that the Italian Government would, on occasion, push for higher price levels.

The conflicting domestic interests over price levels meant that, rather than committing itself one way or the other, the Italian Government was forced to concentrate on increased financial aid towards structural programmes. However, during the policy-forming years, structural policy was not accorded the priority by the other member states that it was by the Italians.

The net result during the 1960s was that Italy did not benefit from the CAP as much as she had originally hoped. However, towards the end of the decade, in recognition of the less favourable (as far as Italy was concerned) aspects of agricultural policy embodied within the CAP, the Italian Government concentrated less on adopting a definitive position on price levels and placed more emphasis on negotiating specific measures in its favour. In this respect it was more successful, strongly influencing the decision to introduce support regimes for wine, tobacco and olive oil, and gaining preferential treatment over the distribution of structural funds.

Netherlands

As the third major exporter of agricultural produce among the Six, the Netherlands, like France and Italy, was expected to benefit from the CAP. However, unlike the French and Italians, the Dutch did not depend so heavily on their agricultural sector as a foreign exchange earner, with Dutch industry during the 1950s and 1960s among the most efficient in western Europe (De Vries, 1975). For this reason, as well as her lack of political power, the Netherlands did not play such a significant role in the formation of agricultural policy within the Community.

The Netherlands was territorially one of the smallest countries in the EC, with its total land area barely 2% of the Community's land area. However, as Table 2.1 shows, while occupying only 3% of the Community's arable area in the mid-1960s, the Netherlands still managed to produce over 6% of total EC food production. With a total population representing only 5% of the Community's, but contributing around 6% of its GDP, these figures give some indication of the country's wealth as well as the intensiveness of its development during the post-war years.

An important feature of the rapid economic development in the Netherlands after the Second World War was the abundance of many natural resources and the existence of a highly efficient agricultural base. With space a fundamental constraint on industrial expansion, careful Government planning and a series of reclamation programmes provided the impetus for the Dutch economy to exploit structural changes in agriculture (encouraged by the CAP) and become increasingly competitive within the Community's industrial sector. However, as Parker (1979) argues, the factor of most fundamental importance was (and remains) the country's geographic centrality in relation to the rest of the Community, in particular its position at the mouth of the Rhine. With Rotterdam, the main Dutch port, leading into the heart of western Europe, the extension of intra-Community trade was a major priority for the Dutch Government during negotiations over the Common Market.

The same reasoning applied to the CAP, with the Dutch Government eager to accelerate trade liberalisation and establish further (and more profitable) markets for her agricultural exports. The proposed variable levy system was acceptable to the Dutch, but they argued against the French proposal for minimum prices. Indeed, as far as common prices were concerned, the Dutch could see nothing to be gained from basing the common level on a compromise between national price levels and preferred an autonomous decision on a desirable level of support prices. The likelihood of higher EC support prices would, as far as the Dutch could see, merely lead to surpluses and problems with third countries over trade arrangements.

The chief objective of the Dutch Government was clearly to force her partners (particularly Germany) to open up their markets to Dutch food exports, particularly of dairy produce which formed the largest proportion of Dutch agricultural production. To this end, they were willing to accept the disadvantages of higher prices and limitations on cheap third country imports of grain, a product in which the Netherlands was particularly deficient.

From the Dutch farm lobby came the most favourable reaction to the Commission's proposals on the CAP (Lindberg, 1963). The main farm union, the Landbouwschap, was in favour of accelerating the adoption of the CAP, provided that all protective devices — import quotas, export aids and minimum import prices were eliminated. The efficient Dutch farmers wanted to exploit their comparative advantage within the Community but were also seeking to maintain long-established trade links with countries outside the Community. Thus, they favoured low support price levels, particularly for grains, which were of crucial importance to the milk and livestock sectors, and criticised the Commission for being too protectionist against third countries.

The problem of cereal prices was particularly complicated in the Netherlands. As Table 2.3 illustrates, the chronic deficiency in wheat during the 1950s meant it was essential for the Government and Dutch dairy/livestock producers alike, (a) to keep common prices low; and (b) to maintain links with cheap grain sources outside the EC. However, given the relatively high level of cereal prices established, Dutch farmers responded by rapidly and substantially increasing wheat production, thus doubling the self-sufficiency level by 1970 and reducing the net import requirements for grain.

This is just one example of the benefits which accrued to Dutch farmers directly as a result of the adoption of the CAP. Dairy and livestock producers were to benefit even more through increased exports and higher price levels, and the Netherlands was to become one of only two net beneficiaries from the farm fund (the major one being France) during the first few years of effective common organisation of the Community's agricultural sector.

Table 2.3: Self-sufficiency levels (%) in wheat for the Six (1955/56–1966/67).

	Average for 1955/56 to 1958/59	Average for 1963/64 to 1966/67
Germany	63	79
France	105	127
Italy	103	95
Netherlands	28	57
Bel-Lux	65	72
EC(6)	89	101

Source: Raup (1970, p.153).

The Federal Republic of Germany

The position adopted by the German Government during negotiations on the CAP is perhaps the most interesting of all the attitudes taken by the Six countries during the policy-forming years. The provision for agriculture in the Treaty of Rome was seen by many commentators (see for example, Marsh and Ritson, 1971, Clerc, 1979, Swann, 1981) as the result of a trade-off between France and Germany, with access to an industrial common market for the Germans to be compensated by access to an agricultural one for the French. This might suggest that Germany's interest in the CAP and agricultural problems in general was relatively unimportant. However, as any analysis of the German agricultural sector reveals, agriculture was of considerable economic and political significance during the 1950s and Germany subsequently took an active role throughout negotiations, making substantial concessions (in the interests of the European Movement) but ensuring some degree of continuity in the nature of support accorded to her farmers.

As Table 2.1 shows, in terms of area and employment, the German agricultural sector was substantial in the mid-1960s. However, immediately after the War the major problem facing German agriculture was inefficiency. The vast majority of farms were very small in size, over-manned with little mechanisation and thus operated at relatively low productivity levels (per hectare and per man). The land consolidation act of 1953 went some way to easing the problem, by encouraging the merger of small farm units, and between 1959 and 1971 some 780,000 farms were merged into larger, more economic holdings.

The main reasons behind Germany's relatively backward agriculture lie in the tradition of protectionism in the agricultural sector, dating back to Bismark's tariffs on grain imports in the 1870s (Tangermann, 1979). While other west European countries chose to liberalise agricultural trade, the Germans embarked on a century of import levies and high national support prices. The proliferation of small inefficient farms put pressure on farm incomes, but by responding to the problem through high domestic support levels, the German Government merely further inhibited the structural adjustments necessary to help the industry become competitive. This, coupled with a growing population, increased

Germany's food import requirement and necessitated higher levels of border protection in order to avoid depressing domestic farm price levels. Thus, as a nation accustomed to regulating agricultural trade, yet deficient in many key products, notably grain, the German Government discovered that its views on the form and objectives of the CAP differed substantially from the member states with considerable export potential.

As a major food importer, Germany's interests were contrary to those shared by France, Italy and the Netherlands, but with an inefficient agricultural sector representing some 12% of the total population, the German Government also had to consider the support of its farmers' incomes. Not surprisingly, the nature of the policy which was eventually agreed, reflected these contrasting interests, not only between Germany and the other member states, but also within the German economy itself.

The German farm organisations, aware of the potential threat which the CAP posed to farm income levels in Germany, took a particularly aggressive stance towards the negotiations right from the start. Enjoying the highest price level in the EC, as well as the benevolence of the German Government, the Deutche Bauernverband (DBV), the main farm union, insisted upon maintaining these advantages within the Community (Lindberg, 1963). It criticised the Commission's proposals for lower common support levels on the basis that consumers (in Germany if not elsewhere) enjoyed a higher standard of living than farmers. Structural policy was not considered to be necessary within the framework of the CAP, as the DBV saw this area of support as predominantly of national concern, to be operated at a domestic rather than Community level. The concept of a common market would only be accepted if trade liberalisation came gradually and was preceded by the removal of unfair competitive practices. As far as the DBV was concerned, the problem was not the high level of German prices but the fact that costs of production elsewhere in the EC (particularly in the Netherlands) were kept artificially low, through the use of export subsidies and price controls. They could not, therefore, agree to the approximation of national prices until domestic policies and production costs had been harmonised throughout the EC. Not surprisingly, the DBV did agree with the Commission on the principle of Community Preference, a theme with which it had grown happily accustomed.

The political power of the DBV at the time of the negotiations on the CAP meant that the German Government had little room in which to manoeuvre. With the need to keep domestic farm prices high (to satisfy farm income demands) and to maintain access to (cheap) non-Community sources of food imports, it could be argued that the German Government did not really want the CAP. However, the CAP's function as the cornerstone of the Community meant that the German Government, rather than overtly expressing its opposition, chose to play for time, pleading for understanding and patience, in the knowledge that the eventual adoption of the CAP was inevitable.

Unlike elsewhere in the Community, pressure on the Government to hold its ground over agricultural policy did not only come from the farm lobby but also from industry. Germany was the largest importer of food (mainly from outside the Community) among the Six and much of these imports were tied to industrial exports by various kinds of barter deals. For example, German grain purchases from Argentina were linked to Argentinian purchases of German manufactured products (Lindberg, 1963). This meant German industry had an interest in maintaining trade links with third countries.

Despite the internal pressures and attempted prolonging of an agreement on the CAP, the German Government was forced to make considerable concessions during the final negotiations of 1962 and, as Neville-Rolfe (1984) suggests, these were largely granted to show that the Germans were now good Europeans. By the end of the marathon Council sessions over the implementation of the CAP, Chancellor Adenauer had clearly accepted the CAP as an integral part of European unity.

It is important to remember that the Treaty of Rome was signed little more than a decade after the cessation of hostilities. In addition to the economic advantages to be gained from an extended, tariff-free Common Market, the German Government was also seeking political refuge through the establishment of a truly united Europe, both economically and politically. It therefore considered that the domestic losses (both economic and political) incurred from the CAP were worth bearing in the interests of progressing towards the broader objective of European integration.

Belgium and Luxembourg

The combined arable area of these two small countries represented less than 3% of the total EC arable area and agricultural production prior to the establishment of the Community represented less than 5% of the total output of the Six (see Table 2.1). As the two smallest countries in the EC, Belgium and Luxembourg have traditionally lacked political and economic influence within western Europe. Indeed the existence (or at least independence) of Luxembourg was, until 1867 largely in the hands of its more powerful neighbours. However, with the two countries at the very heart of the Community, often acting as the driving force within the European Movement and with the capital of the Common Market established as Brussels, they have enjoyed a degree of political recognition which reflects the symbolic importance of Belgium and Luxembourg as integral members of the EC.

With both countries constrained by physical limitations, and Belgium further inhibited by a population density exceeded only by that of the Netherlands, they entered negotiations over the CAP with considerable food import requirements. As a result, facing similar problems to Germany, but lacking the industrial potential, Belgium and Luxembourg often found themselves caught between promoting the success of European integration and protesting against the financial consequences of the CAP.

If they did have an ally among the other members of the EC, then Germany was the most likely candidate. Indeed, the Belgian Boerenbond (the chief Belgian farm union) aligned itself totally with the position adopted by the Germans (DBV), seeking to suppress unfair competitive practices and extend the accession period to allow for the harmonisation of national policies. A similar alliance was established at the governmental level, with the Belgian Government closest to the German position of harmonisation before liberalisation (Lindberg, 1963). Belgian agriculture was relatively backward, which meant higher prices would help to support farm incomes. But substantial import requirements suggested rather lower support price levels within the EC. In the end, the Belgians were less committed to high prices than the Germans and generally went along with the Commission's proposals for an approximation to the existing national price levels.

The position of Luxembourg more closely reflected the requirements of a relatively small economy. More attention was given to the support of individual farm incomes and the Commission was considered to be too concerned about trade with third countries. The Luxembourg Government also felt the Community needed a longer period of time to adopt common policies, and an agreement over the price level should, it believed, have been preceded by measures to reduce costs and compensate farmers for the structural adjustments which they would have to make.

Given the relative inefficiency of agriculture in both countries and their low levels of self-sufficiency in foodstuffs, it was almost inevitable that neither country would gain any net benefits from the CAP. During the 1960s Luxembourg came out as a marginal loser from the farm fund, with the small net gain from structural aid insufficient to cover the inevitable deficit from the guarantee section of FEOGA.

Belgium came out of the 1960s with a much bigger deficit, second only to Germany, with a negative balance on both aspects of CAP support. However, on the positive side, both countries had reduced their import requirements for most food products, particularly cereals, and the Benelux alliance had itself become stronger during the period of intensive negotiations between the Six on common prices and the financing of agricultural support. Thus, on balance, what Belgium and Luxembourg lost on the financial side, they compensated for (to some extent) on the social and political side, from the progress made towards European integration and the effective redistribution of political power among the member states and the Commission.

The Commission

The importance of the Commission's role during negotiations on the CAP were twofold. First, as an intermediary between member states, it was able to help break down national barriers; second, using its (albeit limited) powers as an executive authority, it was able to influence the course of events directly on its own initiative.

In establishing the Common Market and implementing the CAP, the Commission undoubtedly succeeded in promoting the mutual benefits which were likely to accrue to member states through the process of economic and political integration. The fact that the Six were able to reach a compromise on the key issues shows that in the final analysis they were willing to gamble with votes at home rather than stand accused of blocking the advance towards closer European unity (Lindberg, 1963). Most commentators (see, for example, Harris *et al.*, 1983, Tangermann, 1980 and Lindberg, 1963) agree that it was the mediating and brokerage activities of the Commission which made this possible, and the fact that the final agreement closely reflected the Commission's original proposals would seem to confirm this.

Despite mutual commitment to the European Movement from the Six, only the Commission, acting on the Community's behalf, was able to place national arguments into the global perspective, subsequently to coerce or encourage member state governments to make concessions (while granting safety clauses and provisions for special circumstances) in order that an agreement might be reached. While one might choose to criticise the Commission for its lack of precision over the stated objectives of the CAP, one cannot deny that without its assistance in crucial areas of national conflict, any agreement would have been extremely difficult if not unlikely.

However, in its quest to accelerate the harmonisation of national policies and cement a firm foundation for the future of the EC, the Commission was perhaps guilty of underestimating the economic (and inevitably political) importance of those problems which caused most debate. The best example of this is illustrated by the impact which the implementation of the CAP had on EC expenditure during the 1960s. As has already been stressed, nobody had anticipated the degree to which the Community's farmers would respond to the adoption of the CAP. In every country farm output expanded, surpluses (notably of sugar and butter) had already appeared by the end of the decade and, as Table 2.4 shows, the cost of the CAP had escalated beyond all expectations, with FEOGA expenditure accounting for some 95% of total Community expenditure in 1969.

Table 2.4: The growth in FEOGA expenditure (1962/3–1968/9) (million ua).

	Agricultural expenditure		Total Community expenditure	Agricultural expenditure as a % of total
	Guarantee	Guidance		
1962/3	24	7	31	100.0
1963/4	42	14	57	98.2
1964/5	136	45	181	100.0
1965/6	200	67	267	100.0
1966/7	308	103	412	99.7
1967/8	1,094	237	1,505	88.4
1968/9	1,677	237	2,031	94.2

Source: Adapted from De la Mahotiere (1970, p145).

With this financial commitment to the CAP established, it would clearly be difficult to progress with the expansion of the Community to other areas of common organisation. The Commission, like the governments of the Six, had recognised the importance of establishing the CAP, but it had not considered how the Policy (and the sectors to which it related) would develop. As a result, the Commission was forced to look again at the objectives of the CAP and the means by which they were to be respected. In effect, it was obliged to take a closer look at those issues which had originally caused so many problems, not only for the creation of the Community as a whole, but particularly with the adoption of the CAP.

Structural reform and the end of common prices

The implementation of the CAP required, over a decade of intense negotiations, the resolution (albeit partial and to varying degrees of satisfaction) of a number of conflicting issues. One (if not the most important) of these concerned the level of common prices. Too high a level would have upset the balance of payments situation in some member states, while too low a level would have adversely affected the farm income situation in others.

The eventual agreement ensured that common prices were initially set at a level closer to the higher price levels among the Six than the lower ones. This increased the financial cost of the CAP to the net food importing countries, but guaranteed a satisfactory level of price support for the Community's farmers. Inevitably, the resolution of this fundamental problem was to lead to an even greater one — the growth of agricultural surpluses, which became the dominant influence on the development of the CAP from the early 1970s onwards.

Agriculture 80 and the structural directives

By the late 1960s, it became apparent to the Commission that, due to the productive capacity of European agriculture and the incentives provided by a system of guaranteed prices without limitations on production, specific measures would have to be taken to alter the overall structure of the industry. Price cuts on their own, it appeared, were not the answer. First, farm ministers were unlikely to agree to significant price cuts, and second, there was some evidence to suggest that small price cuts would result in a perverse supply response, with farmers (on aggregate) increasing production in order to maintain their incomes (Broad and Jarrett, 1972). Thus, the Commission (and in particular Sicco Mansholt) took the view that production could only be controlled effectively by reducing the area of agricultural land as well as the number of people working in agriculture, and ensuring that those who remained were efficient enough to earn a living from lower support prices.

Commissioner Mansholt had warned farm ministers from the outset (at the Stresa conference) that the combination of family farms and an open-ended price support system would lead to surplus production, inefficient agriculture and ever-increasing costs. Moreover the benefits (in the main) would go to those producers who could increase production, and who (on the whole) were already better off. Thus, his plan, Agriculture 80, provided for "...a price policy devoted to restoring a more normal relationship between market and price levels, and radical land reform measures to bring farms up to a viable size and enable farmers to live as comfortably as possible." (Mansholt, 1972). It aimed to reduce the farm population by five million between 1970 and 1980 and remove some 12.5 million hectares of land from agricultural use, thus increasing the size of farm units and improving labour and capital productivity. This was, however, to be achieved on a voluntary basis, with a variety of financial incentives to persuade farmers to retire early, take other jobs and amalgamate their holdings. These subsidies would be costly and agricultural expenditure would increase in the short term, but it was anticipated that by the mid-1970s substantial savings would be made via reduced price levels and the reduction of structural food surpluses (Commission, 1968).

Not surprisingly, the plan was violently opposed by the Community's farmers and none of the member states were prepared openly and fully to support the measures proposed. With the exception of Italy and Benelux (where structural policy had been largely neglected) most of the member states had pursued their own structural policies for agriculture since before the adoption of the CAP. France had begun consolidating traditionally small holdings before the Second World War and Germany followed suit just after it. During the 1960s, the French Government introduced a series of measures to aid modernisation and improve access for young farmers to farm ownership. In the Netherlands, land consolidation was linked with incentives to early retirement by farmers in their fifties, and by 1967 these three countries were between them spending nearly 1.25 billion ua on structural programmes (Neville-Rolfe, 1984).

Because of the political sensitivity of structural reform, the manner in which the Mansholt plan linked structural change to reduced price support levels and the increased financial commitment which the plan would entail (at least in the short run), the Six were, overall (with the notable exception of France) in agreement with their respective farm organisations in their opposition to the reform package.

The final agreement, outlined in 1971 and ratified under the structural directives agreement in 1972, fell short of the Commission's demands and well short of the original Mansholt plan. The compromise reform plan consisted of five basic points: (a) the encouragement of farmers to leave the land and the provision of a retirement pension not less than 500 ua per year for farmers between the ages of fifty-five and sixty-five; (b) the introduction of farm development schemes providing low-interest loans and loan guarantees; (c) the setting-up of information and advisory services; (d) the encouragement of producer groups and co-operatives, to improve marketing; and (e) the prevention of new land coming into agricultural production.

These in turn were watered down and condensed into the three socio-structural directives which were agreed in 1972, concerning: (a) the modernisation of farms; (b) encouraging the cessation of farming and the re-allocation of utilised agricultural area for the purposes of structural improvement; and (c) the provision of socio-economic guidance for and the acquisition of occupational skills by persons engaged in agriculture.

The year in which the agreement on the structural directives was reached was Sicco Mansholt's last as agricultural commissioner, but while his plan for structural reform stands out as the highlight of internal pressure for concentration of the CAP on structural change, the initiatives which were eventually agreed confirmed that price policies would remain at the core of agricultural support within the Community. While the farm ministers of the Six recognised the need for the reform of the Community's agricultural structure, they were reluctant to accept the political consequences of agreeing to the appropriate reform of its agricultural policy. As Mansholt (1972) himself protested, the price policy was based on consensus politics rather than economic rationale and the failure of the Six to respond to the challenge of structural reform was a shirking of responsibility.

Green money and the MCA — a new concept of common prices

The period over which the CAP was established was one of relative currency stability, with fixed (though in principal adjustable) exchange rates operating throughout western Europe. Indeed, as Tracy (1994) points out, the whole system of common prices had been based on the implicit assumption of exchange rate stability. Thus, few of the Community's policy-makers envisaged the problems which emerged in the early 1970s following the effective breakdown of the gold standard and the floating of individual currencies against the dollar.

However, only two years after the establishment of common prices within the CAP, parity changes were implemented in France and Germany. In August 1969 the French franc was devalued by 11.11% and in October of the same year the deutchmark was revalued by 9.29%. These parity changes marked the end of the only period when common support prices for products under the CAP applied throughout the Community until the establishment of the Single Market in 1993.

The devaluation of the French franc was the first parity change among the Six for eight years. Largely a result of the rising inflation, following the riots and subsequent wage increases of 1968, the devaluation was against Community regulations and with no prior consultation sought with the Commission, it took everyone by surprise. Agricultural and finance ministers met to discuss the impending problems of trade distortion resulting from the devaluation, but ruled out any offsetting change in the value of the ua. Instead, it was agreed that French farm prices would be aligned gradually to the new exchange rate, and that for a *limited* period (not later than the end of the 1970/71 marketing year) the French Government should be obliged to place subsidies on intra-EC food imports and levy taxes on intra-EC exports to remove the unfair price advantage. Thus, the Monetary Compensatory Amount (MCA) was born and, like so many of the temporary measures envisaged by the Commission, it was to become a long standing feature.

The German Government's decision to revalue the deutchmark, following months of upward pressure, was preceded by consultation and agreement from the Commission. However, its decision to impose an import levy of 11% on all CAP products, to compensate German farmers for the downward effect on their prices was not. Instead the Commission requested a total ban on agricultural imports and this for three reasons: first, the Commission wanted to highlight the undesirability of fluctuating exchange rates in the context of the CAP; second, if a currency had to be floated it should be done for a day or two only, as it would not be able to last without importing foodstuffs for a longer period; and third, the Commission did not want the imposition of MCAs to become a permanent feature within the CAP, as this threatened the principle of a single market. Although upheld by the European Court, the Commission's view was not enforced and it was eventually agreed that a positive MCA should be applied to intervention products only and that it should be revised weekly in the light of currency movements.

Following the general election in Germany in 1970, the incoming minister for agriculture, Josef Ertl, agreed to the removal of the MCA on the condition that German farmers would receive compensation from the resultant decline in farm prices. This amounted to a reduction in the farmers' liability for VAT payments, to be spread over three years. The Commission defended the compensatory measures on the grounds that it would have been politically unacceptable for German farmers to face price cuts without compensations, and the alternative of maintaining MCAs would have encouraged the proliferation of MCAs elsewhere in the Community.

However, the reluctance of the French (fearful of the inflationary consequences of raising food prices) and the Germans (wary of the political consequences of lowering prices to farmers) to allow farm prices to adjust following the divergence in market exchange rates meant that the Commission was obliged to allow agricultural support prices, in national currencies, to diverge.

The complicated system for sustaining different levels of support prices in the member states became known as 'Europe's green money' and Chapter 6 explores in more detail the role of green money in the development of the CAP.

The Six become Nine — policy implications of the first enlargement

Despite the problems of monetary instability and the growth of agricultural expenditure within the Community, the potential significance of the EC as an economic and political body of equal power to the US and the USSR, resulted in a growing interest among west European non-member countries to become more directly involved and to share the apparent benefits to be gained from internal trade liberalisation within the protective common tariff barrier. Such interest was reflected in Britain's repeated attempts to gain favour with the Six during the 1960s. British efforts to join the Community failed on two occasions, first in 1962 and again in 1967, but in the accession negotiations of 1970, she was joined by Ireland, Denmark and Norway. By 1973 the terms of entry were agreed, with only Norway rejecting membership at the last hurdle, following a referendum in 1972, and on January 1, 1973, the Six became Nine.

The enlargement promised to have significant economic consequences for the EC as a whole, particularly as Britain provided a substantial export market for the Community's agricultural surpluses. However, while Denmark and Ireland were both heavily dependent on the agricultural sector as an employer and an earner of foreign exchange, Britain, like Germany, was much less interested in the agricultural aspects of accession (apart from the financial burden of the CAP) and was more concerned with exploiting the liberalisation of industrial markets within the EC. Thus, while for Ireland and Denmark, membership was likely to result in substantial (net) benefits accruing to their economies (via the CAP), the immediate prospects for the UK, a major food importer accustomed to relatively low food prices, were less favourable. Indeed, the implications of the CAP for the British economy had been the major factor inhibiting the commitment of successive British Governments to the EC.

Britain's initial departure from the Six, during the early negotiations at the Messina conference, was founded on: (a) its overwhelming scepticism over the ability of the Six to achieve anything of either economic or political note, and (b) on the belief that its own economic strength and industrial competitiveness at the heart of the world economy could be maintained, a contributory factor stressed by Edward Heath (1982), the Prime Minister who eventually took Britain into Europe, in 1973.

The implications for Britain of adopting the CAP were twofold: first, the implementation of tariffs and a higher level of common support prices would increase the cost of food to the British consumer and ultimately have an inflationary effect on the UK economy; and second, as a substantial food importer (though maintaining concessionary agreements with the Commonwealth) the UK would also lose out through the higher cost of food imports and the budgetary contributions to the Community, with most of the benefits going to the agricultural exporting countries of the Nine (Marsh and Ritson, 1971). It would be an over-simplification to suggest that the UK Government accepted this prospect out of hand — the British insisted on a transitional period of at least five years (to ease the process of adjustment), gained considerable concessions for her Commonwealth partners (notably New Zealand) and forced an agreement on a level of budgetary contributions lower than that proposed by the Commission (Swann, 1981). But many of the fundamental differences of opinion (in particular those relating to agriculture and the budgetary implications of the CAP) were swept aside (albeit temporarily) in the effort to secure an agreement.

While British entry clearly presented a number of potential pitfalls for the future of the Community, it also provided a potential short-term solution to the growing problem of surplus food production among the Six: with 40% of temperate food requirements in the UK being imported, the British market could have absorbed all the EC surplus of butter, cheese, grain and sugar, but at a per unit cost twice that which it faced prior to accession (De la Mahotiere, 1970). The fact that this potential was not fully realised was due, in part, to the maintenance of New Zealand butter supplies, the agreements established with the African, Caribbean and Pacific (ACP) countries over sugar exports to the EC, the continued UK preference for North American bread-making wheat, and, finally, expansion of EC production, which offset any increased demand resulting from enlargement.

In the UK, from their original hostility towards UK membership of the EC, the opinion of the farm lobby swung around to its total support of accession by 1973. The deficiency payments system had, until the mid-1960s, worked favourably for the farming population as a whole. But at a time when British farmers were being encouraged to expand production selectively (due to the rising exchequer cost of product subsidies) Community farmers appeared to enjoy more security in high guaranteed prices and an open-ended market.

Prior to accession, British farmers believed that producer prices would increase as a result of the CAP, and given the relatively efficient structure of British agriculture, that they would be well placed to compete. However, as

Marsh (1979) argues, the Government's desire to maintain a relatively high green rate (necessitating a high negative MCA), in order to protect consumer prices, deprived British farmers of many of the expected benefits. As it was, high world prices raised UK prices substantially, which largely offset the efforts to keep prices down (Capstick, 1991).

During the first few years of membership, any potential criticism of the CAP from the British was dampened by the crisis on the world raw materials and agricultural markets. Inclusion in the EC kept prices relatively stable and for some products (notably sugar and cereals) below world market levels. This reduced the direct burden of the CAP, helped control the rising level of inflation in the UK and, as Tangermann (1980) argues, played a significant part in the positive outcome of the British referendum on continued EC membership, in 1975.

However, this period of comparative harmony over agricultural policy was short-lived. In 1977 the Consumers Association published a study calling for reduction in agricultural prices (particularly of surplus products), the National Farmers Union (NFU) began to voice its discontent over the maintenance of high negative MCAs, and the notion of the *juste retour* became a prominent argument in favour of CAP reform. Moreover, the adoption of a price-fixing procedure similar to the annual review practised in the UK meant that the Commission was obliged to take account of "trends in prices and costs, employment, productivity and farm incomes", which opened the door for substantial price concession to the farming lobby (Tracy, 1994).

In contrast to the economic and political significance of Britain's accession to the EC, Ireland, like the other political lightweights in the Community (that is, Belgium and Luxembourg), could expect to do little itself to influence policy developments. However, unlike its small country allies, Ireland, with its agriculture becoming increasingly competitive, was able to sit back and reap the rewards from the decisions influenced most heavily by the more powerful member states. As long as agriculture remained the major beneficiary from the Community budget, the Irish Government had no reason to complain. Not only has Ireland maintained a modest but steady agricultural trade surplus among the Nine, it is also the second largest net beneficiary from the Community budget.

The participation of Denmark in the EC and the process of economic and political integration is characterised by Haagerup and Thune (1983) as being largely non-committal. Sharing many geographical similarities with the Netherlands, a country of about the same size, low-lying with maritime traditions, an intensive agricultural sector and for the most part prosperous in relation to her European neighbours, Denmark's main contrasting feature is her reluctance to become involved with Community affairs. Such a stance towards the Community is in many ways ambiguous: while the Danish economy, particularly the agricultural sector, clearly stood to gain from EC accession, there was a marked reluctance, particularly among the left wing political parties, towards this commitment. This was partly due to the difficulty of reconciling EC membership with a long-standing association with the other Nordic countries; and partly the result of Denmark's dependency on the UK as a trading partner.

The Danish economy is highly dependent on international trade — its exports being worth around one quarter of GDP, and the agricultural sector in particular enjoys a substantial trade surplus. With the UK being her main trading partner (accounting for around 20% of total exports and about 50% of agricultural exports) Denmark was not in a position to contemplate joining the EC from the outset, but instead was obliged to follow Britain in joining the European Free Trade Area (EFTA) in 1959, along with the other Scandinavian countries. Similarly, it was not until the early 1970s, when the UK was clearly set on joining the Community, that the Danish Government adopted a similar policy and entered into negotiations with the Six.

The importance of agriculture in the Danish economy was a major factor in favour of Danish accession: up to 25% of the working population was involved in agriculture during the fifties, but a series of government development programmes after the Second World War helped to rationalise Danish agriculture and render it one of the most efficient in Europe. This was achieved by exploiting the gaps in the agricultural markets of her trading partners, in particular Britain and Germany, and replacing cereal production with pastoral produce, aided by a policy of favourable feed grain prices.

However, as Tangermann (1980) points out, the stimulus for Community membership resulted from the decline in the growth of agricultural production, food exports, farm incomes and agricultural investment during the 1960s, due to the increasing levels of protection and high rates of agricultural expansion in other countries. In this situation, EC entry and the associated unrestricted access to the traditional markets of Britain and Germany, were seen to be extremely attractive. Indeed, the relative economic success of the Danish economy as a whole, and the agricultural sector in particular, under the Community umbrella, has resulted in Danish ministers of agriculture adopting a cautious attitude towards price increases in the Council of Ministers. This is not out of a desire to reduce the benefits accruing to Danish farmers from the Community budget, but rather a pragmatic approach to ensuring the maintenance of the CAP in (more or less) its current form.

Nationalism and institutional initiatives

As anticipated, the first enlargement of the EC in 1973 marked the beginning of a new era for the development of the Community in general and the consideration of the CAP in particular. The accession of Britain posed a significant threat to the CAP in its present form, but more important, British attitudes towards the function of the EC and the role of member states within the Community framework encouraged each member state to look more closely at the budgetary transfers and resource flows which resulted from the adoption of common policies, jointly financed and under the corporate management of the Commission.

The results of this growth in nationalism within the Community, directly related to the financial implications of the CAP, were twofold: first, the pro-CAP member states (notably France, Ireland, the Netherlands and Denmark) sought to protect the national benefits accruing from the CAP, by pushing for higher support prices and maintaining the production of surplus commodities (particularly cereals and milk) which in turn exacerbated the budgetary problems of the CAP. Second, those member states whose interests remained in CAP reform (in particular Britain and Germany) pushed for a prudent price policy with direct compensation for the perceived unfair distribution of the costs of the CAP. For the UK, this took the form of budgetary rebates, while for Germany, the Commission took a lenient position over the removal of positive MCAs.

Such polarisation among the member states over the operation of the CAP made any rational discussion of policy reform at best difficult and more often than not, impossible. Thus, it was left to the Commission to seek a solution to the conflicting problems facing the Nine during the 1970s. However, while agricultural expenditure grew, along with surplus production, the Commission provided few alternatives in a series of reform documents which served merely to reiterate the basic objectives of the CAP, increase the financial pressure on the Community budget and intensify the conflicting interests between member states.

Features of nationalism within the CAP

Following the implementation of the basic mechanisms of the CAP and the establishment of support price levels, decisions on policy developments were to be taken by majority voting in the Council of Ministers. However, the Luxembourg Compromise (see Chapter 3) effectively established the power of veto over decisions of vital national interest and if the Six found this objective undesirable, the possibility of implementing a majority voting system with the Nine was remote. Thus, as Marsh and Swanney (1980) conclude, while the EC was founded on the principle of a pooling of sovereignty over Community decisions, in practice CAP-related issues have been regarded as being of such vital interest that the unanimity rule has been maintained, with the result that major decisions require a compromise of national interests rather than an expression of common interest.

Two aspects of the preoccupation with national interests are examined here: first, the manner in which the economic implications of the CAP are expressed between the member states — income transfers from net importing countries to net exporting countries; and second, the range of stances adopted by member states, in defence of national interests (the maintenance of benefits or the reduction of costs), within the corporate framework of the Community's decision-making process.

Ritson (1979, 1982) points out the significance of the defence of national interests within the Community in relation to the economic analysis of the CAP. As far as member states are concerned, interests lie not in the overall performance of the CAP in attaining basic objectives (as defined in the Treaty of Rome), but in the impact of the CAP on financial transfers between member states (see Buckwell *et al.*, 1982). The importance of these transfers are in turn largely a function of the importance of the agricultural sector in each member state, or more explicitly, the political influence of the farm lobby. Some indication of the relative political significance of agriculture in the Nine is given in Table 2.5, which suggests that agricultural policy should be highest on the political agenda in France, Ireland and Italy and lowest in the UK. However, national costs of the common financing of the CAP diverge between member states and collective price decisions influence individual member states unevenly, with agricultural exporting countries gaining (on the whole) at the expense of agricultural importing countries.

Table 2.5: Indicators of political and economic weight, EC(9).

	Germany	France	Neth.	Den.	Italy	Ireland	UK	Bel-Lux
GNP/head ($) (1976)								
	7510	6730	6650	7690	3220	2620	4180	7020
Average farm size (ha) (1976)								
	14.0	25.0	14.7	23.1	7.5	20.6	4.7	14.2
Ag. pop. as a % of total (1979)								
	6.2	8.9	6.0	8.3	14.9	21.0	2.6	3.3
Salaried agr. labour as % of Total (1979)								
	17.0	20.0	25.0	25.0	37.0	11.0	58.0	10.0
National expenditure on Agr. (1977)[1]								
	13.8	23.7	4.8	9.2	7.5	12.9	21.3	9.3

1) As a proportion of the value of final agricultural production.

Source: Ritson and Tangermann (1979, p.126); Harvey (1982, p.183).

Table 2.6 is one illustration of the extent to which the burdens of common price increases have been distributed among the nine (for alternative measures see Koester, 1977 and Buckwell *et al.*, 1982). The UK, Germany and Italy (the major food importers) suffer the highest welfare losses as a result of the distribution of budgetary contributions. This perception of inequality becomes all the more striking when one considers that Britain and Italy, two of the major contributors to the Community budget, also rank among the poorest members of the EC.

While it is not the objective here to justify or criticise the attitudes adopted by member states during the inflationary period of the 1970s, it is evident that the problems of an unequal distribution of costs and benefits emanating from the CAP were of crucial importance as far as the attitudes of member states towards policy changes were concerned. However, rather than concentrating on the longer term rationalisation of the CAP, there was an overriding tendency to seek the resolution of national conflict via the central policy instrument, the support price.

Table 2.6: Costs and benefits of the CAP with respect to common price changes.

	Ratio of total welfare loss to total gains from 10% price increase	Average cost per unit increase in farm incomes for:			
		FEOGA expend.	VAT contr	User welfare	Economic welfare
EC(9)	1.79	0.50	0.76	0.89	0.66
Germany	2.20	0.39	0.96	1.09	1.05
France	1.61	0.64	0.72	0.77	0.49
Italy	1.98	0.73	0.65	1.25	0.89
Netherlands	1.05	0.62	0.50	0.47	-0.03
Bel/Lux	1.78	0.33	0.66	0.69	0.35
United Kingdom	2.51	0.31	1.12	1.21	1.34
Ireland	0.51	0.90	0.21	0.31	-0.48
Denmark	0.79	0.99	0.37	0.30	-0.34

Source: Harvey (1982, p.178).

It is with regard to the positions adopted by member states over support price levels that the importance of the agri-monetary system is again highlighted. The very fact that national attitudes towards agricultural policy differed among the Nine, along with the perceived role of the CAP in regulating farm prices meant that MCAs served a vital function during the 1970s in allowing member states a degree of freedom over domestic farm prices, when the imposition of common prices (and the resultant increase in income transfers) would have increased the tension between member states as well as the political strain on the Community as a whole. Moreover, the use of MCAs returned some sovereignty over agricultural policy to member state governments, a factor which, as Ritson and Tangermann (1979) illustrate, resulted in most member states pursuing national policy objectives (at least with regard to price support) within the framework of the CAP.

The way in which the MCA system allowed member states a degree of freedom over the level of domestic agricultural support prices is explained in Chapter 6. Britain was able to maintain a (relatively) cheap food policy within the CAP by refusing to devalue the green pound in line with the market rate, thereby sustaining large negative MCAs. A similar position was adopted by Italy, a large importer of northern products which progressively sought to satisfy the Italian farm lobby via structural measures and direct aid from the Community budget. The ambiguity of Germany's maintenance of relatively large positive MCAs, as a net food importer, reflects the importance of the farm vote in Germany and the relative prosperity of Germany among the Nine (see Table 2.5). Denmark, as noted in the previous section, adopted a policy of directly linking the green krone with the market rate, in order to satisfy domestic farmers and exploit the benefits of jointly financed export refunds on high priced CAP products. A similar (low MCA) policy was adopted by Ireland and the Netherlands and among the pro-CAP countries, only France chose to maintain high negative MCAs, although as Ritson and Tangermann (1979) point out, the use of MCAs in France during this period more closely reflected the use for which they were originally designed, to aid the adjustment of the French economy to successive devaluations of the franc.

While the proliferation of MCAs undoubtedly smoothed the path for the Community over the issue of CAP reform, it also staved off the day when the basic operation and underlying principles of the CAP would be reviewed. The Commission had effectively lost much of its managerial control over the implementation of the CAP, and the resolution of conflicting interests became increasingly difficult as national attitudes became further entrenched. However, rather than responding to the challenge, as the Community's policy formulator, of seeking a resolution not only to the conflicting interests of member states but to the growing problems facing the Community's agricultural sector as a whole, the Commission published a series of initiatives which reiterated the basic objectives of the CAP, the economic irrationality upon which they were based and the political tension which they had led to, but offered little in the way of possible solutions.

The Commission dilemma over CAP reform

Since accession in 1973, the UK has been at the head of those countries (albeit a conspicuous minority) in favour of the wholesale reform of the CAP. In recognition of Britain's stance towards agricultural policy and in an early attempt to address the problems of market imbalance and excessive expenditure in the agricultural sector, the Commission produced the first (following the Mansholt Plan) of a periodic review of the policy entitled 'Improvements of the Common Agricultural Policy' (Commission, 1973). Part of the blame for the problems resulting from the CAP was placed on the lack of progress in other areas (notably monetary union and regional policy), but the Commission recognised the need to reduce the disequilibria in certain markets and cut back the level of guarantee spending. Ironically, the one thing that was keeping the Common Market functioning, the MCA system, was considered a threat to market unity and the Commission wanted its removal by 1977.

There was little in the way of new ideas presented in the document and in any case the surge in world commodity prices around this time took some of the pressure off CAP reform. However, in 1974, following the request of the newly elected British (Labour) Government to renegotiate the terms of entry, the Commission was asked to prepare a 'Stocktaking of the CAP' (Commission, 1975) which attempted more firmly than before to comment on the CAP relative to the objectives laid down by the Treaty of Rome. A careful reiteration of the five basic objectives of the CAP was supported by evidence to show how the CAP had (more or less) stuck to the these. However, four main problems were noted: (a) the failure of price policy to reflect the market situation; (b) the failure of structural policy to increase productivity and reduce regional disparity; (c) the continued threat to market unity posed by the MCA system; and (d) the growth in budgetary expenditure under the CAP.

Once again the problems had been noted but no solutions were offered and despite Britain's preoccupation with the CAP, agriculture was not a key issue in the final renegotiation terms agreed in Dublin in 1975. High world prices for cereals, sugar and animal feed protein made the CAP appear less protectionist and again blunted criticism, condemning the issues raised by the Commission to scant recognition within the Community (Harris *et al.,* 1983).

It was not until 1979, when reform proposals were triggered by the dangers resulting from CAP expenditure growing faster than own resources, that the Commission produced a further communiqué — 'Changes in the Common Agricultural Policy to help balance the markets and streamline expenditure' (Commission, 1979), which effectively marked the beginning of a number of serious attempts to modify the CAP in order to counter the growth of structural surpluses and save it from the pressure to reduce expenditure. Thus, two main proposals were made: (a) to move towards closer market balance, especially for milk and sugar; and (b) to enforce producer participation in the cost of surplus disposal.

What followed was a number of Commission documents concentrating explicitly on the method and nature of reform, which served to encourage a concerted effort among the Community's policy-makers to consider an alternative method of agricultural support in the EC. The issue of CAP reform is analysed and interpreted in Chapter 4, but it is important to note here that it was not the initiatives of the Commission which ultimately provoked the consideration of policy reform by the member states, but the real and tangible threat of bankruptcy as a result of unabated CAP expenditure.

All of the Commission's early reform documents recognised the budgetary and market problems resulting from the CAP and stressed the need to reduce market imbalances and cut back on agricultural expenditure. On the other hand, the Commission took great pains to reiterate the goals of the CAP and its support of price guarantees as the central policy. Thus, the initiatives emanating from the Commission, during this period, reflected an (inevitable) ambivalence in attempting to accommodate a diverse range of national perspectives towards agricultural policy and the dichotomy of seeking reform without changing the basic structure of agricultural support accorded to the Nine and the sanctity of the three principles of the CAP.

The Nine become Twelve — agricultural problems of Mediterranean enlargement

On January 1, 1986, five years after the official entry of Greece into the EC and two years after the end of the Greek accession period, Spain and Portugal became full members of the Community. Unlike negotiations with Greece, neither party was particularly anxious to reach an agreement, with the Treaty relating to Spain and Portugal being finally agreed eight years after the start of negotiations. Similarly, while for most products Greece fully adopted the CAP within five years, the transition period for Spain and Portugal was extended to ten years (although eventually accelerated a little).

This contrast in negotiations over the entry of the Three is partly explained by the particular political circumstances dominating Greece's attitude towards European integration and also by the fact that, while Greek accession was not expected to rock the Community boat too much, the further expansion of the Community towards the Mediterranean, in particular to include the greater political and economic power of Spain, threatened significantly to alter the course of European integration in the future. This factor alone necessitated the more careful consideration of the implications of accession, as well as a more structured and gradual process of integration of the Spanish and Portuguese economies within the Community.

Greek involvement with the Community began as early as 1962, with an association agreement, which established favourable arrangements (reduced tariffs, increased import quotas, and so on) for Greek exports to the Community. However, it was in 1972, following the Paris Summit, when the Six agreed a global Mediterranean policy, that the increased involvement of Mediterranean countries was provided for, with a view to creating a free trade area in industrial goods between the Community and each of the Mediterranean countries, with reciprocal tariff reductions, yet maintaining Community preference for agricultural products, with only limited *mutual* concessions

Following the events of July 1974 in Cyprus which led to the fall of the Greek military government, the new (civilian) Greek Government, headed by Constantine Karamanlis, declared its firm intention of joining the EC, on the grounds that accession would help the Greek economy, consolidate the country's democracy and (arguably most important at the time) strengthen its defences and its position against Turkey (Tsakaloyannis, 1980).

Negotiations on Greek accession began in the mid-1970s, with the Greek Government prepared to take the economic risks of speeding up the transitional period in order to profit from the political security which membership of the EC would, it hoped, ensure. From 1975, Greek exports to the EC were exempt from tariffs (apart from compensatory taxes on agricultural products), as were two thirds of exports from the EC to Greece (the other third accounting for 56% of customs duties), and negotiations were enthusiastically pursued during the late 1970s, with the Greek Government making substantial economic concessions (Tsakaloyannis, 1983). Full membership was finally achieved in 1981 with the alignment of prices, the reduction of duties and the application of the common external tariff established just three years later, in 1984.

The net impact of the adoption of the CAP on the Greek economy in general and the agricultural sector in particular is unclear. Before accession, some commentators (for example, Christou and Sarris, 1980, Sarris, 1984) argued that Greek farmers would lose out from the adoption of the CAP, as the prices of many agricultural products would need to be reduced. More recent studies (Georgakopoulos, 1988, 1990), conducted after accession, indicate that farmers have not been so adversely affected, but that higher prices for many agricultural products have contributed significantly to the inflationary pressures in the Greek economy and had a negative impact on consumption of meat, milk and sugar.

The apparent lack of enthusiasm of the Nine towards Greek entry into the EC was largely due to the potential threat which the Mediterranean countries in general and Greece in particular, posed for the future of the CAP and the development of agricultural expenditure under the Community's financial umbrella. Table 2.7 illustrates the relatively important role which agriculture plays in the Greek economy, with over a fifth of the working population involved in agriculture, 14% of GDP generated within the agricultural sector and agricultural exports accounting for 28% of the total value of Greek exports. However, production costs (even in the most efficient areas) are higher in Greece than in most of the member states and the high labour/land ratio reflects the proliferation of small holdings (Siotis, 1983).

Table 2.7: The importance of agriculture in Greece, Spain and Portugal.

	Greece	Spain	Portugal
Percentage of total employment in agriculture (1991)	21.6	10.7	17.5
Agricultural area as a percentage of the EC(12) (1991)	4.3	20.4	3.4
Agricultural population as a percentage of the EC(12) (1991)	9.5	16.3	10.2
Agricultural exports as a percentage of total exports (1991)	28.3	15.6	12.3
Agricultural imports as a percentage of total imports (1991)	12.3	18.3	26.2
Agriculture as a percentage of GDP (1990)	13.9	4.6	5.3

Source: Commission, 1993.

Perhaps the biggest fears within the Community over the impact of Greek accession were voiced by the French and Italian Governments who anticipated Greek competition over particular products, such as fruit and vegetables. This was a fast growing sector during the 1970s, with its share in total agricultural production rising from 15% in 1972 to 25% in 1978. However, while labour costs may be lower in Greece than elsewhere in the Community, the level of production per worker is also lower, making labour costs per unit of output comparable to those in the rest of the EC (Pepelasis, 1983).

While Greek accession added 7% to the total utilised area of the community and 13% to the labour force, it only contributed a further 4% to the value of agricultural production. The relatively small size of Greek agriculture and the degree of complementarity between agricultural production in Greece and that in the rest of the EC as well as the lack of political power within the Council of Ministers, leaves Greece playing a relatively small role, individually, in the decision-making process of the Community. However, as one of a growing number of Mediterranean countries incorporated within the EC, the *collective* action of Greece, Spain and Portugal has had a more significant impact, by shifting the geographical and political balance of Community policies.

Portuguese involvement in the European Movement is a relatively recent phenomenon, with the African colonies generally regarded, up to the 1950s, as a sufficient outlet for future economic expansion (Cravinho, 1983). However, the first step towards economic integration with the rest of Europe came with Portugal's membership of EFTA which if nothing else highlighted the importance of European markets to the Portuguese economy. EFTA's share of Portuguese export markets increased from 18% in 1958 to 35% in 1972 and with the importance of colonial markets declining from around 25% to 19% over the same period, the ruling elite was forced to accept the inevitability of increased involvement with the EC.

This became even more necessary after the first enlargement, which took Portugal's biggest customer, Britain, out of EFTA and into the EC. Prior to British accession, the UK represented 52% of Portugal's EFTA markets and along with Denmark over 21% of total Portuguese trade. Thus, overnight, Portugal had to adjust her trading pattern or risk a dramatic loss of export earnings. The initial result was the association agreement, signed in 1972, which enabled Portugal to maintain trade with the former EFTA members, at preferential tariff rates, progressively over a five year period, to 1977, when negotiations on full membership began.

The Nine played an important role in facilitating Portuguese involvement in the Community by offering financial aid during the mid-1970s, to help avoid the eruption of civil war and to support the pluralistic democracy (Cravinho, 1983). The first emergency loan of 180 million ua was granted in 1975, and as a result of the increased co-operation between Portugal and its European neighbours, the first constitutional (socialist) Government announced in its programme, in 1976, the intention to begin negotiations on EC accession immediately.

With the exception of the Communists, most of the Portuguese political parties were in favour of accession, largely due to the potential economic benefits which were expected to accrue to the Community's poorest member state. The Commission was less concerned with the economic aspects of Portuguese accession, but rather more interested in aiding the democratisation process, although the development of a global Mediterranean policy was facilitated by the inclusion of Portugal, along with Greece and Spain, *within* the Community boundary.

A major factor contributing to Portugal's state of relative poverty is her dependence on a largely stagnant agricultural sector. As in other Mediterranean countries, climatic conditions have contributed to poor agricultural performance, but more significantly, low productivity levels (of both land and labour), a lack of entrepreneurial talent and poor cropping systems, have "transformed Portuguese agriculture from a dynamic sector into a static pool of resources" (De Abreu, 1983).

Production growth rates were more or less static during the 1970s, with the notable exception of the fruit sector, which grew at an average annual rate of 6%. Yields of most crops, particularly cereals, have similarly stayed more or less constant, while the arable area has declined, largely due to out-migration and the limitations imposed by increased (albeit gradual) mechanisation. Traditional crop rotations have failed to meet the pattern of demand, and the level of overall food production has steadily declined from 1960, turning a small food surplus into a significant food trade deficit of over one billion ecu in 1987, with agricultural imports representing 26% of the total value of imports (see Table 2.7).

Despite limited land reform, aimed at encouraging the growth of co-operatives and self-managed farm units, the more recent development of part-time farming (particularly in areas of industrialisation) and the lack of financial aid and co-operation between the Government and the farm organisations has left the structure of Portuguese agriculture dictated by very small units — of the 350,000 holdings, over 75% have less than five hectares.

It is evident that Portugal has faced a number of problems following accession which have many similarities with those experienced by Greece. The agricultural sectors of the two countries have much in common, with low productivity levels, a proliferation of very small units, and concentration on the production of traditional Mediterranean products. However, as with Greece, the agricultural picture is not all gloomy. There is an element of complementarity between the pattern of production in Portugal and the Nine, with potential for exports in wine, nuts, tomatoes and some fruits.

Thus, as a political lightweight, of little economic significance, Portugal, like its Mediterranean neighbour Greece, has had a negligible impact on the nature of Community policies or the decision-making process thereof. However, the inclusion of Portuguese agriculture in the CAP has increased the need for structural reform and necessitated a broader 'Communautaire' attitude amongst the Nine, with the recognition of the need to channel considerable financial aid, through the Community budget, to the poorer areas of the Community.

With Greece and Portugal the two smallest members of the Twelve (with the exception of Luxembourg), the Spanish economy is by far the strongest of the Three, being the fifth largest in the Community, with significant export potential in certain agricultural products (notably fruit and vegetables, wine and olive oil) despite being a net food importer.

Jordan (1985) highlights the extent to which agricultural factors contributed to the difficulty of negotiations over Spanish entry, particularly the impact on the Community's self-sufficiency (and thus the level of agricultural expenditure) in certain agricultural products — Spain has a heavy citrus fruit surplus (250% self-sufficiency) and is more than self-sufficient in most other fruit and vegetables. With the EC as the biggest single fruit and vegetable importer in the world and Spain the largest single exporter, these high levels of self-sufficiency contributed to the rise in the Community's self-sufficiency in citrus fruits from 70% in 1980 to 76% in 1986/7 and from 145% to 182% in processed tomatoes over the same period. Significantly, subsequent reforms to the market regulation of fruit and vegetables (discussed below) has resulted in a return to pre-accession self-sufficiency levels of 71% for citrus and 119% for processed tomatoes (Commission, 1993)

Of all the potential benefits which accession offered to Spain, the financial transfers and expansion of agricultural trade through the adoption of the CAP offered the greatest attraction. Table 2.7 illustrates the importance of agriculture in the Spanish economy (though less than in Greece and Portugal), with some 11% of the active population employed in agriculture and 5% of GDP emanating from that sector. In relation to the Community, Spanish entry has increased the total agricultural population by 17%, the number of holdings by 22% and total agricultural output by around 11% (Commission, 1993).

While large farm units predominate in the south, Spanish agriculture in general is typical of the Mediterranean region in its structural imbalance, with 55% of holdings less than five hectares in size. Moreover, income disparities are particularly marked in certain agricultural areas — the 'poverty pockets' as described by Rodriguez (1983).

However, price incentives for the more efficient fruit and vegetable producers of the south has led to increased production in this sector, which has been reflected by a further improvement in the agricultural trade balance. Prior to accession, despite tariff barriers, less than 60% of Spanish agricultural exports went to EC countries. With the removal of trade controls and the introduction of higher levels of support, the proportion of intra-EC exports has risen to almost two thirds.

This inevitability was a major factor in prolonging the progress of negotiations on Spanish accession, with several member states, particularly those threatened by Spanish competition in fruit, vegetables and wine (France, Italy and the Netherlands) warning of the damaging consequences of opening the Community doors to the heart of the Mediterranean. The position of those countries opposed to Spanish accession is aptly summed up by the concluding remarks of a report published by the Conseil National des Jeunes Agriculteurs (CNJA), the French young farmers organisation, which stated: "Spanish accession is a mistake for all...the conditions that would make the EC enlargement a benefit for Europe are not present...when a ship sinks, it is better not to take on board more passengers" cit Rodriguez (1983).

So great was the concern of the Nine for the impact of the Iberian accession that a special monitoring system — the Supplementary Trade Mechanism — was introduced to monitor and control the trade between Spain and Portugal and the Nine in those products identified as 'sensitive' (Commission, 1986; Ritson and Swinbank, 1994). It is therefore perhaps not surprising that over the period 1984/5 to 1988 agricultural exports from the Ten to Spain increased by 245% whilst Spanish agricultural exports to the Ten increased by just 57% (Hine, 1989). Moreover, within three years of accession, Spanish growers saw the potentially lucrative EC markets in citrus and wine come under further threat, as the Commission launched a new attack on the recurring problems of surplus production and excessive budgetary expenditure.

1986–1990: 'Budgetary stabilisers' and the start of a new era of CAP reform

The Community's inability to control the growth of agricultural surpluses was evident long before Iberian accession began to impact on the Community budget. In 1984, the introduction of milk quotas signalled the beginning of a sustained effort by the Commission to come to grips with the most serious budgetary crisis in the EC's history.

The principle of producer co-responsibility had been long established in the sugar regime, and was extended to milk and cereals towards the end of the decade, with levies being deducted from milk, grain and sugar returns to help finance the cost of surplus production. Yet, low world prices and a weak dollar accentuated the cost of export refunds, at a time when the EC was tabling a reductionist position during the early discussion of the Uruguay GATT round.

The crisis reached a head at the summit meeting of December 1987, when the European Council failed to agree on measures necessary to balance the 1988 budget. This resulted in the application of budgetary twelfths from January 1988, whereby Community spending each month was limited to one twelfth of that in the 1987 budget. Under these arrangements support expenditure could only have been maintained until mid-year, when funds would have begun to run out. The issue was eventually resolved in February 1988, when a compromise was reached which involved the introduction of supply controls across all sectors and major changes in the way the Community would finance the CAP in the future.

The application of stabilisers led to the adoption of Maximum Guarantee Quantities (MGQs) for the majority of agricultural sectors, with automatic price adjustment mechanisms should the MGQs be exceeded. The details of the measures varied slightly from sector to sector (see Commission, 1990), but essentially, MGQs were agreed for the period 1988/9 to 1991/2, based on average production over the period 1984/5 to 1986/7. If actual production exceeded the MGQ, support prices in the following year were to be automatically reduced in proportion to the overshoot.

In addition to the stabilisers, the Council agreed a series of structural measures, including voluntary set-aside, extensification and diversification schemes, designed to facilitate the adjustment of supply to demand whilst compensating farmers for the loss of income, and raising the profile of environmental protection. On the revenue side, the Council agreed to increase the Community budget by enlarging the resource base to include a proportion of the gross national product of each member state. This fourth resource was calculated as the difference between the standard 1.4% of the assessment basis of VAT and 1.3% of gross national product and added 6,000 million ecu to the CAP budget.

In return for the increase in resources, the Council imposed a budgetary guideline on FEOGA (Guarantee) expenditure. This was set at 25,500 million ecu for 1988 and for subsequent years the rate of increase in the guideline was limited to 74% of the annual GNP growth rate in the Community. Two exceptional provisions were made outside the guideline: the first involved 5,600 million ecu for the disposal of accumulated stocks; the second involved the creation of a reserve of 1,000 million ecu to cover any additional costs of surplus disposal arising from unexpected movements in the ecu/dollar exchange rate.

Two important influences on this agreement are noteworthy. First, Commissioner Andriessen, like his predecessor Dalsager (who introduced milk quotas in 1984), was determined to have his own reform of the CAP. His ability to gain an increase in the Community's own resources at a time when member states were adopting a particularly nationalistic stance bears testimony to his skill as an arbitrator. Second, the British, who under Margaret Thatcher were determined to curb the cost of the CAP were instrumental in the decision to adopt the financial guideline and had earlier (1986) been the first to suggest the introduction of set-aside and the 'greening' of the CAP with the introduction of Environmentally Sensitive Areas (ESAs) in 1985 (Tracy, 1995).

The extent to which the stabilisers package was successful in curbing surplus production and budgetary expenditure in the short term is difficult to establish. Whilst production of certain commodities, notably oilseeds, was substantially affected, that of others, notably cereals continued to rise, as agricultural ministers consistently refused to implement price cuts of the required magnitude. Indeed, as the Commission itself acknowledged, the reduction in budgetary pressure at the end of the decade was probably due more to favourable world prices and a strengthening dollar than the effective implementation of the stabilisers (Commission, 1991).

In the longer term, the stabilisers were doomed to failure as they did not attack the underlying problem of support payments being linked to the quantity produced, which represented a permanent incentive to greater (and more intensive) production. With farm ministers lacking the political will to cut support prices sufficiently to establish market balance it was not surprising that by the turn of the decade the Community was lurching back into a surplus-induced budgetary crisis, exacerbated by the urgency of a GATT agreement.

Yet the tone of the much quoted 'Reflections' paper of the Commission (1991) suggests that the experience gained through the attempted implementation of the stabilisers did much to strengthen the Commission's resolve in seeking a more radical reform of the CAP, over which they would have much greater influence and control. The following extract illustrates very clearly the frustration and exasperation which the Commission felt over the failure of the stabiliser package and the perceived need for a fundamental review of the principles and objectives of the CAP, the outcome of which is discussed in the following chapter:

> The reforms of the years 85/88 have not been implemented and are themselves incomplete. It is not surprising that under these conditions the CAP finds itself once again confronted with a serious crisis...It appears in these conditions that the Community's agricultural policy cannot avoid a succession of increasingly serious crises unless its mechanisms are fundamentally reviewed so as to adapt them to a situation different from that of the sixties. The Commission considers therefore that the time has come to stimulate a reflection on the objectives of the Community's Agricultural Policy and on the principles that should guide the future development of the CAP.
>
> (Commission, 1991).

The outcome of this reflection was the so-called Mac Sharry reform of the CAP, discussed in Chapter 5, and characterised in this book as 'the new CAP'.

Conclusion

The development of the CAP from the post war period to the early nineties reflects the emergence of a wide range of contrasting and often conflicting pressures on the process of European integration. These pressures have reflected both the political constraints facing member states, in particular the ceding of sovereignty and the defence of national interests and the economic consequences of adopting a common policy of open-ended price support in a sector which has responded beyond all expectations.

Given the disparity between the agricultural sectors of six, nine and twelve member states, it is perhaps surprising that the basic principles and mechanisms of the CAP remained largely intact during the first thirty years of its evolution. The period following Iberian accession saw a further growth in the pressure for CAP reform, both within the EC as the cost of agricultural support has continued to rise, and from outside the Community, as the struggle to maintain exports markets in the face of subsidised production intensified. Yet the CAP remains as the single most important common policy, still accounting for the lion's share of the EC budget and symbolic of the overwhelming desire to achieve common solutions to the Community's problems.

That there was still a common policy for agriculture in the Community of twelve is as much the result of the importance which the CAP attained as a symbol of European unity as it was of the desire of member states to respect the objectives for agricultural policy as outlined in the Treaty of Rome. In this respect the developments of recent years, namely the completion of the Single Market, the quest for economic and monetary union and the key role of agricultural policy reform in the Uruguay round of the GATT, have all served to reduce the symbolic importance of the CAP and thereby facilitate, if not necessitate, effective rather than cosmetic changes to the way in which the Community — and now the European Union — chooses to support its farming population.

References

Broad, R. and Jarrett, R. (1972) *Community Europe Today*, Oswald Wolf, London, 161–171.

Buckwell, A., Harvey, D., Thomson, K. and Parton, K. (1982) *The Costs of the Common Agricultural Policy,* Croom Helm, London, 25–66.

Capstick, C. (1991) British agricultural policy under the CAP. In: Ritson, C. and Harvey, D. (eds) *The Common Agricultural Policy and the World Economy*, CAB International, Wallingford, 71–88.

Christou, C. and Sarris, A. (1980) The impact on Greek agriculture from membership in the European Economic Community, *European Economic Review*, 14, 159–188.

Clerc, F. (1979) Attitudes Francaises vis-a-vis de la Politique Agricole Commune. In: Tracy, M. and Hodac, I. (eds) *Prospects for Agriculture in the European Economic Community,* College of Europe, Brussels, 353–363.

Comite Intergouvernemental (1956) *Rapport des chefs de delegation aux Ministres des affaires etrangeres* (The Spaak report), Brussels.

Commission of the European Communities (1958) *First General Report on the Activities of the Community,* Brussels.

Commission of the European Communities (1958a) *Receuil des documents de la conference agricole des etats membres de la communaute economique europeenne a Stresa au 12 Juillet 1958*, Brussels.

Commission of the European Communities (1968) *Memorandum on the reform of agriculture in the European Community,* COM (68) 1000, Brussels.

Commission of the European Communities (1973) *Improvements of the Common Agricultural Policy,* COM (73) 1850, Brussels.

Commission of the European Communities (1975) *Stocktaking of the Common Agricultural Policy,* COM (75) 100, Brussels.

Commission of the European Communities (1979) *Changes in the Common Agricultural Policy to help balance the markets and streamline expenditure,* COM (79) 710, Brussels.

Commission of the European Communities (1986) *Agricultural aspects of Community enlargement to include Portugal and Spain,* Green Europe, No.1, 1986, Brussels.

Commission of the European Communities (1990) *Restoring equilibrium on the agricultural markets*, Green Europe, No. 3, 1990, Brussels.

Commission of the European Communities (1991) *The development and future of the CAP – Reflections paper of the Commission*, COM (91) 100, Brussels.

Commission of the European Communities (1993) *Annual situation in the Community: 1992 Report*, Brussels.

Cravinho, J. (1983) Portugal: Characteristics and motives for entry. In: Sampedro, J. and Payno, J. (eds) *The Second Enlargement of the European Community: A Case Study of Greece, Portugal and Spain,* Macmillan, London, 131–147.

De Abreu, A.T. (1983) Portugal: The Agricultural Sector. In Sampedro, J. and Payno, J. (eds) *The Second Enlargement of the European Community: A Case Study of Greece, Portugal and Spain,* Macmillan, London, 149–165.

De La Mahotiere, S. (1970) *Towards one Europe,* Pelican, Harmondsworth, 139–175.

De Vries, J. (1975) Benelux 1920–1970. In: Cipolla, C. (ed) 1978. *The Fontana Economic History of Europe,* 6(1), Fontana, London, 55–63.

European Communities (1987) *Treaties establishing the European Communities, Treaties amending these Treaties, Single European Act, Resolution-Declarations,* Office for Offical Publications, 1, 207–257 (Articles 1–43 of the Treaty of Rome), Luxembourg.

Fennell, R. (1987) *The Common Agricultural Policy of the European Community,* BSP Professional Books, Oxford, 1–15.

Georgakopoulos, T. (1988) The impact of accession on agricultural incomes in Greece, *European Review of Agricultural Economics*, 15(1), 79–88.

Georgakopoulos, T. (1990) The impact of accession on food prices, inflation and food consumption in Greece, *European Review of Agricultural Economics*, 17(4), 485–493.

Haagerup, N. and Thune, C. (1983) Denmark: The European Pragmatist. In: Hill, C. (ed) *National Foreign Policies and European Political Co-operation,* Allen and Unwin, London, 106–118.

Harris, S., Swinbank, A. and Wilkinson, G. (1983) *The Food and Farm Policies of the European Community,* Wiley, Chichester.

Harvey, D.R. (1982) National interests and the CAP, *Food Policy,* 7(3), 174–190.

Heath, E. (1982) Interview in: *Europe 82,* No.4, 19.

Hine, R. (1989). Customs union enlargement and adjustment: Spain's accession to the European Community, *Journal of Common Market Studies*, 28(1), 1–27.

Jordan, J. (1985) *The implications of Spain's accession to the EEC and the citrus fruit sector,* unpublished MA Dissertation, University of Reading, 3–10.

Kindleberger, C. (1965) The Post-War resurgence of the French economy. In: Hoffman, S. (ed.) *In search of France,* Harper and Row, London.

Kitzinger, U. (1967) *The European Common Market and Community,* Routledge and Kegan Paul, London, 33–198.

Koester, U. (1977) The redistribution effects of the common agricultural financial system, *European Review of Agricultural Economics,* 4(4), 321–345.

Lindberg, L. (1963) *The Political Dynamics of European Economic Integration*, Oxford University Press, Oxford, 261–283.

Mansholt, S. (1972) The Promised Land for a Community. In: Barber, J. and Reed, B. (eds) 1973. *European Community: Vision and Reality,* Croom Helm, London, 315–318 .

Marsh, J. (1979) United Kingdom Attitudes to the CAP. In: Tracy, M. and Hodac, I. (eds) *Prospects for Agriculture in the European Economic Comunity*, College of Europe, Brussels, 364-377.

Marsh, J. and Ritson, C. (1971) *Agricultural Policy and the Common Market.* Chatham House, London.

Marsh, J. and Swanney, P. (1980) *Agriculture and the European Community,* Allen and Unwin, London, 11–37.

Neville-Rolfe, E. (1984) *The Politics of Agriculture in the European Community,* European Centre for Political Studies.

Parker, G. (1979) *The Countries of Contemporary Europe,* Macmillan, London.

Pearce, J. (1983) The Common Agricultural Policy: The accumulation of special interests. In: Wallace, H., Wallace, W. and Webb, C. (eds) *Policy-making in the European Community,* Wiley, Chichester, 143–159.

Pepelasis, A. (1983) The implications of accession for the Greek agricultural sector. In: Sampedro, J. and Payno, J. (eds) *The Second Enlargement of the European Community: A Case Study of Greece, Portugal and Spain,* Macmillan, London, 70–84.

Raup, P. (1970) Constraints and potentials in agriculture. In: Beck, R. *et al. The changing structure of Europe,* University of Minnesota Press, Ann Arbor, 126–170.

Ritson, C. (1979) An economic interpretation of national attitudes to CAP prices, Discussion Paper prepared for the second Wageningen seminar on

The role of the economist in policy formulation with respect to the European Community's Common Agricultural Policy.

Ritson, C. (1982) Impact on Agriculture. In: Seers, D. and Vaitsos, C. (eds) *The Second Enlargement of the EEC: The Integration of Unequal Partners,* St. Martin's Press, New York, 93–106.

Ritson, C. and Fearne, A.P. (1984) Long term goals for the CAP, *European Review of Agricultural Economics*, 11(2), 207-216.

Ritson, C. and Swinbank, A. (1994) *Prospects for exports of fruit and vegetables to the European Community after 1992*, FAO, Rome.

Ritson, C. and Tangermann, S. (1979) The economics and politics of Monetary Compensatory Amounts, *European Review of Agricultural Economics,* 6(2), 119–130.

Rodriguez, J. (1983) Spain: The agricultural sector. In: Sampedro, J. and Payno, J. (eds) *The Second Enlargement of the European Community: A Case Study of Greece, Portugal and Spain,* Macmillan, London, 210–221.

Sarris, A. (1984) Agricultural problems in EEC enlargement, *European Review of Agricultural Economics*, 11(2), 195–205.

Siotis, J. (1983) Greece: Characteristics and motives for entry. In: Sampedro, J. and Payno, J. (eds) *The Second Enlargement of the European Community: A Case Study of Greece, Portugal and Spain,* Macmillan, London, 57–67.

Swann, D. (1981) *The economics of the Common Market,* Penguin, Harmondsworth.

Tangermann, S. (1979) Germany's position on the CAP — is it all the German's fault? In: Tracy, M. and Hodac, I. (eds) *Prospects for Agriculture in the European Economic Comunity*, College of Europe, Brussels, 395–404.

Tangermann, S. (1980) National attitudes to the CAP, unofficial translation by Rollo J. of German article *'Agrarpolitische Positionen in den Mitgliedslanden der EG und den Institutionen'*.

Tracy, M. (1984) Issues of agricultural policy in a historical perspective, *Journal of Agricultural Economics,* 35(3), 307–317.

Tracy, M. (1989) *Government and Agriculture in Western Europe — 1880–1988,* Harvester Wheatsheaf, Brighton, 243–356.

Tracy, M. (1994) The spirit of Stresa, *European Review of Agricultural Economics*, 21, 357–374.

Tracy, M. (1995) *Major influences on the CAP — Attitudes of the various member states and the international context*, Summary paper for the Department of Environment Seminar, London, January 26th, 1995.

Tsakaloyannis, P. (1980) The European Community and the Greek-Turkish dispute, *Journal of Common Market Studies,* 19(1), 35–54.

Tsakaloyannis, P. (1983) Greece: old problems, new prospects. In: Hill, C. (ed) *National Foreign Policies and European Political Co-operation*, Allen and Unwin, London, 121–134.

---

[1]The role of this and various other committees is discussed in Chapter 3 on the CAP decision-making process.

Chapter 3

The CAP Decision-making Process

Alan Swinbank

Introduction

As explained in the previous chapter, the Common Agricultural Policy (CAP) was expressly provided for in the Treaty of Rome, and although the European Economic Community (EEC) of the late 1950s has evolved into the European Union (EU) of the mid-1990s, those Treaty provisions of 1957 remain the legal foundation of the CAP today. Furthermore, the formal process of CAP decision-making remains unaltered, even though in many other respects the powers and processes of the EU are very different from those of the EEC. The role of this chapter is to explain the formal process of CAP decision-making, and to outline the institutional framework in which those decisions are taken. In a later chapter, Harvey explores the political and economic pressures which are brought to bear on the decision-making process, and outlines an economic framework for examining *rent-seeking* behaviour.

From EEC to EU

During the course of the 1950s six west European nations came together to form three European communities:

- the European Coal and Steel Community (ECSC), which began life on 1 January 1952;
- the European Economic Community (EEC); and
- the European Atomic Energy Community (EURATOM), both of which came in to force on 1 January 1958.

EURATOM was concerned with the peaceful use of atomic energy, whilst the EEC, in "establishing a common market and progressively approximating the economic policies" of its member states, had as its task "to promote throughout the Community a harmonious development of economic activities, a continuous and balanced expansion, an increase in stability, and accelerated raising of the standard of living, and closer relations between the states belonging to it". In its early days, the emphasis was placed upon establishing a common market, in particular by the creation of a customs union and the elimination of technical barriers to trade between the member states. The Treaty establishing the EEC specified that the common market was to include agriculture and trade in agricultural products, and provided for the creation of a common agricultural policy. Fearne's chapter has explained how those Treaty provisions were implemented and how the CAP evolved.

All three communities had the same six states as members; Belgium, West Germany,[1] Italy, France, Luxembourg and The Netherlands, and each had similar institutions:

- an assembly parliament;
- a council;
- a commission (or in the case of the ECSC, a high authority); and
- a court of justice.

From the outset a single Parliament and Court of Justice served all three Communities, and the EEC and EURATOM also shared an Economic and Social Committee (Ecosoc). The merger Treaty of 1965 created a single Commission of the European Communities and a single Council of the European Communities, and thus in effect the EC was born. The European Communities, often referred to in the singular as the European Community (EC), derived its authority from the ECSC, the EEC and the EURATOM treaties, and still does.

These three treaties have been amended on a number of occasions. Fairly technical changes were made for the accession of Denmark, Ireland and the UK in 1973, Greece in 1981, Spain and Portugal in 1986, and Austria, Finland and Sweden in 1995. A court of auditors was established. More fundamental changes came in 1987 with the implementation of the Single European Act. This:

- extended the competencies of the Community,[2] in particular into the sphere of R&D and the environment;
- extended the range of topics on which decisions could be reached by the Council by qualified majority vote;
- enhanced the power of the Parliament in certain spheres of policy; and
- committed the member states to a new deadline of 31 December 1992 for the creation of an internal market: "an area without internal frontiers in which the free movement of goods, persons, services and capital is ensured" in accordance with the provisions of the EEC Treaty.

Decision-making on the CAP was not directly affected by the Single European Act, though closely associated areas of policy were, such as measures concerned with environmental protection, and the food law harmonisation programme which became subject to greater parliamentary control and qualified majority voting. Policy changes were, however, prompted. Thus, the 1992 programme meant that the old 'green money' system (see Chapter 6), involving border taxes and subsidies (monetary compensatory amounts) on intra-Community trade, had to be revamped.

The EU is the creation of the Maastricht Treaty on European Union signed at Maastricht in The Netherlands. It is the product of an intergovernmental conference which had tried to reconcile the conflicting views of those who wanted the Community to adopt more federalist characteristics and move towards a united states of Europe, and those who saw it, essentially, as a free trade agreement among sovereign states. In Denmark, Germany, France and the UK, significant opposition had to be overcome before those countries could ratify the Treaty and it could come into force.

The Maastricht Treaty modified, but did not displace or abolish, the existing treaties that had established the European Communities, and created two new 'pillars' of joint endeavour between the member states. These two new pillars are concerned with a "common foreign and security policy", and "co-operation in the fields of justice and home affairs". In these two respects, the member states work together, but the institutions created by the treaties establishing the European Communities have no formal role.

The title of the Treaty establishing an EEC was amended to delete the word 'Economic',[3] the Treaty was amended to embrace the rather ill-defined concept of *subsidiarity*,[4] and nationals of the member states also became citizens of the Union. There was a further extension of the EC's competence — in particular in relation to economic and monetary policy, and the objective of achieving economic and monetary union, but only eleven of the then member states were willing to be bound by an agreement on social policy. Thus the social chapter lies outside the EC Treaty. The European Parliament was given additional powers, but not with respect to the CAP, and there was a further extension of qualified majority voting.

When the Treaty on European Union came into force on 1 November 1993, the Council assumed the title 'Council of the European Union', and the Commission the title 'European Commission'. However, it is as the Council of the European Communities, and the Commission of the European Communities that they legislate on and implement the CAP and the other policies referred to in this book. Furthermore, it is still the Treaty establishing the EEC, signed in 1957, that remains the legal basis for the CAP, the customs union and the EU's trade policy, its food law harmonisation programme, and its legislation on environmental matters. The acronyms can be confusing! CAP legislation up to 31 October 1993, is in the form of *EEC* regulations and directives, whereas that put in place since 1 November 1993 is expressed in terms of *EC* regulations and directives, as a consequence of dropping the word 'economic' from the title of the Treaty.

As will be appreciated, the EU has evolved over the years, and the political debates that have accompanied that constitutional change have usually been intense. Already the preliminary skirmishes of the next intergovernmental conference, are underway; and many people will read this text during the course of those negotiations, or when the next Treaty has been approved. To date, however, the decision-making provisions for the CAP remain unchanged from those laid down in 1957. But it is not inconceivable that they will be different following the intergovernmental conference.

At the time of writing a potential formal, and informal, agenda would include the following interlinked items:

- *The federal character of the EU*, and its commitment to policies such as economic and monetary union.
- *The policy coverage of the EU.* Is it the intent that the same policies should apply throughout the EU, or might there be an inner core with a looser grouping of countries on the periphery, or a Europe of 'variable geometry' in which a member state need not participate in all of the EU's policies? Taking the precedent of the social chapter, could a member state opt out of the Common Fisheries Policy, or the CAP?
- *The size of the EU.* Several countries in central Europe, and the Mediterranean have been promised membership after the 1996 intergovernmental conference, but this may well necessitate policy change.[5] The conventional wisdom is that there will have to be further reform of the CAP before the next enlargement, as extension of the present policy to central Europe will be impossibly costly to the EU's budget, and incompatible with the EU's GATT/WTO obligations. This issue is explored further in Chapter 14.
- *The EU's institutional structure and its decision-making processes.* As will be shown below, the EU's institutional structures and decision-making procedures were barely adequate for a community of 12. The alpine and Nordic enlargement of 1995 is expected to exhaust the present arrangements, and a larger EU cannot readily be contemplated with the present decision-making structures. But small member states will be reluctant to give up their council seats and commissioners, and others will be reluctant to accept a weakening of their power to block decisions in the Council under qualified majority voting, or see the powers of the European Parliament further extended.

The institutions

As amended by the Maastricht Treaty, the EU has five institutions:

- the European Parliament,
- the Council,
- the Commission,
- the Court of Auditors, and
- the Court of Justice.

The Council and Commission are assisted by an economic and social committee, and a Committee of the Regions, which both act in an advisory capacity.

The overarching provisions of the Treaty on European Union also provide for a European council. The European Council brings together the Heads of State or of Government,[6] and it discusses matters relating to the wider interests of the EU. As such, it is not a legally constituted meeting of the Council of Ministers authorised under the three treaties establishing the European Communities. The European Council has, *de facto*, existed for many years and in fact it has proved to be an important political arena in which the member states have forged new policies, and brokered agreements. Many key developments in the CAP have been agreed in the European Council, to be subsequently enacted into law in the Council of the European Communities. ('The Council of Europe', and the 'European Commission and Court of Human Rights', are separate entities and not part of the EU, even though press coverage often fails recognise this fact.)

All EU legislation relating to the CAP — but not all EU legislation of relevance to agriculture, the food industries and the environment — is enacted on the basis of article 43(2) of the Treaty of Rome which states:

> The Council shall, on a proposal from the Commission and after consulting the European Parliament...by a qualified majority...make regulations, issue directives, or take decisions, without prejudice to any recommendations it may also make.

It is clear from this text that the Council and Commission are the two important institutions in creating CAP legislation, and that the formal role of the Parliament is slight. To use the phraseology often deployed in EU circles: the Commission *proposes* and the Council *disposes*. Note that the Council can only legislate on the basis of a proposal from the Commission. It can, however, adopt an amended version of a Commission proposal if it is able to secure unanimity amongst its members.

The Council

The Council, which is served by a permanent secretariat, is made up of a ministerial representative of each of the member states. Thus, fairly frequently, 15 ministers meet as the Council of the European Union, to deal with matters that the EU is competent to discuss, as provided for in the treaties. The items on the agenda of a particular Council will tend to reflect one arena of EU policy, such as the CAP. The ministers that attend that meeting will usually be ministers of agriculture, whereas a Council meeting discussing EU transport policy will be attended by ministers of transport. Thus, *de facto*, we have a Council of Agriculture Ministers[7] and a Council of Transport Ministers; though all meetings of the Council of the European Union have equal decision-making powers. First among equals, in the sense that it is usually assumed it will provide some sort of policy lead, is the Council composed of ministers of foreign affairs, and usually referred to as the General Affairs Council.

The Presidency

The chairmanship (Presidency) of the Council rotates on a six-monthly basis, as the delegations move clockwise round the Council table. The cycle is illustrated in Table 3.1. Column 1 is fairly clear. With 12 member states there was a six-year cycle, with Belgium taking the Presidency for the first six months of 1987, and the UK the last six months of 1992. The countries are listed in alphabetical order, depending upon the spelling of the country's name in its own language. This had been the pattern since the formation of the EC.

However, because of holiday periods, the first six months of the year is more important than the latter part of the year in the EU's legislative programme; and with an even number of member states, the burden or prestige of the Presidency is unevenly spread. Thus, in January 1993 a new cycle began in which, within the year, the Presidency was switched. Consequently Denmark served before Belgium, and Greece before Germany. This new cycle has now been interrupted by the accession of three new member states, but will continue until the end of Luxembourg's Presidency in December 1997 when on the basis of the alphabet Austria (Österreich) would have been due to take the chair. The Council's current plan is that from then on the Presidency should rotate between large and small member states.

Table 3.1: Presidency of the Council of Ministers.

1987	I	Belgium	1993	I	Denmark
	II	Denmark		II	Belgium
1988	I	Germany (Deutschland)	1994	I	Greece
	II	Greece (Ellas)		II	Germany
1989	I	Spain (España)	1995	I	France
	II	France		II	Spain
1990	I	Ireland	1996	I	Italy
	II	Italy		II	Ireland
1991	I	Luxembourg	1997	I	Netherlands
	II	Netherlands (Nederland)		II	Luxembourg
1992	I	Portugal			
	II	United Kingdom			

There are actually 17 seats around the Council table: the President-in-Office at one end, a commissioner at the other, and 15 ministerial representatives of the 15 member states. On assuming the Presidency, the minister's deputy takes over the task of representing the interests of the member state. The Presidency imposes a considerable burden on a member state, for although the Council secretariat services the meetings, a Council President's home ministry must provide additional briefing papers and help its minister in the round of bilateral negotiations with each of the other member states which are a frequent feature of deliberation on contentious issues. Some member states have a better reputation than do others in terms of their institutional capacity to service the Presidency.

Associated with the Presidency is some international and national prestige — the country holding the Presidency speaks on behalf of the EU for example — and some influence. A good deal can depend upon the competence of an individual minister to chair a meeting, but beyond that, and despite the impartial role that the Presidency assumes, the Presidency can influence the progression of legislation through the Council. On assuming the Presidency, the country's foreign minister will outline to the European Parliament the matters that it is hoped will be resolved during the country's six month tenure. With the Council secretariat the Presidency draws up the agenda for Council meetings (though other member states can insist on the inclusion of an item) and from the chair it can push a contentious item to conclusion by, for example, keeping the Council in session overnight, or insisting on a vote. A Presidency less committed to resolving a particular issue might be tempted to defer further discussion to the next meeting. A former UK permanent representative to the EU, Sir Michael Butler, has commented:

> It is almost impossible to get a conclusion out of a Council against the wishes of its President
>
> (Butler, 1986, p. 26).

In the same text he wrote:

> With a good brisk chairman and proper preparation, a lot can get settled at a Council. It is best when the Presidency and the key delegations concerned with the difficult items have done a lot of quiet work behind the scenes on compromise formulae. But without these favourable conditions, or if the problems involved are particularly intractable, the meetings can drag on far into the night. If the agenda is heavy, the Council will in any case have to continue on the following day. It is alleged that some delegations find it easier to settle in the middle of the night because it is not then possible to ring up their own President or Prime Minister. This may sometimes be a consideration, but not often. More usually the meetings go on because the President hopes to wear the other ministers down. This can be an incredibly boring and tiresome process.
>
> (Butler, 1986, pp. 79–80)

Most CAP measures are so pressing they cannot easily be set aside by a reluctant Presidency, but in other spheres the progression of legislation through the Council can be very slow. As Table 3.2 shows, the agriculture Council meets frequently, usually on a monthly basis, except in August, but more frequently during the annual farm price review. Thus ministers of agriculture have more opportunity to get to know their colleagues in the Council, than do — say — ministers of transport. Furthermore, ministers of agriculture tend to serve for long periods. Josef Ertle served as Germany's minister of agriculture from October 1969 to March 1983; and his successor, Ignaz Kiechle, from March 1983 to January 1993. Both men derived much of their political authority from farming interests in Bavaria, and represented a minority party in a succession of coalition governments. To a large extent, they determined Germany's farm policy, and were powerful exponents of the farm interest in the agriculture Council.

Farm ministers, of course, are answerable to their cabinet colleagues at home, but it is easy to see why many observers believe that the Council of Agriculture Ministers frequently acts as a Council *for* agriculture, rather than just another meeting of the Council which happens to be attended by ministers of agriculture. In its club-like atmosphere, with its frequent late night sittings during the annual farm price review, many suspect that the outcome of the agriculture Council's deliberations will tend to err in favour of the farm lobby, whereas if finance ministers were asked to deliberate on the same issue a rather different outcome would result.

Table 3.2: Meetings of the Council, 1993.

Composition	Number of Meetings
General Affairs*	20
Agriculture*	12
Economic and financial questions	11
Internal market	6
Environment**	6
Industry	5
Fisheries	5
Transport**	5
Research	4
Labour and social affairs	4
Telecommunications	3
Energy**	3
Education	2
Budget	2
Development co-operation	2
Consumer protection and information	2
Health	2
Cultural affairs	2
Justice and home affairs	1
Total	94

* Including a joint meeting on general affairs and agriculture.
** Involving one joint meeting on environment and transport, and one on environment and energy.

Source: General Secretariat, 1994, p. 52.

Qualified majority voting

As we saw above, as far as the CAP is concerned the Council is authorised to legislate by qualified majority on the basis of a Commission proposal. In some areas of policy, a unanimous decision would be required by the treaties, but many areas of EU policy are now subject to qualified majority voting. This means that the votes of the member states are weighted as laid down in Table 3.3, and for a measure to be adopted a *qualified majority* means that 62 votes must now be cast in favour of the measure. Thus a *blocking minority* amounts to 26 votes.

It cannot be claimed that the weights attached to the member states' votes reflect with any precision the number of electors, or the economic importance, of the member states. On whatever criteria cited, Luxembourg with a voting strength of 2 is over-represented, and the unified Germany with a voting strength of 10 is under-represented. However, these provisions have evolved over a number of years and represent an uneasy compromise between the member states. Debate on this issue will be heated at the 1996 intergovernmental conference. In the 1970s, the system was perceived of as a mechanism which stopped the four 'large' member states 'ganging up' to impose their will on the 'small' member states: the four needed the support of at least one other member state before they could muster a qualified majority. More recently, the 'low' blocking minority has been seen as a mechanism for minimising the opportunity of the less federally-minded countries, such as the UK, being outvoted. It seems unlikely that the present British government in the context of the 1996 intergovernmental conference will agree to a change in the Treaty — which itself would require the assent of all member states — to reduce the number of votes required for a qualified majority.

Table 3.3: Qualified majority votes.

	Votes in a Community of:					Population
	6	9	10	12	15	(Million)
	1958	1973	1981	1986	1995	1993
Belgium	2	5	5	5	5	10.1
Denmark	-	3	3	3	3	5.2
Germany (Deutschland)	4	10	10	10	10	81.2
Greece (Ellas)	-	-	5	5	5	10.4
Spain (España)	-	-	-	8	8	39.1
France	4	10	10	10	10	57.3
Ireland	-	3	3	3	3	3.6
Italy	4	10	10	10	10	58.1
Luxembourg	1	2	2	2	2	0.4
Netherlands	2	5	5	5	5	15.3
Austria (Österreich)	-	-	-	-	4	8.0
Portugal	-	-	-	5	5	9.9
Finland (Suomi)	-	-	-	-	3	5.1
Sweden (Sverige)	-	-	-	-	4	8.7
United Kingdom	-	10	10	10	10	58.2
Qualified majority	12	41	45	54	62	
Blocking minority	6	18	19	23	26	

For those who advocate CAP reform, qualified majority voting is a major impediment to change for it favours inertia and the maintenance of the status quo. Most CAP legislation has an unlimited life. This means that, unless amended, it continues for ever. It can only be amended (on the basis of a proposal from the Commission) if a qualified majority can be mustered in favour of change. One might speculate on how different the evolution of policy might have been if the Treaty of Rome had insisted that CAP legislation had to be renewed every five years, on the basis of a qualified majority; or if fewer votes had been required to muster a qualified majority.

In practice, during the CAP's formative years the situation was worse than this, and qualified majority voting although provided for by the Treaty of Rome was not applied. Although largely of historical interest, an appreciation of the so-called Luxembourg Compromise can still help our understanding of CAP decision-making in the mid-1990s. The Luxembourg Compromise (which is outlined in the previous chapter) stems from the mid-1960s, and French concerns about being outvoted on matters of vital national interest. In January 1966 the member states meeting in Luxembourg thrashed out an agreement to disagree, thus resolving the crisis which had paralysed decision-making over the preceding six months. As a result, the other member states took care not to outvote France even though they could have done so at any time.

The habit grew of adopting all measures by consensus, except in the budget Council, even when the treaties provided for qualified majority voting. Entry of Denmark and the UK into the Community in 1973 confirmed the veto, for they too shared French distaste for qualified majority voting, and with three member states willing to act together they were able legally to enforce a veto. It is not that the treaties now provided for a veto, but as Table 3.3 shows, in 1973 had France and the UK jointly refused to participate in a vote, a qualified majority could not have been achieved and the measure could not have passed. All that was required was for one country to declare 'a vital national interest', and the other as a matter of principle would refuse to vote. This ethos pervaded the Council. Thus decision-making rested on consensus (or rather, the slightly less onerous requirement that no-one should object), and it was in this environment that the annual farm price marathon developed. Discrete decisions could not be taken on individual measures, because member states would unscrupulously threaten to veto that decision unless their particular pet project was approved. Thus package deals had to be concocted that satisfied all the member states; and constructing such package deals usually involved several rounds of late night sittings, and seldom resulted in a significant cutback to the CAP.

In May 1982, distracted by the Falklands War, the UK was outvoted in the Farm Council. The then agriculture minister, Peter Walker, had invoked the Luxembourg Compromise to block the 1982/83 farm price package until such time as Britain had secured a satisfactory settlement on the UK's net contribution to the EU budget. When, without precedent, the Belgian Presidency asked for a vote, the UK, Denmark and Greece refused to take part; but this did not comprise a blocking minority and so the measure was passed. France, in participating in the vote and supporting the measure, claimed that the Luxembourg Compromise could only be invoked when a country declared that a matter of vital national interest was *directly* involved and, as the UK had indicated satisfaction with the farm price settlement, the attempt to link the issue with Britain's budget contribution was invalid (for further details see Butler, 1986, p. 100). Butler, rather dramatically, comments that this was "our worst defeat", but it marked a shift in EC decision-making and gradually qualified majority voting, when provided for by the treaties, became the norm rather than the exception in the Council of Ministers. Indeed, shortly afterwards in the Single European Act, and subsequently in the Maastricht Treaty on European Union, qualified majority voting was extended to other spheres of Community competence.

The Luxembourg Compromise was formally invoked by Germany in 1985, in an attempt to block price cuts for cereals, and by Greece in 1988 in relation to Greek demands for a devaluation of that country's green conversion rate (see Swinbank, 1989).[8] Although many people would argue that the Luxembourg Compromise is now a relic of the past, it is not quite dead and its spirit lives on. In May 1992, in a debate in the French Senate over the Maastricht Treaty, the French government declared that it still had the "right to impose a veto in defence of vital national interests, by virtue of the so-called Luxembourg Compromise" (*Financial Times*, 14 May 1992), and later the same year the French Prime Minister threatened to veto the Blair House Accord (see Chapters 16 and 17). Indeed, during the major part of 1993, there was much speculation about the threat of a French veto of the evolving GATT agreement, despite the provisions for qualified majority voting, and it was evident that the EU was unwilling to isolate a member state and hence had to proceed on the basis of consensus (for further discussion see Swinbank and Tanner (1996) Chapter 6). EU Commissioner Sir Leon Brittan, in defending the principle of qualified majority voting, points out that:

> the Union always seeks, through consensus, to avoid isolating any of its members anyway
>
> (Brittan, 1994, p. 231).

The Special Committee on Agriculture (SCA)

The Council is served by a vast committee structure, manned by civil servants from the member states and the Commission, and chaired by the country holding the Presidency. At the apex of this committee structure is the Committee of Permanent Representatives (COREPER), dealing with all Community matters other than the CAP, and the Special Committee on Agriculture (SCA) dealing with CAP.

COREPER and the SCA, usually meeting weekly, prepare the material for Council meetings, identifying areas of agreement and disagreement between the member states. Where the member states are agreed, and no further discussion at Council level is deemed necessary, the matter would go to the next meeting of the Council, whatever its composition, as an 'A' point for automatic agreement. Thus the agriculture Council on 18–21 May 1992, as well as thrashing out CAP reform, agreed two measures under the ECSC Treaty, antidumping duties on imports of video cassettes from Hong Kong, the establishment of an EU–Japan centre for industrial co-operation, and various environmental measures previously agreed in principle by the environment Council on 12 December 1991 (Council, 1992).

The SCA and COREPER themselves set up a whole series of specialist committees, as necessary, to examine the detailed technical issues that inevitably arise in examining Commission proposals. Thus dozens, if not hundreds, of national civil servants regularly commute from their national capitals to Brussels to serve this committee structure.[9] Many of the civil servants sitting on Council committees will also sit on the Commission's management committees, discussed below. Despite this common membership, Council committees and the Commission's management committees serve different functions in CAP policy making.

The European Commission

The Commission is made up of 20 members who collectively serve a term of 5 years. They must be nationals of member states, and in practice Germany, Spain, France, Italy and the UK each nominate two Commissioners, and the others one. The Commissioners' independence should be 'beyond doubt', and their job is to serve the EU and not act as representatives of the member states from whence they came. However, some will want to serve more than one term, and hence will be dependent upon the government of their home country for nominating them for a subsequent period of office, and others will want to return to political life back home. Even if these pressures failed to colour the perceptions of individual Commissioners, they are inevitably imbued with the preferences and prejudices of their home countries, and are often looked upon as representatives of the member states.

Each Commissioner is responsible for a discrete area of policy. In truth, there are barely enough jobs to keep 20 Commissioners fully employed; but some of the key jobs are very demanding. These include the President of the Commission (Jacques Santer from Luxembourg for the 5-year term 1995–9) and the Commissioner with responsibility for agriculture (Franz Fischler from Austria). Over the years a number of powerful personalities have stamped their mark on EU farm policy. They include the former Dutch minister of agriculture, Sicco Mansholt who presided over the CAP in its formative years and only gave up the mantle in 1972 when he briefly assumed the role of President of the Commission; Petrus Lardinois another Dutchman; the Dane Finn Gundelach, who died at the beginning of his second term as Farm Commissioner; and Ray Mac Sharry from Ireland (see Table 3.4).

The Commission is a collegiate body which determines policy by a simple majority vote at its weekly meetings. As well as having the sole responsibility for initiating policy proposals, as outlined above, it acts as the guardian of the treaties and as such ensures that the member states implement legislation, if necessary arraigning them in the European Court if they fail to comply. Under the competition rules of the treaties, it has the responsibility for implementing EU competition policy, subject to arbitration in the European Court. Furthermore, in a number of arenas of policy, notably the CAP, the Council has delegated to the Commission day-to-day management authority with respect to EU policies. The Commission is serviced by a permanent staff of 17,000+ officials, many of whom are involved in interpreting and translating between the EU's eleven official languages.

Table 3.4: Commissioners of agriculture.

1958–1972	Sicco Mansholt, became President April 1972
1972	Carlo Scarascia–Mugnozza
1973–1976	Petrus Lardinois
1977–1980	Finn Olav Gundelach, died 13 January 1981
1981–1984	Poul Dalsager
1985–1988	Frans Andriessen
1989–1992	Ray Mac Sharry
1993–1994	René Steichen
1995–1999	Franz Fischler

Directorates-General

The Commission services are organised into a number of Directorates-General, known cryptically as DGs, in rather the same way that a national civil service would be arranged into ministries. DG VI, the Directorate-General with responsibility for agriculture, is the service that implements the CAP. DGs tend to have strong vertical links, reflecting their very hierarchical structure and the limited delegation of responsibilities; and historically they have had difficulty fostering and sustaining working arrangements between DGs at anything less than the highest level. DGs are staffed by EU civil servants, and headed by a Director General (and possibly several Deputy Director Generals). Next in the hierarchy come Directorates, headed by Directors; and then Divisions led by a Head of Division, and staffed with administrators and support staff.

DG VI is a large, well resourced, Directorate-General, which is often referred to as the Directorate-General for Agriculture. Although less beholden to the farm sector than it was in the 1970s and early 1980s, DG VI is still very protective of the farm interest. By contrast, the food industry's sponsoring body is the 'Food Production and Biotechnology' Division located within DG III, the Directorate-General for Industry. Here, a small staff seeks to protect the interests of the food industries from the consequences of DG VI's actions. A sister division enacts and implements the EU's major programme of food law.

Council and Commission

The annual farm price review forms a useful case study of the flow of policy making under the CAP. Over the years the Council has adopted CAP price support regimes for most agricultural products of importance in European agriculture. In EU jargon, these support policies are known as CMOs (Common Market Organisation). There will be a Council regulation setting out the basic principles of the CMO, supported by a whole host of Council and Commission regulations laying down more detailed rules.[10] Thus, following the May 1992 reforms, the basic Council regulation setting out the support arrangements for cereals was Council regulation (EEC) No 1765/92.

The Council regulation for each commodity regime established, for example, intervention arrangements; but intervention and other prices had to be fixed by the Council on an annual basis. Thus, each year the Commission must propose to the Council support prices for the following year. In the new cereals regime, following the 1992 Mac Sharry Reforms, the Council has fixed support prices until such time as they are changed. This has reduced somewhat the importance of the annual farm price review, and shifted decision-making power from the Council to the Commission; but the Council still has to fix intervention standards, the monthly increases in intervention prices, and other details, which engender heated debate in the Council.

The price proposals are formulated by the Commissioner for agriculture in liaison with members of the Commissioner's private office, or *cabinet*, and the Director General and other senior officials of DG VI. A mixture of political and objective criteria would feed into the package, for it could be counterproductive to propose something that stood little prospect of acceptance by the Council of Ministers.[11] Depending upon the circumstances of the year, and the personalities involved, those initiating and participating in this discussion would vary; but it would be a small inner group. Lowly officials in DG VI would only be involved on a 'need-to-know' basis, and discussions with other DGs would be minimised. These discussions would typically take place before Christmas, but when a new college of Commissioners takes office — as in 1995 — the proposal would normally be formulated by the new Commission in the New Year. When the package is ready it is presented to the Commission at their weekly meeting for discussion, and hopefully approval. The Commissioners' discussion would have been preceded by that of their *chefs de Cabinet*.

The Commission, having debated the proposal at one or more of its meetings, would ultimately adopt it as its formal position by simple majority vote, as was the case on 15 February 1995 in the context of the 1995/96 price fixing (see Table 3.5). The proposal should then officially enter the public domain, as a Commission working (or COM) document; though typically its contents would already have been well and truly leaked to the press. It now has the status of a formal proposal from the Commission to the Council; and as such will also be dispatched to the European Parliament.

On receipt by the Council, formal appraisal can begin. The Council Secretariat will officially dispatch the document to the governments of the member states; and they in turn can begin such domestic parliamentary scrutiny as is provided for in their legislative systems. Typically the Commissioner will talk to the proposal at the next meeting of the agriculture Council, and ministers give their initial reactions. The Special Committee on Agriculture will begin its work.

Meanwhile the European Parliament undertakes its review of the proposals, finally adopting its view in plenary session. This is a 'green light' for the Council. In the absence of an opinion from the Parliament it is not empowered to conclude its deliberations; but the Council does not have to take any notice whatsoever of the Parliament's opinion. It should be noted that in other spheres of Community competence, but not the CAP, the Parliament has powers of co-determination which considerably enhance its role in decision-making.

The role of the Presidency can be crucial, both inside and outside the Council chamber. Frequently, bilateral discussions (sometimes referred to as 'confessionals') will be held with other ministers in an attempt to identify potential compromise solutions. The President-in-Office will usually draw up an alternative package that it believes will be acceptable to the Council. This alternative proposal from the Presidency can, however, only be adopted by qualified majority vote if it secures the endorsement of the Commission, because the Treaty states that a modification of the Commission proposal can only be adopted unanimously. Thus, the Commissioner of necessity takes an active part in all the negotiations in and out of the Council chamber. Final decisions on the 1995–96 farm price review were reached on 22 June 1995, after a four-day negotiating session; and probably almost 12 months from the time when the Commission services first began working on the price package.

Table 3.5: The 1995/96 Farm price fixing.

23–24 January 1995: First meeting of the agriculture Council with the new French President-in-Office (Jean Puech), the new Farm Commissioner (Franz Fischler) and the farm ministers of the three new member states. Items on the agenda included vexed animal welfare questions relating to the use of veal crates and maximum journey times for animal transport, reform of the sugar regime, and support for Austria's durum wheat producers.

10 February 1995: The journal Agra Europe carried a comprehensive account of the farm price proposals that Franz Fischler was to present to his fellow Commissioners the following week.

Commission Meeting: proposals adopted at the Commissioners' weekly meeting on Wednesday 15 February 1995. The proposals were published in COM (95)34, dated 17 February 1995.

1st Round: Agriculture Council, 20–21 February 1995. Ministers voiced their initial reactions to the price proposals. Several member states expressed their opposition to further cuts in support to cereals, over and above the reductions embodied in the Mac Sharry reforms of 1992.

1 March 1995: Commission tabled reform proposals for the cotton regime, which subsequently became part of the price package.

13 March 1995: An informal agricultural Council was held at Toulouse in France. The main agenda item was a review of the long-term future of agriculture, but animal welfare issues and the overdue reform of the wine regime were also discussed. The emerging crisis over suspended revaluations of green conversion rates was beginning to dominate the policy-making agenda.

18 March 1995: Five member states, facing green rate revaluations, met in Aachen, Germany, to discuss strategy.

2nd Round: Agriculture Council, 27–28 March 1995. There was a brief round-table discussion on the price proposals, with ministers repeating the positions outlined a month earlier. The 1994/95 marketing years for milk and beef were extended to 30 June 1995. Agrimonetary matters dominated discussion over

Table 3.5 (Cont.)

dinner on the Monday evening; and the Commission lodged a new proposal on animal journey times.

No Council had been planned for April, because of the French presidential elections, but because of the agrimonetary crisis a special meeting of the Council was held on 10 April 1995. Following the presidential election, and the change in government, Philippe Vasseur was named French minister of agriculture.

European Parliament: Adopted its opinion on the package in Strasbourg in May 1995. It rejected its agriculture committee's proposal to increase all CAP prices by 1% in 1995/96.

3rd Round: Agriculture Council, 29 May 1995. By now the SCA had finished its work on the price package, and the dossier was back on the Council table; but with a new French President-in-Office this Council was a brief, low-key affair, concluded on the Monday evening. Agrimonetary and animal welfare issues dominated the formal discussions, with much of the Tuesday devoted to bilateral meetings between the Presidency and the other member states.

The 'Marathon': Agriculture Council, 19–22 June 1995. Brought forward by a week, this was the ritual marathon session to settle the price package and all the other dossiers that had become entangled with it. Thus the final deal included a controversial package to resolve the agrimonetary crisis, a resolution of the animal transport saga, a new regime for cotton, and the 1995/96 price package. The Austrian minister secured increased aid for his durum wheat producers. During the course of the four days the Presidency had tabled three compromise papers, and held two full days of bilateral discussions with other delegations. Various delegations expressed their opposition to particular items in the package; indeed the Swedish minister was opposed to all its parts. However, the Presidency declared that all the different elements of the global compromise that it had presented to the Council with the Commission's assent had achieved the qualified majority required for adoption. The negotiations had been completed — just — under the French Presidency, and new support prices were applied from 1 July 1995. Amongst the other decisions taken, the Council unanimously agreed that France could give additional national aids to its wine producers.

Management committees and delegated powers

The annual farm price marathon attracts considerable attention in the farming press; and ministers usually wish to be seen to be listening, if not responding, to the entreaties of the farm lobby. If pressure on farm incomes is particularly acute, big set-piece demonstrations can be expected on the streets of Brussels. The food industries, of course, have an interest in the level of prices set; but their main concerns, and lobbying activities, will lie in the day-to-day management of the market by the Commission. The impact of the CAP on the food industries is discussed in Chapter 12.

It would be quite impossible for the Council to be involved with the daily business of running the policy: fixing export refunds for example. Thus, in setting up the CMO, the Council will have delegated such market management activities to the Commission; and the *Official Journal of the European Communities* is full of Commission regulations exercising this delegated responsibility under the CAP.

Some responsibilities are delegated to the Commission entirely on its own authority. Thus it was the Commission that fixed import levies without having to consult any other body, though this responsibility is considerably reduced as a consequence of the Uruguay Round GATT agreement. In other areas of delegated responsibility, its activities are more constrained. Before fixing export refunds, it must normally consult the appropriate management committee.

For each CMO the Council has established a management committee, each with identical powers, to oversee the Commission's management of the market. They are chaired and serviced by the Commission, and manned by national civil servants, and meet on a frequent basis. For commodity sectors such as cereals or sugar, the weekly meeting of the management committee is a dominant feature of the work schedule of the division.

As noted above, in a number of areas of delegated responsibility such as the fixing of export refunds, the Commission must *consult* the management committee. The management committee itself does not take decisions, it merely gives opinions on the Commission's proposed actions. In counting the votes of the national delegations, the same weights apply as in the Council, as tabulated earlier. A total of 62 or more in favour of a Commission proposal amounts to a favourable opinion; 62 or more against, as an unfavourable opinion; and neither 62 for nor against, as no opinion. In practice the Commission tries very hard to present proposals which will attract a favourable opinion, and unfavourable opinions are very rare.

Having received a favourable opinion, or no opinion, the Commission can implement the measure; but is not obliged to do so, and sometimes does not, which can catch the trade on the hop. Even with an unfavourable opinion the Commission is entitled to enact its proposed legislation, but the Council has the right to overturn this within a month.

The deliberations of management committees are supposed to be secret, but in practice their business is soon in the public domain; and a mini information industry based in Brussels transmits this material to its clients around the world. *Agra Europe* regularly reports on the business conducted in management committees.

Commercially sensitive information becomes available in the management committee: if for example the Commission declares its intent to change export refunds in a day or two, this is of considerable importance to the trade, and any company which has privileged access to such information can gain a competitive edge over its rivals. Thus delicate questions are bound to be raised about the conduct of the Commission (or more particularly Commission officials), and national delegations, in their formal and informal 'debriefing' sessions after management committees, particularly if it is suspected that some interested parties have preferential access to information. Even greater concerns are raised if, as has been suggested, information 'leaks' from the management committee whilst still in session. Management committees, for example frequently review bids received in tenders to determine sales out of intervention, or the level of export refunds determined under the tender mechanism. Although all bids are anonymous, an early indication of the likely outcome could allow a successful bidder to consolidate its position, and an unsuccessful bidder to unwind its exposure, prior to the formal announcement of the outcome.

The food industries have in the past expressed considerable disquiet about DG VI's powers, and the lack of transparency in management committee deliberations. For example, in 1987 the Food and Drink Federation (FDF) said:

> The FDF believe there should be:
> (a) improved consultation procedures between the Commission and the food and drink industry, thus reflecting the importance of the industry to the economy of the Community;
> (b) greater democratic accountability of the work of the Commission, particularly with the usage of Committee procedures.
>
> The industry's experience of the way in which the Commission's devolved powers are used is less than satisfactory. The management committee system...make(s) many decisions of considerable financial and commercial importance to the industry. They are administered by DG VI of the Commission: this DG is responsible for agricultural policy. The committees work in such a way that agendas are tardy or not known at all in advance by the industry, there is frequently little or no consultation with the industry and the committees deliberations are subject to a disproportionate agricultural interest.
>
> (FDF memorandum, in House of Lords, 1987, p. 42).

Implementation in the member states

Although the Council and the Commission have crafted the CAP legislation, and have a major role in its implementation, it is in fact the member states that deal with farmers and traders, provide intervention stores, and ensure that charges are collected and that payments are legally made. In the UK for example, HM Customs and Excise implements all the import regulations under the CAP, collecting import duties on behalf of the EU. The Intervention Board, an executive agency responsible to ministers, administers most of the other CAP regulations relating to price and income support, although other organisations such as the Home-Grown Cereals Authority and the Meat and Livestock Commission do act as the Board's agents. The Intervention Board employs just over a thousand individuals with its main offices in Reading and Newcastle upon Tyne (Intervention Board, 1994). Structural policy is the responsibility of the agriculture departments.

The European Parliament

The 626 members of the European Parliament have a very limited role in CAP law making, although in other areas of policy making their importance is greater. Under the CAP, their opinion must be sought before the Council can act upon a Commission proposal; but there is no obligation on the Council to heed Parliament's advice. The Court has insisted upon the niceties of these arrangements, and in 1980 rejected as invalid a Council regulation on isoglucose on the grounds that it had been adopted unlawfully *before* the opinion of the European Parliament had been given.

The Parliament shares budgetary responsibilities with the Council, and together in a highly complex exercise they must jointly adopt the Community's annual budget. However, even here the CAP is subject to rather special rules; for expenditure arising from the market support mechanisms is deemed to be 'compulsory', and the power of the European Parliament to amend such budget lines is limited.

Article 144 EU does confer upon the Parliament a rather awesome power: by a two-thirds majority of the votes cast it can dismiss the Commissioners as a body. This is, however, such a draconian power that it has never been used. Nor has there ever been a sufficient threat that a two-thirds majority could be mustered, and so it has not been an effective bargaining counter *vis-à-vis* the Commission. After implementation of the Maastricht Treaty, Parliament acquired new powers relating to the appointment of the Commission President and the college of Commissioners. Thus, in January 1995 Parliament held a series of scrutiny hearings before approving the new Commissioners.

The Parliament's main role is that of exposing Community policy to wider scrutiny. First, in coming to an opinion on draft CAP legislation, the Parliament's agriculture committee, and perhaps others, will publicly scrutinise the draft with Commission officials in attendance, and take evidence. The agriculture committee's report will then be debated in a public plenary session, which a Commissioner will attend and at which s/he has the right to speak. These events attract media coverage; and although the Parliament has no *formal* power to adopt or amend CAP legislation, the Commission will be anxious to appease Parliament in order to secure favourable publicity.

Second, Parliament can pose written and oral questions to the Commission. As in Westminster, a carefully drafted question can elicit information that would otherwise not readily be disclosed; and the Commission would usually take great care in framing its responses. Although many questions are probably frivolous, and others 'planted' by lobbyists, the system does give Parliamentarians some power over the Commission.

In other areas of Community policy the Single European Act gave Parliament an enhanced role in law making. This 'co-operation procedure' involves a review by Parliament of the Council's 'common position', adopted by the Council on first examination of the Commission's proposal, before final adoption of legislation by the Council. The Maastricht Treaty further enhanced Parliament's role; but not to the extent originally sought by many MEPs.

The Court of Auditors

The Court of Auditors was established to examine the revenue and expenditure accounts of all the Community's institutions, to report to the Council and Parliament on the reliability of the accounts and the legality of the underlying expenditures, and thus to assist the Council and Parliament in the exercise of their financial responsibilities. In practice this has led the Court of Auditors to examine the efficacy of policies, and not just the implementation of policy, and a number of unflattering reviews of the CAP have emerged (see for example the review of sugar policy in Court of Auditors, 1991).

Court of Justice of the European Communities

The EU regulations which set out CAP price policy are legally binding in all member states as soon as adopted by the Council (or Commission) and published in the *Official Journal of the European Communities*. If national legislation contradicts EU regulations, then EU law prevails. In the main it is the national courts which must apply EU law on a day to day basis; but the European Court remains the final arbiter. Thus the European Court *implements* and *interprets* EU law, and acts as a supreme court in ruling upon the legality of regulations adopted by Council and Commission. A fairly arbitrary selection of cases will illustrate its role.

In the *isoglucose* case (Case 138/79 of October 1980) cited above, the Court struck down the offending legislation in defending the privileges of the European Parliament. Earlier, in Case 114/76 (of July 1977) the Court had ruled that the Council had illegally insisted upon the compulsory incorporation of skimmed milk powder in animal feed, in that it imposed an unwarranted financial burden upon the purchasers of such feed; and in Case 77/86 (of February 1988) Britain's National Dried Fruit Trade Association won its point that Commission legislation on minimum import prices for dried grapes imposed an unwarranted burden on importers trading under the minimum import price. More recently, in a whole series of cases, various provisions relating to the allocation of milk quotas in 1984 have been unravelled.

Free trade, under article 30 of the EEC Treaty, has been a dominant concern of the Court. Stemming from the *Charmasson* case of 1974 (Case 48/74 of December 1974), relating to bananas, the need to introduce a Common Market Organisation for sheepmeat was recognised; and *Cassis de Dijon* (Case 120/78 of February 1979) is widely credited as being the critical ruling that led the Commission to formulate its principle of mutual recognition of food legislation in force in other member states.

Literally hundreds of cases concerning the CAP, and trade in food and agricultural products, go before the Court; and lawyers specialising in the CAP are in heavy demand. A number of cases before the Court relate to litigation between the institutions of the EU, or between the institutions and the member states; but many cases are referred to the European Court from the national courts for rulings on an interpretation of EU law. Such Court action can be protracted, and very expensive for the parties concerned; but if a company believes that it is suffering an injustice as a consequence of wrongly applied EU law then it may well have to fight its case through to the European Court.

Lobbying and the lobbyists

As this book makes clear, the CAP has a considerable impact on the EU and the wider world economy. In attempting to raise farm incomes, it changes asset prices and resource use in the farm sector and thus has an impact on the environment; it alters the raw material costs faced by the food industries resulting in increased food prices; and it tends to depress and de-stabilise world market prices. Under the circumstances it is hardly surprising that individuals and organisations, acting either individually or collectively, seek to inform and influence the decision-makers. The lobbyists are frequently engaged in a two-way flow of information, for their paymasters need to extract from Brussels timely information on developments in policy, and advance warning of possible policy change. For traders in particular, the day-by-day decisions of the Commission as it manages the markets can be particularly important.

Given the limited role of the European Parliament in implementing the CAP (and even more so in the case of the Ecosoc) relatively little lobbying activity tends to be directed at MEPs, though this has been growing in importance; and in spheres of EU policy where Parliament can have a more pronounced impact on the legislative process, lobbying can be intense. Thus, Parliament was a focus for lobbying on the sweeteners directive, on which there was a major clash of national and industrial interests (see Earnshaw and Judge, 1993). Nonetheless, the Parliament's role in CAP policy making should not be overlooked: it was, for example, as a result of Parliamentary pressure that the use of beef hormones were banned.

Ministers, in the Council of ministers, represent a legitimate target for the lobbyists; but the focus of lobbying tends to be the national capital. Ministers arrive in Brussels or Luxembourg for a Council meeting, and depart as soon as it is over. In the context of the annual farm price review, the Council of Agriculture Ministers does meet with a delegation from the *Comité des Organisations Professionnelles Agricoles* (COPA), the main EU-wide lobby group representing European farmers, but it is difficult to believe that this meeting has more than symbolic importance.

It is the Commission, both Commissioners and the Commission services, which is the main focus of the Brussels-based lobbyists. Once inside the Commission buildings, visitors find that Commission officials are surprisingly free with their time and frank with their views. On the whole they welcome the two-way flow of information and ideas, as desk-bound in Brussels they tend to feel isolated from the sector they administer. Thus a steady flow of people, representing individual companies, regions, third countries, or trade associations, pass through their offices. Most companies will maintain an office in Brussels, and foreign countries an embassy with an ambassador accredited to the EU.

The trade associations reflect those established at national level, and on the whole act as umbrella organisations. Thus COPA's membership includes Britain's National Farmers' Union and similar organisations from other member states. A full listing of the hundreds of organisations cannot be attempted here. The food industries are represented by the *Conféderation des Industries Agro-alimentaires* (CIAA) and the consumers by the *Bureau Européen des Unions de Consommateurs* (BEUC). These umbrella bodies tend to perform three roles: (a) they lobby on behalf of their constituent members; (b) they collect information to pass back to their members; and (c) they provide a forum in which their members can co-ordinate their lobbying strategies for use on ministers and civil servants back home. More specialist associations, such as the *Confederation of Importers and Marketing Organisations in Europe of Fresh Fruit and Vegetables* (CIMO), will only warrant a small staff in servicing the particular needs of their members.

In addition to these informal contacts, the Commission also has a more structured — but perhaps less useful — network of advisory committees which advise it on the machinations of the CAP. Thus, for each major CAP commodity, an advisory committee made up of farmer, trader and consumer representatives will meet periodically with Commission officials to discuss various aspects of policy. Similarly, the original intent had been that the Ecosoc, grouping together the social partners of workers, producers and consumers, would bring a wider expertise to bear on EU policy deliberation. There is little evidence to suggest that Ecosoc has had such an impact, though it is frequently asked to comment on CAP proposals and can also report on pertinent matters on its own initiative. Whether or not the recently formed Committee of the Regions will have any more impact than Ecosoc remains to be seen.

Concluding comments

In this chapter I have attempted to explain the formal institutional framework in which CAP policy is made. It is only within this framework that the CAP can be changed, though the institutional framework itself could be changed by treaty at the next intergovernmental conference. The Luxembourg Compromise, qualified majority voting, and the balance of power within the Council and between the Council and Commission, have tempered the evolution of policy over the last three decades. Those impatient for CAP reform need to understand the framework in which reform could be effected, and perhaps need to ask whether or not institutional reform might be a prerequisite for CAP reform.

References

Brittan, Sir L. (1994) *Europe. The Europe we need*, Hamish Hamilton, London.

Butler, Sir M. (1986) *Europe: More than a Continent*, Heinemann, London.

Council of the European Communities, General Secretariat (1992) *Press Release* 6539/92, 1579th Council Meeting – Agriculture – Brussels, 18, 19, 20 and 21 May 1992, Brussels.

Court of Auditors (1991) Special Report No 4/91 on the operation of the common organisation of the market in the sugar and isoglucose sector accompanied by the replies of the Commission, *Official Journal of the European Communities*, C290, 7 November 1991.

Earnshaw, D. and Judge, D. (1993) The European Parliament and the sweeteners directive: From footnote to inter-institutional conflict, *Journal of Common Market Studies*, 31(1), 103–116.

General Secretariat of the Council of the European Union (1994) *Forty-First Review of the Council's Work. 1 January to 31 December 1993 (The Secretary-General's Report)*, Office for Official Publications of the European Communities, Luxembourg.

House of Lords Select Committee on the European Communities (1987) 3rd Report Session 1986–87, *Delegation of Powers to the Commission (Final Report)*, HL38, HMSO, London.

Intervention Board (1994) *1993–94 Annual Report and Accounts*, HMSO, London.

Swinbank, A. (1979) The Objective Method: A Critique, *European Review of Agricultural Economics*, 6(3), 303–317.

Swinbank, A. (1989) The Common Agricultural Policy and the Politics of European Decision Making, *Journal of Common Market Studies*, 27(4), 303–322.

Swinbank, A. and Tanner, C. (1996), *Farm Policy and Trade Conflict: The Uruguay Round and CAP Reform*, The University of Michigan Press for the Trade Policy Research Centre, Ann Arbor.

---

[1]In October 1990 the German Democratic Republic (East Germany) was absorbed by the Federal Republic of Germany (West Germany), and thus became part of the EC.

[2]Although the EC/EU has certain federal characteristics, it remains a group of independent nation states bound by Treaty. It can only act as a Community if its Treaties permit it to do so.

[3]Thus, until 1 November 1993, we referred to the Treaty Establishing the European Economic Community, and subsequently to the Treaty Establishing the European Community. Here, and elsewhere, this Treaty is often referred to as the Treaty of Rome.

[4]Article 3b of the EC Treaty reads:

"The Community shall act within the limits of the powers conferred upon it by this Treaty and of the objectives assigned to it therein.

In areas which do not fall within its exclusive competence, the Community shall take action, in accordance with the principle of subsidiarity, only if and in so far as the objectives of the proposed action cannot be sufficiently achieved by the member states and can therefore, by reason of scale or effects of the proposed action, be better achieved by the Community."

"Any action by the Community shall not go beyond what is necessary to achieve the objectives of this Treaty."

[5]Six central European states (Bulgaria, the Czech Republic, Hungary, Poland, Romania and Slovakia) have *Europe Agreements* with the EU which hold out the promise of eventual membership; and similar agreements have been negotiated with the Baltic States and Slovenia. The EU's trade relations with these countries are outlined in Chapter 14. The four Visegrad countries (the Czech Republic, Hungary, Poland and Slovakia) will probably prove to be earlier entrants than the other six. Cyprus and Malta are the other potential entrants, although Turkey has also asked to be allowed to join.

[6]The President of France, a head of state, represents France. Otherwise, prime ministers or chancellors (heads of government) participate.

[7]Alternative descriptive expressions include 'the Agriculture Council' or 'the Farm Council'.

[8]'Green money' is discussed in Chapter 6.

[9]In 1993 the Council sat for 119 days; COREPER, the SCA and similar for 115 days, and various other Council Committees and working parties for 2,105 days (General Secretariat, 1994, p. 255).

[10]Both Council and Commission regulations have direct force of law in all member states, and displace any contradictory national legislation in force. EC regulations are published in the *Official Journal of the European Communities*.

[11]In the 1970s, but now long forgotten, the Commission developed an 'objective method' of determining support prices on a cost-plus pricing formula. See Swinbank (1979).

Chapter 4

Reform of the CAP: From Mansholt to Mac Sharry

Lionel Hubbard and Christopher Ritson

Introduction

Virtually since its inception, the Common Agricultural Policy has been subject to proposals for reform. Among the academic community, the debate rapidly achieved something of a consensus and subsequently evolved remarkably little. Academics and other CAP specialists, either as individuals or in groups, have produced numerous reform proposals. Notable in this context were the Wageningen Memorandum of 1973, the Sienna Memorandum of 1984 (Ritson, 1984), and a report commissioned by a cross-party group of members of the European Parliament on "The Changing Role of the Common Agricultural Policy" (Marsh, 1991). Reading these documents one is struck by how little seemed to change – even several of the people are the same! A typical form of this kind of argument is outlined in the next section.

The academic reform argument

The argument would begin as follows. When the agricultural ministers of the original six Common Market countries launched the CAP, they made a fatal mistake. Partly because of a failure to appreciate the potential growth in production, and partly because of the overriding necessity to achieve a political agreement in agriculture to cement the establishment of the European Economic Community, they set support prices for cereals at too high a level (more than 50% in excess of, what was then, a very stable world market price). As a consequence, because most other agricultural products are related to cereals, either as competitive arable crops or as users of cereal-based feeding-stuffs, most other agricultural product prices had similarly to be set at relatively high levels.

From this single set of decisions there followed, so it was argued, a number of undesirable consequences — undesirable, that is, when viewed against a set of criteria, widely accepted (though often implicitly) among academics, concerning agricultural policies. This approach judges the success of agricultural policies relative to certain fundamental goals in society, such as the efficient use of resources and equity in income distribution (in each case, both within agriculture and between agriculture and other sectors) and good international relations. Because support prices were so high, it was argued that the CAP encouraged inefficient, high-cost, production; impeded structural adjustment; disadvantaged low-income consumers (because of high food prices); benefited large farmers greatly and small farmers very little (because the benefit was distributed pro rata to the amount produced); and damaged trading relations with both rich and poor countries alike. All would be well, however, if agricultural product support prices were reduced — and the debate, as such, was really about how to achieve this.

The problem, of course, was the damage to farm incomes that would result from lower prices. The key, in most reform proposals, was the introduction of some form of direct income supplement for low-income farmers. Such a switch, from price support to direct income support for agriculture, seemed highly desirable when judged by the criteria listed above. Greater equity could be achieved, on account of the benefit of lower food prices to low-income consumers and the ability to target support to those farmers most in need; efficiency could be improved, in that the Policy would no longer underwrite high cost marginal output; and international trading relations would be improved. Such a development seemed also to favour the prospects of low-income country exporters to the EC, and so the cause of reform of the CAP has typically been embraced by the development lobby — though, as pointed out in Chapter 15, the issue is far from clear-cut.

The famous Mansholt Plan of 1968 included substantial price cuts linked to structural reform, but these proposals were rejected by the Council of Ministers. Subsequently, despite various policy initiatives CAP support prices remained stubbornly well in excess of international levels (Tables 1.1 and 1.2) and only with the Mac Sharry reforms of 1992–95 has a significant step been taken towards switching from price support to more direct income measures. This chapter provides a theoretical framework for understanding the pressure for CAP reform and the kind of policy initiatives which have followed.

The formal approach

Most of the literature on CAP reform, although written by academics, was directed towards a more general audience with the aim of influencing policy-makers. Underpinning it, however, was other more professional analytical work which, with the benefit of hindsight, can be seen to originate with the publication in 1969 of Josling's article "A Formal Approach to Agricultural Policy". It is a peculiar feature of post-war agricultural economics that, whereas an analytical approach to farm production economics was well advanced by 1969, agricultural policy had tended to be the preserve of a more 'literary' type of agricultural economist. Josling applied public choice theory to agricultural policy

— attempting to analyse the appropriate choice of policy in terms of objectives, constraints and instruments.

In Fig.4.1 we illustrate, first, the idea of instruments and complementary objectives — that is, where the instruments all have a positive relationship with the objectives — which are here, by way of example, to raise farm output (perhaps for security of supply reasons) and to raise farm incomes. For the sake of argument, the instruments might be product subsidies, import controls and investment subsidies — and Instrument 3, to illustrate the various possibilities, is drawn so that, past a certain level, the objectives are no longer complementary.

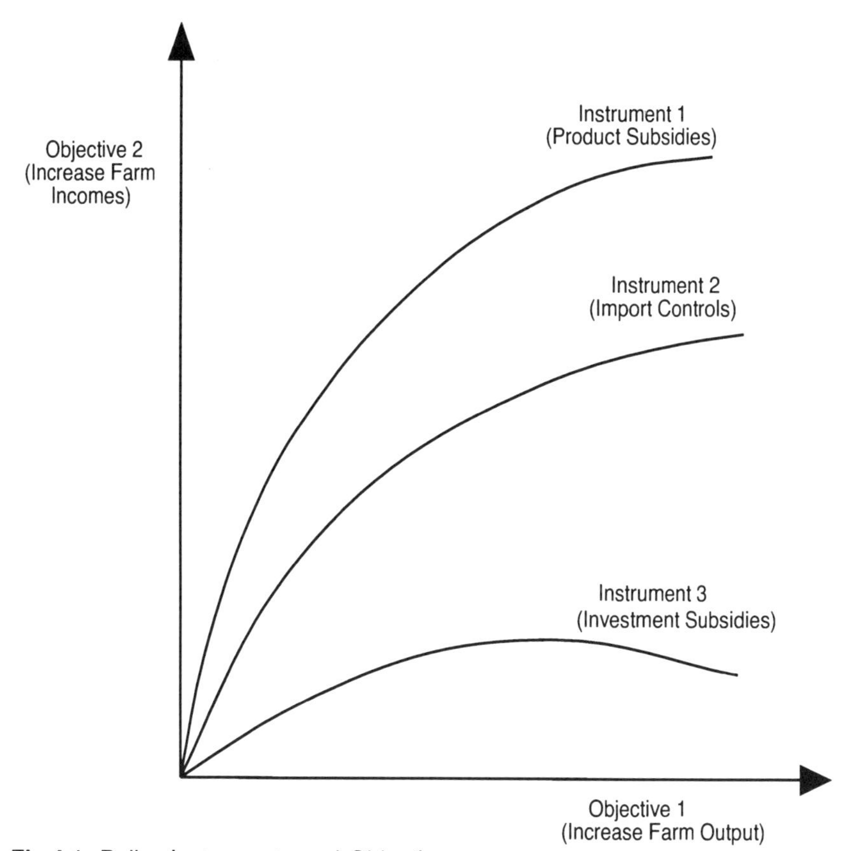

Fig.4.1: Policy Instruments and Objectives.

Now the important message from the diagram is that: "...a necessary (though not sufficient) condition for the reaching of a number of quantitative objectives is that one employs a similar number of policy instruments" (Josling, 1969). In other words it is extremely unlikely that the choice of any one of the instruments in Fig.4.1 would allow the achievement of target levels of both objectives.

Thus, the more 'academic' version of the argument conveyed in the previous section was that "the CAP is unsuccessful because it is attempting to use one instrument (raising market prices) in order to fulfil a number of objectives".

Official reform plans

Meanwhile, it was not just academics who were calling for reform of the CAP — but also successive European Commissions. Indeed, perhaps the most famous reform plan of all — the Mansholt Plan referred to earlier — came from the Commission of the European Communities (1968), and was in general consistent with (and indeed largely pre-dated) the academic work. It advocated reduced prices for surplus commodities, but linked various forms of financial compensation to structural reform. It was either 'get bigger or get out', whereas other reform proposals were willing to contemplate 'stay small — here is a bit extra to supplement your income from farming'.[1]

The Mansholt Plan was far from popular with the agricultural interests in western Europe, but subsequently there have been a succession of Commission documents proposing change to the CAP — though after Mansholt the term 'reform' became infra dig. — euphemisms such as 'adjustment', 'development', 'guidelines' and 'improvements' have been used.[2] Gradually, however, as we moved into the 1980s, the preoccupation of the European Commission, and some other CAP commentators, diverged from that of what one might call 'academic orthodoxy', and a parallel debate developed in which the central concern was the budget cost of the Policy. This came to the fore because of the failure of the Community's automatic system of generating revenue (its 'own resources') to match the budgetary cost of its policies — mainly the cost of the CAP. And this became the real debate and the driving force behind change.

For an academic, it was somewhat sobering to realise that the arguments involved in the consensus over CAP reform mentioned earlier were almost wholly irrelevant to the actual reform of the CAP. Herein lies a paradox, for it was the peculiarity of having to forge a common agricultural policy for six countries which was partly responsible for the original mistakes with respect to the CAP; and it has been the fact that the Policy has had to meet six (and then nine, then twelve and now fifteen) member state interests which has partly been responsible for the failure of the Policy to reform in the way advocated by many. Given this impasse, reform became dominated by preoccupations with the budget.

It was of no great surprise then that the most recent reform package, agreed by the Council of Ministers in 1992, was initiated in response to an impending budgetary crisis. Export refund expenditure had soared due to weakening world markets in cereals and dairy products, coupled with large increases in purchases of beef into intervention. In addition, there was mounting international pressure, within the Uruguay GATT Round, for countries to lower their levels of support for agriculture. Against this background, a set of draft proposals was tabled by the then Agriculture Commissioner, Ray Mac Sharry, in January 1991 (Commission of the European Communities, 1991a) and a set of formal proposals tabled in July 1991 (Commission of the European Communities, 1991b). The final package was agreed in May of the following year and phased implementation began almost immediately, to be completed by 1996.

The Commission's intention with the 1992 reform package was to combine substantial price reductions for the major agricultural commodities with annual compensation payments which would be modulated to favour smaller farms.

Other measures were aimed at reducing the amount, or intensity of use, of land in production, through 'set-aside' and reduced stocking densities, again compensated by annual payments. When the draft proposals became known there was widespread objection to both the severity of price cuts and the notion of modulation, with the UK particularly critical of the latter. The final outcome reflects the compromises necessary for agreement. The general principles of the reform package have been retained and it remains radical, but only just (Harvey, 1994). Details of the measures contained in the package are reported in Chapter 5.

Real CAP reform and public choice theory

Does this emphasis on budgetary considerations mean that the theory of public choice is of no relevance in practice to understanding the development of the CAP? We think not; rather it has been a failure of the academic work to catch up with the changed policy environment.

First, the 1980s saw a proliferation of policy instruments under the CAP, which can be viewed as coming to terms with the need for more than one instrument to attain more than one objective. Second, the priority of objectives and the relation between objectives and constraints are largely a political issue. Figure 4.2 considers the second case, where the instruments are associated with conflict between the objectives. We have chosen to insert a new objective 'economic efficiency', but could also have introduced a new instrument (say quotas) to Fig.4.2. Product subsidies and import controls both adversely affect efficiency (but import controls more so, because they raise consumer prices).

In these circumstances choosing an optimum policy is best seen as maximising one objective subject to viewing the other objective as a constraint. One then chooses the instrument which achieves the highest value of the objective (say farm incomes) subject to the constraint (efficiency). In this example product subsidies would be preferred.

In the 1970s farm incomes were usually taken as the prime objective, with equity, import saving and efficiency as subsidiary objectives. Public expenditure was seen, if at all, as a constraint. What seemed to happen to the CAP, during the 1980s, is that what was once best viewed as a constraint became a major objective. Why this should have become so is now discussed.

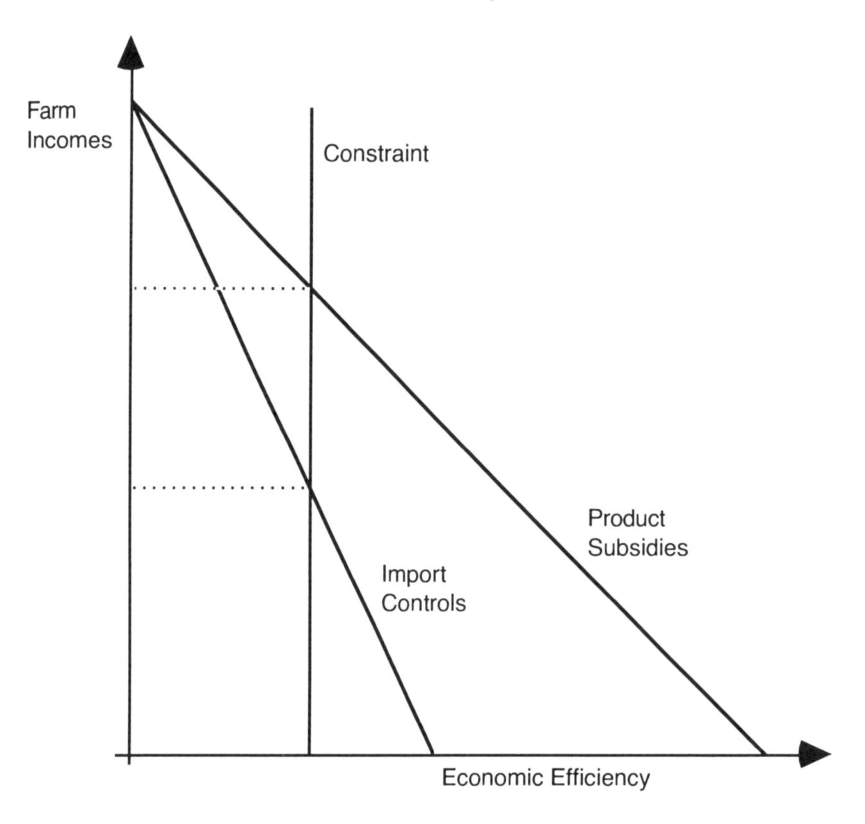

Fig.4.2: Policy objectives: Trade-offs.

The budget problem

Over the years, the EU has moved from a net importer to a net exporter of several major agricultural commodities. To the academic CAP specialist the point at which the EU moves from 100% to, say, 105% self-sufficiency in a commodity was seen as of no greater significance than when it moves from 95% to 100%. High-cost, inefficient output is still just that, whether or not the cost is expressed as an export subsidy or as a levy which displaces cheaper imported produce; high prices are still high prices to consumers and producers; and the world trading system is affected in much the same way by the elimination of a previous import requirement, as it is by the addition of a new export availability. But, from the point of view of the administration of the CAP, the movement into export surplus matters a great deal.

This is because it represents the point at which the Community budget begins to have to 'top up' the contribution of food consumers to enhancing the revenue of farming. When the Community is less than 100% self-sufficient, market prices can be supported solely by taxing imports. When EU production exceeds EU consumption, however, prices can only continue to be supported if the Community finances the purchase and disposal of surplus production; and, as surpluses grow, so does the budgetary cost of their disposal. By the early 1980s, the cost of surplus disposal had taken EC expenditure to the ceiling imposed by 'own resources', and the Community was forced to resort to various devices to balance the books. The most significant of these was to allow stocks to accumulate, resulting in the infamous commodity mountains and lakes.

Member states bear the cost of intervention until stocks are disposed of. By manipulating the level of export refunds, the Commission was able to control the timing of exports and effectively transfer intervention expenditure forward into the next financial year. But accumulation of stocks only delays the time at which the EU budget has to bear the cost of surpluses. Consequently, to understand much of the reform of the CAP over the past 15 years, it is necessary to appreciate that there has been really only one criterion at work, epitomised by the question 'Does the change make a positive contribution to the Community budget?' In attempting to classify the various policy changes, it is helpful to categorise them according to the main way in which they might be expected to contribute positively to the EU budget. This is done in Table 4.1. Broadly, there are two groupings — those which seek to reduce expenditure and those which seek to increase revenue.

Table 4.1: CAP reform measures classified according to budgetary effect.

Expenditure-reducing		
1)		a) price cuts
		b) stabilisers
		c) intervention criteria
2)	limit production	
		a) marketing quotas
		b) production quotas
		c) 'set-aside'
Revenue-increasing		
1)	product taxes	
		a) co-responsibility levies
		b) vegetable oil tax
2)	input taxes	
3)	national financing of CAP policies	
4)	new sources of revenue	
		a) related to agricultural production
		b) related to GNP

In choosing between alternative policy measures which have similar budgetary implications, a second criterion will be apparent. This is that the reform will be favoured which involves least change in the current balance of benefit for the interest groups affected by the CAP — particularly when one interest group is biased towards some member states — which is nearly always the case.

Expenditure-reducing measures

There are two dimensions to this particular group of measures: a) the quantity of surplus produce, and b) the unit cost of disposal (essentially the difference between EU intervention prices and international trading or 'world' prices). The first possible reform measure cited is that of *reducing support prices*. This is the measure favoured most by 'market economists'. Lowering the intervention price for a commodity will secure an immediate reduction in the unit cost of disposal. Subsequently, it should reduce (or at least impede the growth in) production and therefore the size of any surplus. (This assumes, fairly rationally, that the supply of a commodity is positively related to its price.) In the longer term such a policy measure may be expected to reduce yet further the unit cost of disposal, with the world price rising as a consequence of reduced EU production. Tyers and Anderson (1988) estimated the price-depressing effects of the CAP to range from 6% for wheat to 33% for dairy products.

Experience tells us, however, that although support prices were sometimes 'frozen' (at least in common ECU terms) and occasionally fell a little in real terms, with the Commission putting pressure on the Council of Ministers to moderate prices, it was simply not possible from a political point of view to cut prices sufficiently to constrain the cost of the CAP. The clash between the economist's solution of price cuts and the need for political acceptability is evidenced by the extreme difficulties that the Council had in 1990 in agreeing the modest price cuts implied by the Commission's proposal to reduce the 'Aggregate Level of Support' by 30% over a ten-year period, as the EU's initial offer under the Uruguay Round of GATT negotiations. Since then some progress has been made in reducing support prices and overall levels of agricultural protection, under the Mac Sharry reforms of 1992 and the final GATT agreement of 1994. Ironically, the compensatory payments and premia that accompany these price cuts may cause budgetary expenditure to *increase*, at least in the short run as the eventual agreement involved compensation for all farmers, rather than limited it to small farmers, as originally envisaged. Nevertheless, these cuts will bring CAP prices closer to world levels.

Since, in the past, straightforward price-cutting proved so difficult a task for the Council of Ministers, it became embodied in the agricultural *'stabilisers'*. These were designed to stabilise markets for the major commodities. Each stabilisation mechanism was tailored to the specific features of the relevant product, but with the same basic idea: whenever production broke through a ceiling set in advance, support was automatically reduced for that product (Commission of the European Communities, 1987). The production ceiling for a commodity was referred to as the 'maximum guaranteed quantity', and when this was exceeded a price reduction followed automatically in the subsequent season. In principle, the system had an advantage over earlier attempts at price-cutting, in that its effects were automatic and did not require approval from the Council each year. In practice, its major shortcoming was that, whilst price cuts were automatic, there was no such restraint on the setting of 'gross' prices. Thus, the cut related to a price which still required the agreement of the Council of Ministers. If there was to be any significant impact on curtailing supply, these 'gross' prices needed to be kept in check. Despite the open-ended nature of this price determination, there were some large reductions in prices for certain commodities in the late 1980s as a result of the stabiliser mechanism (for example, oilseeds).

Over recent years a number of changes have been made in the criteria governing *intervention* buying to make it a less attractive outlet. Minimum quality standards have been raised, the times of the year when intervention buying is undertaken have been shortened, and the prices at which commodities are purchased by intervention agencies have been reduced, that is, the 'buying-in' price and the intervention price have become separated. In bringing about these changes the Commission has sought to persuade farmers to produce for market rather than for intervention, and to restore the intervention agencies to their original role of 'buyer of last resort'. As a result, budgetary savings have been created.

The second set of expenditure-reducing measures is that which physically *limits production*. Viewed from the narrow criteria of budgetary effects and damage limitation to the existing balance of interests, *quotas* have considerable attractions. Production can be reduced very quickly to whatever level is regarded as manageable, and since prices do not need to be reduced as well (indeed they may be increased in compensation) the impact on farm income is likely to be only marginal. In the case of milk quotas, introduced in great haste in April 1984, production has been reduced significantly. As a consequence, intervention stocks of butter and skimmed milk powder have fallen, although with production in the Community still exceeding domestic consumption, milk remains an expensive sector within FEOGA. Quotas create innumerable problems and inequity but, perhaps somewhat surprisingly, have been integrated into Community agriculture more easily than had been originally envisaged. In part this is because, faced with a choice between quotas or a price cut equivalent, most producers will be better off opting for quotas. Marketing quotas (essentially, where a limit is placed on the quantity of output from each producer eligible for price support — as with the arrangements that now apply to milk) are only feasible where the major part of production has to pass through a number of controllable processing units. This is possible with milk (and sugar), but more difficult with many other products, in particular cereals. Thus, there have appeared proposals for restricting what is produced, rather than what is marketed — though of course the response to a marketing quota is probably to reduce production. One method of directly limiting production is to control the amount of land that can be devoted to certain crops — or 'set-aside' as it is commonly known. Set-aside was introduced for cereals in the EU in 1988.

Experience of set-aside in the United States, where it has been employed intermittently since the 1930s, is not encouraging. For a number of reasons, the impact on production is not as great as might be imagined, and any reduction that does occur tends only to be short-lived, as advancing technology continues to improve yields. Indeed, this last point is particularly pertinent in the case of member states of the EU, where production over the last 25 years has risen almost entirely as a result of better yields from virtually the same land area.

Work undertaken to ascertain the impact of set-aside has suggested, therefore, that it needs to be combined with longer-term measures, such as price reductions. [3] Indeed, these have now been introduced, although set-aside still has many critics. From the economist's viewpoint, it is perhaps most appropriately seen as an attempt on the part of the authorities to meddle with one of the inputs, albeit a rather special one, in the production process. In this respect, set-aside has some similarities with input taxes (see below). From a purely budgetary angle, the important question to be answered is whether the cost of the set-aside policy, in terms of the payments that have to be offered to farmers to induce them to idle part of their land, will be less than the cost of surplus grain disposal. If the former outweighs the latter, as some critics claim, set-aside offers no solution to the budgetary problem — though it does reduce production, and thus subsidised exports, which is a requirement of the GATT Agreement (see Chapter 17).

Revenue-increasing measures

Policy measures aimed at increasing the Community's budgetary revenue can seek to raise additional funds from various sources. With product taxes, the intention may be to shift part of the financial burden of surplus disposal on to producers. This is the idea behind the *co-responsibility levy*, a measure based on the so-called 'self-financing' production levies operated in the sugar sector. Introduced for milk in 1977 and for cereals in 1986, before being abolished for both in 1992, the co-responsibility levy is a flat-rate tax on all (or most) of a product passing through marketing channels. Essentially, it means that part of what the consumer pays for the product is diverted from its previous path to the producer, and into the Community budget. Whether the levy acts also as a producer price cut (and thus discourages production) depends on what the ministers do to support prices. Provided the levy is high enough, it is quite possible to make a particular CAP commodity regime 'self-financing', with the tax revenue used to finance surplus disposal. This is the direction in which the sugar regime has gone; the milk and cereal co-responsibility levies were very small in comparison.

Co-responsibility has long been a favourite with the Commission and in fact dates back to the earliest days of the EC. The principle — to make producers bear at least part of the cost of surplus disposal — seems eminently sensible. In practice, co-responsibility has had a chequered history. The main problem is essentially the same as that outlined above for the 'stabilisers', in that, whilst the co-responsibility levy reduces the price received by the producer, no formal restraint operates on the setting of the 'gross' support price by the Council of Ministers. Thus, it is not difficult for the levy to be transformed into a consumer tax, thereby avoiding any real hardship to producers (Hubbard, 1986). Even so, it was an unpopular measure with farmers.

Included in the 1987 annual CAP package from the Commission was a proposal for a tax on vegetable oils (strictly speaking, a *consumer price stabilisation mechanism*) which would operate in a similar way to a co-responsibility levy. The difference with oilseeds is that, because import tariffs are bound in the GATT, EC support has taken the form of deficiency payments, and the oilseeds policy has become increasingly expensive, even in the absence of surplus production. Domestic production has expanded and low market prices have required substantial budgetary payments. The effect of the tax would have been to reduce the deficiency payments on domestically grown oils, and to raise revenue from oils based on imported oilseeds. The Commission claimed that consumption and imports would not be affected — this was doubtful. In the event, the proposal was not accepted, which was probably for the best if American threats of a trade war were to be taken seriously.

An alternative measure of raising revenue, and one which has the added attraction of reducing production at the same time, is that of an *input tax*. This will raise the price of an input and discourage producers from using it. Interest has centred on the possible use of a tax on nitrogen fertiliser. This has yet another attraction — that of lowering what are becoming dangerously high levels of nitrate in ground water. Rickard (1986) argued that reduced nitrogen usage would inflict least harm on the smaller farms — a feature which some may find particularly appealing. However, there is considerable debate about the extent to which production is sensitive to fertiliser use and, more particularly, the extent to which fertiliser use is sensitive to rises in its price. So whether such a tax would have a marked impact on production is unknown but it would, of course, raise revenue. Its main attraction lies in the unusual coincidence of the preoccupation of CAP reform, as discussed here, and environmental concern.

Another area of reform, which in part is already under way, is that of *national financing* of the CAP. Member states have been responsible for partial financing of some regimes — for example, butter disposal measures and various production premiums in the beef sector. However, the main concern here is that a move to national financing of agricultural policies on a significant level would be seen as a regressive step and a departure from the principle of the single market. A similar but more acceptable development might be to link the financing of surplus production to those member states judged (by some criteria) to be most responsible, or to base member states' budgetary contributions on their shares of total Community agricultural production, rather than on value added tax and GNP, as at present (Buckwell *et al.*, 1981).

Traditionally, the EU has obtained revenue from its 'own resources', comprising customs duties, agricultural levies and the VAT-based contributions of member states. A fourth own resource, related to member states' GNP, has been added to these original three. Whilst GNP-based contributions will alter the distribution of net costs (costs minus benefits) between member states on to a more equitable basis, they are unlikely to make it any easier in the future for the Commission to increase the total amount of revenue available. In fact, the 'budgetary discipline' now imposed by the Council of Ministers restricts the annual rate of growth in the Guarantee Section of FEOGA. Whilst this puts a ceiling on the level of agricultural expenditure for any given year, it is unlikely to bring to an end the quest to reduce the budgetary burden and the need for further reform.

The future of reform

This chapter has traced the development of the CAP reform debate from its origins within the academic community to a discussion of the forces which have motivated reform during the 1980s and 1990s. The reform of the CAP in practice has been characterised, in the academic language of formal analysis, as 'what was once best viewed as a constraint (finance) is now a major objective'. Ironically, the 1992 reform package could increase the EU budget cost of the CAP, in that compensation payments and premia may exceed the savings from reductions in support prices, at least in the short term. This suggests that budgetary pressure will remain a major internal force for further reform of the CAP (see, for example, Tangermann and Josling, 1994), although some commentators argue that agricultural spending will be kept within acceptable limits following the Mac Sharry reform (Expert Group, 1994).

Whatever the budgetary outcome, there are other considerations. An unfortunate outcome of the 1992 reform has been an increase in the administration and bureaucracy with which the industry has to contend and there have already been demands for the 'paper work' to be simplified. Moreover, further changes to the CAP are likely by the end of the century to ensure that the EU meets its commitment to the 1994 GATT agreement. And enlargement of the EU, to include other west European countries and some from eastern Europe, will also act as a force for continued reform, as discussed in Chapter 14.

As for long-term reform of the CAP, if the goal is for agricultural prices in the EU to be brought down to world market levels, then eventually this will negate the need for any artificial restrictions on output. That is to say, output restrictions are not necessary at unsupported prices, since any production which is surplus to domestic requirements can be sold in non-EU (world) markets at no cost to the budget. In this situation, the EU would be an exporter as a result of some comparative advantage in agriculture. Quantitative (financial) limits would still need to apply to any existing compensation payments and premia, though whether these remain will be at the discretion of future politicians.

The ultimate aim must be for an agriculture which is market-led. As the world moves towards greater liberalisation in agricultural production and trade, farm support prices in the developed countries have to come down. This is the direction in which agricultural policy in the EU, and elsewhere, is moving, and the change to a market-led policy may be aided by firmer prices for agricultural products on world markets as we move towards 2000.

References

Buckwell, A.E. (1986) *Cereals Set-aside in the European Community,* Paper presented at the Agricultural Economics Society Conference, Reading University, October.

Buckwell, A.E., Harvey, D.R., Parton, K.A. and Thomson, K.J. (1981) Some Development Options for the Common Agricultural Policy, *Journal of Agricultural Economics,* 32(3).

Commission of the European Communities (1968) *Memorandum on the Reform of Agriculture in the European Community* (the Mansholt Plan), COM (68) 1000, Brussels.

Commission of the European Communities (1987) *Implementation of Agricultural Stabilizers,* Vol. 1, COM (87) 452, Brussels.

Commission of the European Communities (1991a) *The Development and Future of the CAP – Reflections Paper of the Commission*, COM(91) 100 Final. Commission of the European Communities, Brussels.

Commission of the European Communities (1991b) *The Development and Future of the CAP – Proposals of the Commission* COM (91) 258 Final. Commission of the European Communities, Brussels.

Expert Group (1994) EC Agricultural Policy for the 21st Century, *European Economy*, Reports and Studies No.4.

Harvey, D.R. (1994) Policy Reform after the Uruguay Round, in: Ingersent, K.A., Rayner, A.J. and Hine, R.C. (eds) *Agriculture in the Uruguay Round*, Macmillan, Basingstoke.

Hope, J. and Lingard, J. (1988) *Set-aside — a Linear Programming Analysis of its Farm Level Effects,* DP8/88, Department of Agricultural Economics and Food Marketing, University of Newcastle upon Tyne.

Hubbard, L.J. (1986) The Co-responsibility Levy — A Misnomer? *Food Policy*, 11(3).

Josling, T. (1969) A Formal Approach to Agricultural Policy, *Journal of Agricultural Economics,* 20(2), 175–191.

Marsh, J. (ed) (1991) The changing role of the common agricultural policy, *The future of farming in Europe*, Belhaven Press, London.

Marsh, J. and Ritson, C. (1971) *Agricultural Policy and the Common Market,* PEP, London

MAFF (Ministry of Agriculture, Fisheries and Food) (1986) *Diverting Land from Cereals (Note by the UK),* Paper presented at the Agricultural Economics Society Conference, Reading University, October.

Rickard, S. (1986) *Nitrogen Limitations: a Way Forward?,* Paper presented at Agricultural Economics Society Conference, Reading University.

Ritson, C. (Rapporteur) (1984) Sienna Memorandum, The Reform of the Common Agricultural Policy, *European Review of Agricultural Economics,* 11(2).

Ritson, C. and Fearne, A. (1984) Long Term Goals for the CAP, *European Review of Agricultural Economics,* 11(2).

Tangermann, S. and Josling, T. (1994) *Pre-accession Agricultural Policies for Central Europe and the European Union*, Study commissioned by DGI of the European Commission, Brussels, mimeo.

Tyers, R. and Anderson, K. (1988) Liberalising OECD Agricultural Policies in the Uruguay Round: Effects on Trade and Welfare, *Journal of Agricultural Economics,* 39(2).

Wageningen Memorandum (1973) Reform of the European Community's Common Agricultural Policy, *European Review of Agricultural Economics,* 1(2).

---

[1]For a full discussion of the Mansholt Plan, see Marsh and Ritson (1971).

[2]The 'evolution' of Commission reform proposals is traced in a paper (Ritson and Fearne, 1984) prepared for the Sienna meeting, which led to the 'Sienna Memorandum' mentioned earlier.
[3]See, for example, MAFF (1986), Buckwell (1986) and Hope and Lingard (1988).

Chapter 5

The New CAP

Alan Swinbank

Introduction

Earlier chapters have discussed the evolving nature of the CAP in the period since 1958, and the continuing pressures for CAP reform. These pressures led to the introduction of quotas on milk production in 1984, and a reduction in the real level of support prices throughout the 1980s, but did not fundamentally change the nature of the CAP. Although considerable differences existed between commodity sectors, a caricature of the CAP of the 1970s and 1980s (illustrated in Figure 1.1) would be of a managed market system with:

- imports restricted by the payment of variable import levies which were designed to ensure that imported products did not undercut a predetermined threshold price;
- subsidies paid on exports to enable excess supplies to be dumped on world markets; and
- intervention buying, and other market disposal mechanisms, to remove excess supplies from the internal market.

All this was designed to ensure that EC market prices remained in excess of, and more stable than, those prevailing on world markets. Farm revenues were thus enhanced — which policy-makers naively assumed would be reflected in a permanent increase in farm incomes — and consumers as well as taxpayers bore the burden of support.

Furthermore, as the Commission reported, this system of income support which depends almost exclusively on price guarantees, is largely proportionate to the volume of production and therefore concentrates the greater part of support on the largest and most intensive farms. So, for example, 6% of cereals farms account for 50% of surface area in cereals and for 60% of production; 15% of dairy farms produce 50% of milk within the Community; 10% of beef farms have 50% of beef cattle. The effect of this is that 80% of support (provided by Community taxpayers via the EU Budget) is devoted to 20% of the farms which account for the greater part of the land used in agriculture (Commission, 1991a, p. 2).

However, in the period 1992 to 1995, significant changes to this system of support were introduced such that the CAP of the latter part of the 1990s might legitimately be labelled the new CAP. The changes which justify the use of the adjective 'new' were:

- the Mac Sharry reforms enacted in May 1992, which partially shifted the burden of support from consumers to taxpayers, but which were by no means as radical as the Commission's reform proposals of 1991;
- changes to the green money system which swept away monetary compensatory amounts (MCAs) as of 1 January 1993 as part of the wider single market programme, and abolished the switchover coefficient from 1 February 1995; and
- the GATT Agreement of December 1993, implemented in 1995, which should turn variable import levies and other non-tariff barriers into conventional tariffs, and which imposes constraints on the level of support that can be provided by the CAP.

This chapter focuses on the Mac Sharry reforms of 1992, but necessarily notes the new import arrangements and other aspects of the GATT Agreement. Chapter 6 deals with green money, and Chapter 17 more specifically with the GATT Agreement. In the following discussions on the new CAP, however, it should not be presumed that the reform process is complete. The Mac Sharry reforms focused on cereals, oilseeds and protein crops, tobacco, and beef and sheepmeat production. Other sectors were barely affected, and await reform. Furthermore, the budgetary problems of the CAP have not gone away: indeed they were exacerbated rather than assuaged by the Mac Sharry reforms, and as a consequence budgetary pressures are bound to lead to further adaptations of the CAP in future years. The GATT Agreement will over time impact more heavily on the CAP. Even if radical change can be avoided during the six-year implementation period of the Agreement, this cannot be so over a longer time horizon. And the prospect of a further enlargement of the EU to embrace countries of central and eastern Europe, as discussed in Chapter 14, raises challenging questions about the future of the CAP. Thus the new CAP of this chapter may well prove to be a fleeting phenomenon, with a radically different policy emerging by the end of the decade.

The Mac Sharry reforms

When the Irish Commissioner Ray Mac Sharry was given responsibility for agriculture in Jacques Delors' 1989–1992 Commission, many commentators presumed that this was an unwise choice. As a result of being a net exporter of agricultural products, the Irish economy has benefited significantly from the CAP, and it was widely presumed that an Irish Commissioner for Agriculture, who was said to have political ambitions on his eventual return to Ireland, would not prove to be a reformist Commissioner. In retrospect it can be seen that Mac Sharry's name is associated with the most significant changes in the CAP to date; and although he had left office before the Uruguay Round GATT negotiations had been concluded, he was the Farm Commissioner at the time of the Blair House Accord of November 1992 which helped settle key components of the GATT Agreement on Agriculture between the EU and the USA. Indeed, as shall be noted in the paragraphs that follow, the GATT negotiations and CAP reform were interlinked, even though the Commission at the time insisted that they were two separate issues.

The proposals

The original timetable for the Uruguay Round of GATT negotiations provided for a formal conclusion in December 1990. In October 1990, however, the EC was still attempting to resolve internal disputes as to the magnitude of any cut in CAP support it was willing to countenance. The debate centred on the Commission's proposal for an EC offer of a 30% cut in support. It is reported that the German Chancellor, Helmut Kohl, intervened directly in this debate, and sought "from the Commission concrete guarantees that losses in farmers' incomes resulting from the GATT price cuts would be compensated by additional production-neutral aids, and by reinforced set-aside and extensification programmes" (*Agra Europe*, 19 October 1990, p. P/3). When agreement on the EC's negotiating stance was finally settled at a joint meeting of agriculture and foreign trade ministers on 6 November 1990, "the Commission undertook to submit in the very near future [proposals for] support measures designed to soften the effect on Community agriculture of the reductions in support which will ensue from the Community offer" (Council, 1990, p. 5). Mac Sharry's reform proposals of February 1991 (Commission, 1991a) met this requirement, but by now the Commission was denying any formal link with the GATT negotiations.[1]

The plan, elaborated more fully in July 1991 (Commission, 1991b), was to reform the support arrangements for cereals (and oilseeds and protein crops), tobacco, milk, beef and sheepmeat, together with an "Agri-Environmental Action Programme" and plans for the afforestation of agricultural land, and to encourage early retirement.

A central feature of the Mac Sharry reforms was a proposal to reduce the support price of cereals by 35%, to something approaching world market price levels, whilst introducing an area payment to cereal producers to make up — in part — the fall in farm revenues. However, beyond a certain farm size, farmers would have to set-aside a portion of their arable land if they were to qualify for the area payments. Set-aside land could not be used to produce crops subject to CAP support, or graze livestock, during the period of set-aside. However, arrangements would be made to allow farmers to grow 'non-food' crops on set-aside land for industrial use, provided the marketing channels were outside the normal CAP support arrangements.

Arable area payments would be 'modulated': that is, small farmers would be compensated in full, but beyond a certain farm size threshold only partial compensation would be paid. The concept of modulation attracted a hostile response from organisations representing the larger farmers, and was subsequently dropped from the package, thereby significantly increasing the taxpayer cost of the new support arrangements. In the February 1991 document, the Commission had suggested that "in the immediate future" the sugar policy should be subject to similar reforms. In practice this did not come about, and in 1995 the existing policy mechanisms were rolled forward for a further five-year period.

For milk, the quota system was to be retained, but there was to be a 5% cut in quota and a 10% cut in support prices. In compensation: the co-responsibility levy was to be abolished, headage payments on dairy cows were to be introduced for less-extensive (i.e. grass-based) producers who would not benefit from lower-priced animal feeds, and there was to be compensation for quota reduction. Bearing in mind the fact that agricultural economists have long advocated the introduction of a transferable bond in compensation for reductions in price support, the Commission's text is reproduced in full:

> Farmers whose [milk] quotas are reduced, will receive an annual compensation of 5 ecu per 100kg over a period of 10 years. Member states can add a national supplement.
> The compensation arrangements will be operated through a bond issued to the farmers concerned, on the basis of which the Community would make annual payments over its life-time (10 years). The farmers could choose to keep the bond and receive the associated annual payments, or could sell it on the private market (Commission, 1991b, p. 21).

The bond scheme is discussed later in this chapter.

In the event, neither the bond scheme nor the headage payments were introduced. Limited reductions to support prices were implemented, and because of the complicated politics of the quota regime in Italy, the total volume of milk quota was actually increased rather than reduced.

For beef, intervention was to be significantly reduced and the revenue loss made up by headage payments; and for sheepmeat the ewe premia (headage payments for female sheep) were to be limited by quota. Support payments for tobacco were to be significantly reduced, and production limited by quota.

The new support arrangements for cereals

The Mac Sharry package was accepted by the agriculture ministers in May 1992, and there were then three annual reductions in support prices for cereals: first for the 1993 harvest, then for 1994, and finally for the 1995 harvest. The 1995/96 intervention price, at the beginning of the marketing year, was 119.19 ecu/tonne as illustrated in Fig.5.1.[2] This new level of price support was fixed for an indefinite period, and thus will apply until such time as it is changed. Fig.5.1 also defines a target price of 113.11 ecu/tonne. It should be noted that this target price is not the same as the target price that was fixed under the pre-1992 cereals regime. The target price is significantly below the intervention price that used to apply for cereals prior to the Mac Sharry reforms.

It is the approximate difference between the old intervention price, and the new target price, which defines the rate of area aid paid to cereal growers, and is discussed more fully below. The intervention and target prices apply for all cereals, whereas under the old regime a different level of price support applied depending upon the cereal (e.g. barley or wheat). Similarly, all cereals receive the same arable area aid — though additional area payments are paid on durum wheat in recognised production regions.

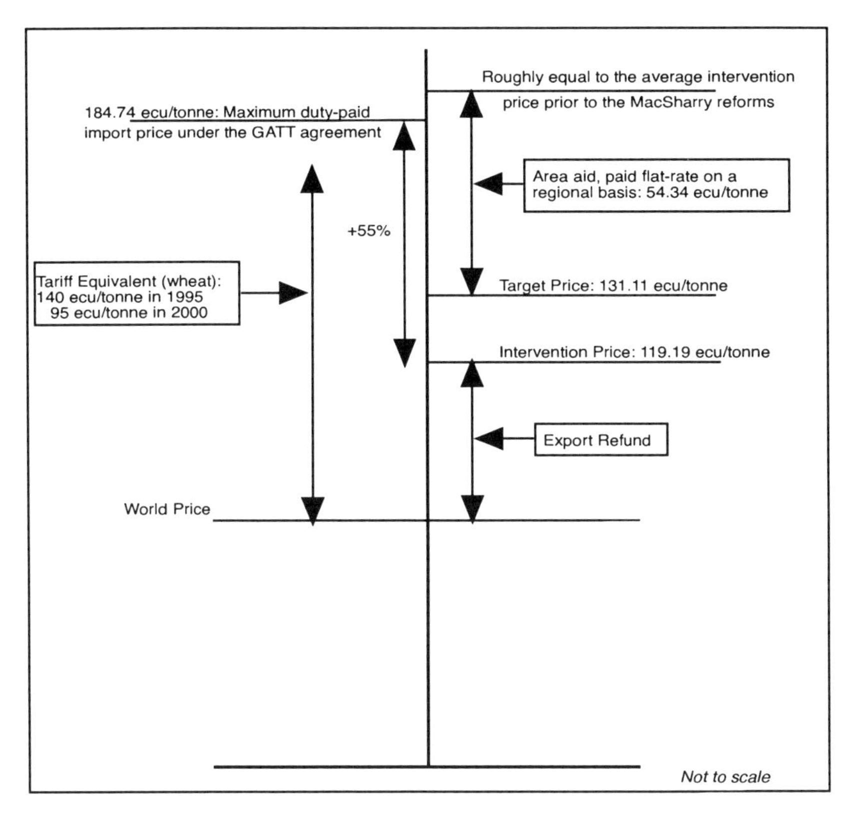

Fig.5.1: Price support for cereals from July 1995.

From 1 July 1995 new import arrangements have applied as a consequence of the GATT Agreement on Agriculture. Thus, under the process of tariffication, the old system of threshold prices and variable import levies has been abolished. Instead, fixed tariffs are to apply on imports. Under this process, the EU determined a new tariff equivalent of 149 ecu/tonne for common wheat, be reduced in six equal steps to a new bound rate of 95 ecu/tonne from the year 2000 onwards. The first of the annual reductions resulted in a tariff on wheat of 140 ecu/tonne from 1 July 1995.

However, recognising that these tariffs legitimately calculated under the tariffication rules would result in punitively high import charges on cereals — for the simple reason that they were calculated on the basis of the price gaps prevailing prior to the Mac Sharry reforms of 1992 — the US insisted that the Community limit itself to applying an import duty "at a level and in a manner so that the duty-paid import price for such cereals will not be greater than the effective intervention price... increased by 55%".[3] In Fig.5.1 this defines a maximum duty-paid import price of 184.74 ecu/tonne, and for much of the time this is expected to impose a binding limit on the import charge that can be applied. It should be noted that the 'intervention price plus 55%' rule, determining the maximum-duty paid import price, results in a price level at the beginning of the marketing year exactly equal to the now abandoned threshold price.

The way the EU honoured this commitment was to fix import charges for six categories of cereals on a fortnightly basis.[4] Prices on specified US grain markets for:

- Common wheat
  - Hard Red Spring No 2 (high quality)
  - Hard Red Winter No 2 (medium quality)
  - Soft Red Winter No 2 (low quality)
- Durum wheat
- Maize
- Barley (which also determines the import charge for rye and sorghum)

are measured, to which transport costs to Rotterdam are added. This determines a reference price for each of the six cereals. The import charge for each cereal, fixed fortnightly from a Thursday, is then equal to the intervention price plus 55% minus the reference price for the cereal concerned. Consequently on 1 July 1995, for the first time under the new arrangements, the import charges were determined at:

	ecu/tonne
—Common Wheat	
—high quality	10.20
—medium quality	47.79
—low quality	60.02
—Durum wheat	zero
—Maize	111.95
—Barley and rye	78.28

Britain's Home-Grown Cereals Authority noted that under the old rules the variable import levy on common wheat — of all qualities — would have been about 51 ecu/tonne (Home-Grown Cereals Authority, 1995, p. 1). It is evident that the new arrangements result in a narrowing of the landed-price gap between the different quality wheats, and results in some reduction in the cost of importing high quality bread-making wheats.

Abatements can be claimed by traders under specified circumstances, but there remains a risk that on certain consignments the maximum duty-paid import price could be exceeded. Indeed, within weeks both Canada and the US had indicated their intent of challenging the EU's new import arrangements within the disputes settlements procedure of the World Trade Organisation. The EU's critics argued that the 'intervention price plus 55%' limitation had to apply on a consignment basis; but the EU has with justification argued that if the import charge were to be determined on a consignment basis there would be a major incentive for fraud. Traders would try to ensure that the invoices they presented to customs officers would minimise the import charge to be paid. The net result, of course, is that the new system of import charges for cereals (and rice) in effect recreates the old variable import levy mechanism that was outlawed under tariffication!

In addition to import charges and intervention purchases, export refunds can also be used to support EU market prices, though the GATT Agreement does place limits on the volume of subsidised exports and the overall level of expenditure on export subsidies (see Swinbank and Tanner, 1996; and Chapter 17). It is these export constraints which threaten to constrain the CAP in coming years.

In the context of Fig.5.1 it might be helpful to note that the EU declared a world market price of 93 ecu/tonne for the base period 1986–88, in determining the tariff equivalent for wheat. However, by the end of 1995, poor harvests and reduced stocks had driven world cereal prices up to EU levels, and consequently EU import charges on cereals fell away to zero and instead export taxes were introduced.

For most other sectors, with the notable exception of fruit and vegetables (see Swinbank and Ritson, 1995), the post-GATT import tariffs differ significantly from the old variable import levies. Fixed import tariffs apply, and consequently the landed price of imports will rise and fall reflecting movements in world market prices. For many products, for example butter and sugar, these MFN (most favoured nation) tariffs will probably be prohibitively high in the early years of the Agreement on Agriculture. Furthermore, under the safeguards clause, additional import taxes can be applied if the invoiced price of a particular consignment falls below a reference level based upon the EU's actual import price in the 1986–88 base period. However, tariff quotas — under which specified quantities can be imported at less than the full MFN tariff rates — have been established for those products for which the EU previously granted tariff concessions (for example on New Zealand butter, and sugar from India and the African, Caribbean and Pacific states under the Lomé Convention), and under the minimum access arrangements where imports claimed less than 5% of the EU's market in the base period (for example for cheese).

Arable area payments and set-aside

As noted above, although the Mac Sharry reforms cut the levels of market price support for cereals, a new arable area aid system was brought into play which was designed to maintain farm revenues. Fig.5.1 depicts the basis for the determination of area aids for cereals: an aid payment of 54.34 ecu/tonne in 1995/96. (Similar, but not identical, schemes apply for oilseeds and protein crops.) Each member state had to specify a number of cereal growing regions, and for each of those regions determine an average annual yield for a period prior to 1990/91. Each member state has determined its regions on a slightly different basis: the United Kingdom, for example, originally declared five regions, with an average cereal yield in England of 5.93 tonnes/hectare. The regional yield (5.93 tonnes/hectare), and the basic rate of aid payment (54.34 ecu/tonne) determine the area aid of 322.24 ecu/hectare paid in England on eligible cereal land in 1995. Assuming unchanged cereal yields and with a market price equal to the target price, the average grower now receives 71% of revenue from growing cereals from the market, and 29% from area aid. Clearly the switch in support has favoured farmers with less than average cereal yields, and penalised those with higher than average yields.

On small claims, producers can receive arable area payments without a set-aside obligation. Small claims are defined on a regional basis. It is that area of land which is capable of producing 92 tonnes of cereals at the regional yield, and averages out at about 20 hectares across the EU. If a farmer wishes to claim arable area payments in excess of the 'small claims' threshold, then land must be set-aside. Originally the set-aside rate was set at 15%, so for example on an arable area of 100 hectares, a farmer could claim arable area payments on 85 hectares of a mixture of cereals, oilseeds and protein plants provided 15 hectares had been set-aside. Detailed rules apply to the management of set-aside land. In addition, it should be noted that farmers also receive area payments on their set-aside land.

Integrated Administration and Control System (IACS)

The EU's clear intent in introducing set-aside was the need to convince the US in the GATT negotiations that the EU was willing and able to control supply. However, it would be naive to presume that a 15% set-aside rate would result in a 15% reduction in output.[5] First, the 'small claims' exemption means that many farmers do not need to set-aside land but they are still entitled to claim area payments on all, or the bulk of, their sown land. Second, even though the original set-aside rate of 15% was for rotational set-aside, there is still scope for farmers to set-aside their worst land. It was originally set for a six-year rotation, and thus covered only 90% of a fixed area of land. Furthermore, a period of fallow was historically used to boost soil fertility and subsequent yields, and some such effect could be expected from rotational set-aside.

A more important potential source of slippage, however, was of farmers *maintaining* their cropped area unchanged, whilst bringing other land under the plough to set aside. There are two controls which attempt to limit these possibilities. First, for the individual farm, any land that was in permanent pasture, permanent crops, forest or non-agricultural use on 31 December 1991 cannot be used for set-aside or to grow crops on which arable area payments are claimed. Even so, it must be recognised that a farmer who had 100 hectares of cereals, and 20 hectares of temporary grass on 31 December 1991 would be entitled to claim area compensation payments on 102 hectares of cereals and 18 hectares of set-aside under the rules.

Second, in any particular region, if the total area of crops and set-aside land on which area payments are claimed exceeds the base period area, then all arable area payments claimed that year will be scaled back by a corresponding percentage, and an additional area of uncompensated set-aside would be imposed on the region's producers in the following year. It must be noted, nonetheless, that there has subsequently been political slippage on this provision, in that ministers have found excuses for not applying the full penalty when regional base areas have been over-shot.

To help combat fraud, and ensure that this new scheme of arable area (and headage) payments could be effectively applied, ministers introduced a new Integrated Administration and Control System (IACS) for the CAP. In effect, if farmers wish to claim arable area and headage payments on their holdings, they must complete detailed IACS forms. These are designed to ensure that only eligible land is entered into the scheme, that the correct set-aside rate and rotation is applied, and that only one claim is made on any individual piece of land. This is particularly important with respect to arable area and headage payments. As will be explained below, headage payments are subject to an extensification rule. That is, there must be a minimum area of forage on the farm for each qualifying animal. Land sown to cereals can count towards this minimum forage area requirement, but the same piece of land cannot be used to claim both arable area payments and headage payments.

Impact on the EU's cereal market

The Mac Sharry package of May 1992 was not achieved without cost. In particular the reform proposals for milk were jettisoned, the need to extend the reform into other sectors — sugar, fruit and vegetables, wine — was forgotten, and the proposal to introduce modulation into the arable area payment scheme was rejected. Farmers receive arable area payments in full, without any hectarage limit. Arable area payments have in part switched support from consumers to taxpayers, and because Mac Sharry's proposal to modulate arable area payments was rejected the taxpayer cost has escalated, threatening a new budget-driven crisis for the CAP in the latter part of the 1990s. In 1996, arable area payments are budgeted to account for 38% of total expenditure on CAP income and price support.

But having secured ministerial approval for CAP reform, Mac Sharry's ambition was now to conclude a GATT agreement. This involved satisfying two sceptical groups: first, the US negotiators, that the CAP reforms were in line with any negotiable GATT agreement based upon Arthur Dunkel's Draft Final Act of December 1991; and second, EU farm ministers, that a GATT agreement would not involve further CAP reform. This was largely achieved in the Blair House Accord of November 1992.

As a consequence of the Blair House and subsequent discussions, the arable area and headage payments introduced into the CAP by the Mac Sharry reforms were deemed to be decoupled. Decoupled is a new word introduced into the language by the GATT negotiations. It is used to describe a support payment to the farm sector that has no discernible impact upon production, and hence upon trade volumes, and thus which need not be subject to any GATT disciplines. We discuss below the extent to which arable area payments can objectively be deemed to be decoupled. But this decision, together with the computation of a base-period Aggregate Measurement of Support (AMS) for the EU's farm sector *prior to* the Mac Sharry reforms, means that the 20% AMS reduction built into the GATT Agreement has no discernible impact on the CAP.[6] The so-called Peace Clause, which extends to the year 2004, ensures that these payments cannot be challenged, provided the overall level of support does not exceed that agreed in 1992.

In response to its EU-based critics who claimed that the export constraints built into the draft GATT agreement implied further CAP reform and in particular cuts in production, the Commission's response was that "the draft GATT agreement and the Reform of the CAP form a coherent whole" (Commission, 1992, p. 10). In particular, with respect to cereals, the Commission argued that yields were likely to stabilise: "the reform of the CAP is based on per hectare payments with a fixed yield...There is no longer any incentive for higher yields as there will be no payments beyond the fixed yield" (Commission, 1992, p. 7). Thus the Commission's "worst case scenario" was for yield increases of 1% a year, even though over the previous five years yields had been growing at 1.5% per year. Lower cereal prices, as a consequence of the Mac Sharry reforms, would also stimulate consumption, first as a consequence of the replacement of cereal substitutes by cereals in animal feed rations, and second as a consequence of increased consumption, and hence production, of pig and poultry meat.

The Commission's many critics argued that the computed impacts on both production and consumption were over-stated, and concluded from this that further controls on production — in particular, in the form of a higher rate of set-aside — would be necessary if the GATT export constraints were to be met. This is not the place to attempt an arbitration between these views. In addition to uncertainties about the size of the relevant price (and cross price) elasticities of supply and demand, and the vagaries of the weather in coming seasons, the way the Commission will manage the EU's cereal market, and thus its impact on EU market prices, will have an important bearing on the outcome. However, some comments on the likely impact of the Mac Sharry reforms on supply response are warranted.

Under normal circumstances, a price cut would have a two-fold impact on cereal production: a yield, and an area response. Josling's 'guesstimate' is that 70% of the normal supply adjustment might be attributed to a yield response, with the remaining 30% an area response (Josling, 1994, p. 517). Thus, in Fig.5.2, whilst a price reduction from P1 to P2 would normally result in a cut in the planned supply from S1 to S2, there would be a smaller response — from S1 to S3 — under the Mac Sharry reforms as land would be retained in crop production in order to qualify for the arable area payments. Consequently economists would predict that the Mac Sharry reforms would result in a reduction of yields in the first instance, as farmers applied lower applications of fertilisers and agro-chemicals in their quest for profit maximisation; but of course over time yields would resume their upward trend as productivity gains were captured by the farm sector. Unless the yet to be adopted technologies are uniquely linked to high fertiliser and agro-chemical application rates, it is difficult to see why future increases in yields would be affected by the Mac Sharry reforms. Furthermore, it is evident from Fig.5.2 that the area payments for arable crops are not fully decoupled, in that the quantity supplied with area payments (S3) exceeds the volume that would be supplied at that price level in the absence of arable area payments (S2).

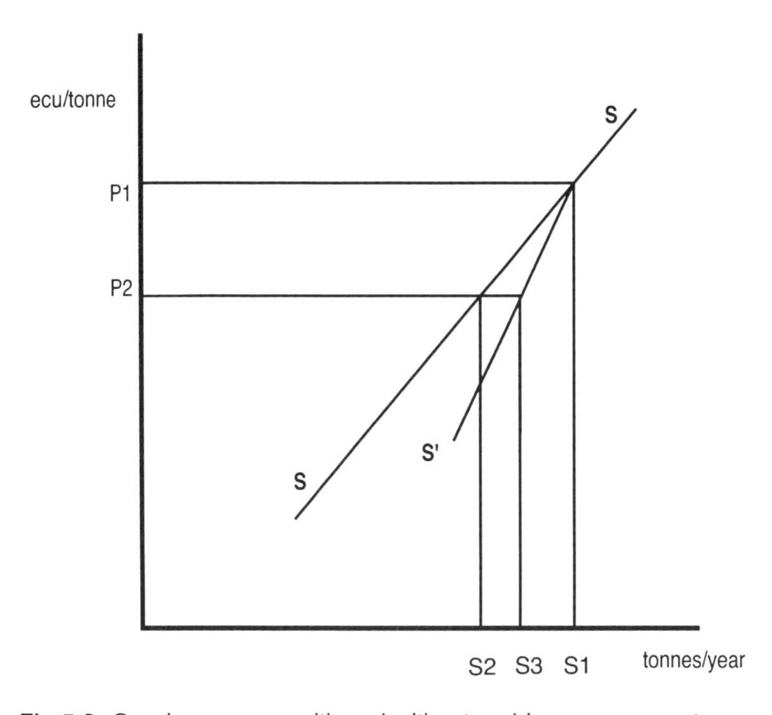

Fig.5.2: Supply response with and without arable area payments.

Arable area payments, the price of farm land and the advantages of a bond

As can be seen from Fig.5.1, we now have two systems of support for cereal producers, both of which have to be administered and both of which are susceptible to fraud. Significant administrative economies could be gained, it is claimed, if the mechanisms for intervention buying and the granting of export subsidies could be removed, and arable area payments correspondingly increased.[7] Furthermore, this would mean that the GATT export constraints on cereals (and indeed on pig and poultry meat) would be redundant as export subsidies would no longer be paid. This in turn would suggest that set-aside would no longer be required. However, there would be a further escalation of budgetary costs, and the switch would do nothing to relieve the distortive impact on the land market induced by arable area payments.

In the Council Regulation of June 1992 establishing the new system of support for arable producers, the Council referred to the need to "*compensate* the loss in income caused by the reduction of the institutional prices by a *compensatory* payment for producers who sow such products" (Council, 1992, p. 12; emphasis added). In this chapter, however, we have eschewed use of the term 'compensatory payments', preferring instead 'arable area payments'.

When the scheme was first introduced it did compensate existing producers for the cut-back in market price support. But payments are not restricted to individuals who were farming when the new scheme was introduced: they are paid to whoever happens to farm the land, and unless revoked or reduced will be paid annually for ever. Nor are the payments related to subsequent movements in market prices: thus they remained unchanged even though world cereal prices soared in the winter of 1995/96. Landlords on letting their arable land, and owners on sale, will be aware of these future income streams attached to arable land; and consequently we must expect that the sale price, and rental value, of arable land will reflect the future level of arable area payments. New entrants into the industry will incur inflated costs for land use, and in turn they will be dependent upon the continued payment of arable area aids to meet their mortgage or rental commitments. Thus the high cost structure of the industry continues.

The Mac Sharry reforms — and in particular the arable area payments including the area payments on set-aside — sustain the price of arable land, and this benefits the landowner rather than the tenant farmer. High land prices discourage the diversion of arable land into alternative uses, such as forestry or bird sanctuaries, and — as we saw in Fig.5.2 — maintains more land in crop production than would otherwise be the case. The system also affects the relative prices of land and land-saving inputs such as fertilisers and agro-chemicals. Economic theory suggests that if the price of land is increased relative to that of fertilisers and agro-chemicals, then heavier applications will be applied.

For the farmer, political uncertainties undermine the credibility of the system. Although the legislation provides for the payment of arable area aids for an unlimited time, many farmers doubt their longer term sustainability. Not only does the high taxpayer cost render the policy vulnerable to budget cuts, but the transparency of the system, with very large payments to some farms, could lead to a popular backlash which would undermine the political consensus in favour of farm income support.

For the EU there is a potential political embarrassment. If arable area payments are part of the normal package of CAP farm income and price support, can they be denied to the farmers of central and eastern Europe if and when these countries join the EU? And if not, what then of the budgetary cost of the CAP, and of the EU's GATT obligation to reduce its AMS by 20% by the year 2000? In short, one of the reasons many commentators have suggested that the CAP will have to be reformed before the next enlargement is because of the difficulty of extending arable area payments to central and eastern Europe.

Some of these difficulties could be overcome by announcing now a time limit for the continued payment of arable area aids (and also headage payments), and possibly limiting payments to individuals who were engaged in farming on a specified date. The problem with the latter proposal is that it would discourage the retirement of individuals who would otherwise have left farming, or ceased growing cereals, as the implicit value of their assets would fall on retirement. Furthermore, these limited changes to the system would do little to further decouple support from production.

A fully decoupled reform would involve the replacement of existing entitlement to area payments by a transferable bond, as advocated for example by Tangermann (1991). The IACS system means that bonds could readily be issued to individuals who have claimed arable area aids, and livestock headage payments, since 1992. Bonds could also be issued to replace milk quotas, and to individuals who hold delivery contracts for sugar beet to the beet processing factories; and ideally the system would be extended throughout all the commodity regimes currently subject to CAP price support. Undoubtedly some rough justice would be involved, as when milk quotas were first allocated in 1984.

In year one, the payment to the bond-holder would be the same as that made under the displaced scheme. Thus an existing arable area aid would be directly replaced by a bond payment. However, unlike the present arrangements, farm ministers (backed by the EU and member state financial authorities) would guarantee the future pattern of payments to the bond-holders. The bonds would have a fixed life, say ten years; and annual payments would be tapered, falling away to zero at the end of ten years.

Payments would be made to the bond-holder, and would not be conditional upon future farming activities. A farmer could retire, but still receive the annual payments. Furthermore, bonds could be bought and sold, and a market would develop. With future payments clearly defined, and guaranteed by the EU and member state financial authorities, these paper assets would offer the same financial security as gilt-edged government stock. The market price of bonds would reflect their future earning potential, prevailing interest rates, etc. Farmers could choose to sell their bonds and retire, or indeed continue farming. Some critics of transferable bonds have misunderstood this proposal: they have assumed that a farmer's ability to sell his/her bond for a capital sum implied an up-front call on the EU's budget, and thus that the scheme was untenable from a budgetary perspective. But this is an invalid criticism: it is the financial markets which would provide the capital sum, whilst the EU's Budget would be called upon to make regular annual payments to the bond-holders — whoever they might be — for the lifetime of the bond.

The creation of this financial asset — the bond issued to farmers — would result in a corresponding fall in the price of other assets: notably in the price of arable land and of milk quotas. This would mean that new entrants could buy into the industry without incurring the excessive entry costs currently associated with milk quotas, the arable area payment and other schemes. Furthermore, agricultural support would be truly decoupled from production, and thus would be compatible with the EU's GATT obligations and would no longer result in distortions in the rural land market.[8] Existing agricultural businesses would however be sheltered from bankruptcies by the annual payments under the bond.

One unresolved problem relates to the allocation of the bond between landlord and tenant, for in practice the replacement of existing support measures by a bond will result in a heavier fall in the value of landlords' capital than in tenants' capital. This is also the case when milk quota or entitlement to ewe premia is sold off the farm. However, for both milk quota and ewe premia, the

British legal system has found mechanisms for the appropriate allocation of quota value between landlord and tenant, and so farm ministers should perhaps be encouraged to press ahead with this long overdue reform without worrying unduly about the eroding asset values of landowners.

Headage payments

Beef and sheep producers are entitled to claim a complex array of premia on the number of animals kept. No attempt will be made here to describe the support arrangements in any detail, and only three of the headage payment schemes will be mentioned specifically:

—the beef special premium, which is payable twice during the lifetime of male cattle reared for beef production;

—the suckler cow premium, which is paid annually on cows of beef-breeds kept to rear beef calves; and

—the ewe premium, which is paid annually on female sheep and goats.

All three schemes share the characteristic that the number of claims is limited to a total entitlement established in the past, and thus in the context of the GATT Agreement all three are deemed to be decoupled payments which do not form part of the overall AMS calculation, and are not subject to reduction commitments.

Two of the schemes — the suckler cow premium and the ewe premium — are limited to the number of claims an individual producer made in 1991 or 1992. Thus they are akin to a quota mechanism, though the quota limits the number of claims made and not the number of animals kept. Additional animals may be reared, but they do not qualify for the premia payments. Furthermore, entitlement to receive premia can be transferred: not just with the land, but from farm to farm. Thus — as with any transferable quota mechanism — the entitlement to receive these premia creates assets with financial values which will reflect the market participants' expectations of future premia payments. As the quota determines only the eligibility to receive premia, but does not restrict supply, we would expect the price of quota to reflect fully the premia payments — provided the market price in itself is sufficient to encourage some producers to keep ewes and suckler cows. In these circumstances, the entitlement to receive premia payments is very similar to a bond scheme. The differences lie in the uncertainty attached to the level of future payments, and in the fact that the entitlement is only of value when linked to sheep or cattle rearing.

The beef special premium differs in that any producer is entitled to claim, but claims are liable to be scaled back if the total number of claims in a particular region exceeds the regional maximum. Thus individual quota rights — and hence assets — are not created. The beef special premium is however limited to a maximum annual claim of 90 animals per holding, for each of the two eligible age groups.

As well as these constraints (entitlement quota and/or headage limit as applicable) on a farmer's claims for headage payments, there is a third limitation. This is that the beef special premium and the suckler cow premium are only paid up to maximum stocking density limits. Once the Mac Sharry reforms have been fully implemented in 1996, the maximum stocking density limit will be 2.0 livestock units (LU) per forage hectare; but with an additional extensification premium paid if the stocking density falls below 1.4 LU per forage hectare. Once again it should be noted that these limits do not control production, but rather they limit the number of claims for premia which can be lodged. The computation of the number of LU per forage hectare involves a count of the number of beef cattle and ewes on which the three headage payments are claimed, and the number of dairy cows deemed necessary to produce the holding's milk quota entitlement. It also involves a delimitation of the holding's forage area, taking care to ensure through the IACS scheme that land in set-aside, and on which arable area aids are claimed, is not counted as part of the forage area.

All of these rules involve farmers in complicated computations as they fill in their annual IACS forms and determine what is the optimum portfolio of animals, arable crops and set-aside on which to claim headage and arable area payments. The bureaucracy has certainly increased but, despite the complaints of some farmers, a day or two in the farm office carefully filling in the IACS form can be more rewarding financially than hours spent on the tractor, or in the lambing shed.

Concluding comments

This chapter has attempted to give a flavour of the CAP in the mid-1990s, particularly the support system for cereals after the implementation of the Mac Sharry reforms in 1992 and the GATT Agreement in 1995. It has been suggested that the CAP of the mid-1990s is sufficiently different from the CAP of 1990 to warrant use of the phrase 'the new CAP'. However, the CAP remains in a state of flux, and policy mechanisms will change. This chapter has not attempted to predict the future evolution of policy; but it has suggested that a preferable alternative would be the replacement of existing support mechanisms by a transferable bond, an issue which appears again in the final chapter which considers the CAP of the future.

References

Commission of the European Communities (1991a) *The Development and Future of the CAP. Reflections Paper of the Commission*, COM(91)100. CEC: Brussels.

Commission of the European Communities (1991b) *The Development and Future of the CAP. Follow-up to the Reflections Paper — Proposals of the Commission*, COM(91)258 as amended. CEC, Brussels.

Commission of the European Communities (1992) *Agriculture in the GATT Negotiations and Reform of the CAP*, SEC(92)2267. CEC, Brussels.

Council of the European Communities General Secretariat (1990) Special Council Meeting. Agriculture — with the participation of the Ministers for Foreign Trade — Brussels, 5 and 6 November 1990. *Press Release* 9721/90 (Presse 173), Brussels.

Council of the European Communities (1992) Council Regulation (EEC) No 1765/92 of 30 June 1992 Establishing a Support System for Producers of Certain Arable Crops, *Official Journal of the European Communities*, L181, 1 July.

Home-Grown Cereals Authority (1995) *Weekly Bulletin*, 30(1), 3 July.

Josling, T. (1994) The Reformed CAP and the Industrial World, *European Review of Agricultural Economics*, 21, 513–527.

Swinbank, A and Ritson, C. (1995) The Impact of the GATT Agreement on EU fruit and vegetable policy, *Food Policy*, August.

Swinbank, A. and Tanner, C. (1996) *Farm Policy and Trade Conflict: The Uruguay Round and Common Agricultural Policy Reform*, The University of Michigan Press, Ann Arbor.

Tangermann, S. (1991) A Bond Scheme for Supporting Farm Incomes. In John Marsh (ed.), *The Changing Role of the Common Agricultural Policy: The Future of Farming in Europe*. Belhaven Press, London.

[1]For a fuller discussion, see Swinbank and Tanner (1996).

[2]The switch from the green ecu, to the commercial ecu, in February 1995 complicates this story. The first four columns of numbers in Table 5.1 below are reported in the green ecu of the time. The final column, reporting the actual support prices in 1995/96, is in the commercial ecu then used in the CAP. In Chapter 6 it will be found that the appropriate coefficient to convert from green to commercial ecu is 1.207509. This is reflected in the last row of the table, but in the first two rows some reduction in green ecu support prices had been implemented since the 1992 agreement, to claw-back one quarter of the inflationary effect of the increasing switchover coefficient. Because of the increase in the switchover coefficient, the actual 1995/96 support prices are 4.1% higher than was agreed in 1992, and the area aid 5.4% higher.

Table 5.1: Cereal prices under the Mac Sharry reforms (July prices).

	1992 Agreement:				Actual:
	1991/92	1993/94	1994/95	1995/96	1995/96
		— green ecu/tonne —			ecu/tonne
Intervention	155	117	108	100	119.19
Target	—	130	120	110	131.11
Threshold	228.67	175	165	155	—
Basis for calculating area aid	—	25	35	45	54.34

[3]The commitment covers wheat, rye, barley, maize and sorghum, and thus excludes oats. A comparable constraint applies to rice.
[4]All of the detailed rules implementing the EU's import and export commitments under the GATT Agreement were introduced in the first instance on a one-year trial period.
[5]Subsequently a permanent set-aside scheme was introduced (at 20%, reduced to 18% in certain circumstances), and the rotational set-aside rate was reduced to 12% for 1995 and 10% for 1996.
[6]Until, that is, the EU is extended to embrace central and eastern European states.
[7]This assumes that the high world cereal prices of the winter of 1995/96 are a temporary phenomenon, and that EU intervention prices lie above long-term world market prices. If this is not the case, then there would be no need to increase arable area payments and the GATT export constraints are irrelevant.
[8]A sleight of hand here. Support would only be fully decoupled if all existing CAP mechanisms were replaced by a bond. In the case of cereals, this would involve not only arable area payments, but also the remaining support offered by high import tariffs, intervention buying and export subsidies. For cereals this could imply higher initial taxpayer payments under the bond scheme than under existing arrangements; but it would also mean that set-aside could be abandoned.

PART II

Mechanisms and Analysis of the CAP

Chapter 6

Europe's Green Money

Christopher Ritson and Alan Swinbank

Introduction

The CAP's agri-monetary system — its 'green money' — is almost certainly the least understood aspect of the policy. In much economic analysis of the CAP the agri-monetary dimension is completely ignored, or referred to obliquely. Yet the system has been of major significance in the development of the CAP: first as a palliative, relieving tension between member states with conflicting price objectives; second as an illusion, allowing average support prices to farmers to increase without explicit Council decisions to raise them; and third as an irritant, to the Commission (as it devised ever more ingenious mechanisms to square a circle), the Council (as it struggled to avoid the unpalatable consequences for the CAP of currency instability), and the trade (in that it confronted policy-induced price uncertainty and an additional administrative burden).

The reason why the importance of the agri-monetary system in the development of the CAP has been overlooked is, of course, its immense complexity. The *detail* is difficult to understand, and has caused headaches for many politicians, civil servants, business executives and students over the years; but there are a number of basic principles underlying the green money system which we outline in this chapter. We do add some of the detail, to give the reader an idea of the flavour of operating the green money system. A number of apocryphal tales add to the mystique of green money: of the German Chancellor emerging from a particularly gruelling Council meeting complaining that he had only one official who understood the green money system but that official could not explain it, and that the official who could explain it did not understand it. Or of the British ministry official responsible for agri-monetary matters who lamented that he had never succeeded in making anyone understand the 'green pound'; the most that he had ever achieved was to convince someone who thought they had understood it that they had not. We aim to do better in this chapter.

First we explain what green money is. Then we show how governments, in the pursuit of national price stability, exploited the green money system to maintain differing national price levels which fragmented the concept of a common level of support prices under the CAP — one of the CAP's basic principles. Differences in national price levels could only be sustained if taxes and subsidies were applied on intra-Community trade. These border taxes and subsidies were known as MCAs (monetary compensatory amounts). Although MCAs were abolished as of 1 January 1993, an explanation of the MCA system helps us understand green money. Furthermore, it is only in the context of the debate over the abolition of MCAs that the switchover coefficient of 1984 can readily be understood; and the switchover coefficient — which survived until 1995 — defined the level of support prices under the CAP. From 1984 to 1993 there was a largely invisible upward creep in support prices which non-CAP specialists could easily have missed, and which generated some interesting twists with respect to the EU's application of the GATT Agreement. Finally, we briefly outline the green money system applicable in the summer of 1995, although for any reader directly involved in trading CAP products we should caution that there have been a number of not insignificant changes in the system since January 1993, and further changes of detail will undoubtedly occur in the future.

What is green money?

Since 1979 all CAP support prices — intervention prices, production aids, export refunds, etc. — have been fixed in ecu. However, in their transactions with farmers and traders in applying the CAP, national intervention agencies pay out and receive national currencies. Consequently a conversion rate is required to translate CAP prices expressed in ecu into national currencies. That conversion rate (e.g. 1ecu = £0.95), fixed under CAP rules, is known as the *agricultural conversion rate* — but is popularly referred to as green money. It is the rules governing changes in the agricultural conversion rate which give rise to most of the complexities of the green money system.

Prior to the Maastricht Treaty on European Union, the letters 'ecu' formed the acronym for European Currency Unit, but now seem to form a word in its own right. The ecu is in fact a basket of European currencies, as listed in column 1 of Table 6.1, and thus its value in any currency can be determined at any time simply by adding up the value of its constituent parts.

If and when the ecu — or its successor unit — becomes the single currency of Europe, the green money system will disappear because all CAP transactions will then be conducted in ecu. On a number of occasions it has been suggested that CAP transactions should already be conducted in ecu, similarly removing the need for a green money system. EU research contracts are denominated and paid in ecu for example, and in most member states, ecu bank accounts can be held. Indeed, British universities were urged to open ecu bank accounts to receive payments for EU research contracts and student exchanges from the European Commission. Whilst the ecu remains a fledgling currency, rather than the

Table 6.1: The ecu, and its value at the last readjustments of central rates within the Exchange Rate Mechanism (ERM).

Currency	Quantity in basket	Value on 6 March 1995 central rate 1ecu =	% Weight in value of ecu
Belgian/Luxembourg franc	3.431	39.3960	8.71
Danish kroner	0.1976	7.28580	2.71
German mark	0.6242	1.91007	32.68
Greek drachma	1.44	292.867*	0.49
Spanish peseta	6.885	162.493	4.24
French franc	1.332	6.40608	20.79
Irish punt	0.008552	0.792214	1.08
Italian lira	151.8	2,106.15	7.21
Dutch florin	0.2198	2.15214	10.21
Portuguese escudo	1.393	195.792	0.71
British pound	0.08784	0.786652*	11.17

* nominal values, as the Greek drachma and British pound do not participate in the ERM.
In the past the composition of the ecu (3.431 Belgian or Luxembourg francs, etc.) had been changed on a number of occasions. However the Maastricht Treaty ruled that "The currency composition of the ecu basket shall not be changed". Consequently there is no provision to include the Austrian, Finnish or Swedish currencies.

accepted money of day-to-day commerce, any recipient of an ecu payment must incur bank charges as ecus are changed into national currencies, and anyone who holds ecus (in an ecu denominated bank account) incurs an exchange rate risk, for the conversion rate between ecus and national currencies could change at any time. However, this proposal — that all CAP transactions should be conducted in ecu, with farmers and traders incurring the cost of currency conversion and the exchange rate risk — has been rejected by the European Commission, ostensibly because not all member states allow private citizens to hold ecu bank accounts.

We noted above, however, that the value of the ecu can be determined at any moment simply by adding together the value of its constituent parts. Thus at 11.15 am, say, on any particular morning, the pound sterling value of the ecu is simply the sterling value of all the currency units listed in column 1 of Table 6.1. Could not the national intervention agencies conduct CAP transactions in national currencies using the market rate of the hour, or of the day, and thus avoid the need for the periodic fixing of agricultural conversion rates specific to the farm sector? Again, the Commission has rejected this proposal:

> Application of the real exchange rate, which fluctuates daily, would destabilise the market and lead to constant movements in institutional prices and other amounts expressed in national currencies, which could

place the objectives of CAP (as stated in Article 39 of the Treaty) in jeopardy.
(Leenders, 1992)[1]

Thus, because of the need to maintain price stability as provided for in the CAP's objectives, a green money system has been preserved.

Exchange rate movements

The pursuit of national price stability in the face of currency movements inevitably leads to conflict. Under these circumstances, the law of one price and national price stability cannot both be sustained. A non-agricultural example may illustrate the difficulties. Suppose the exchange rate between the pound sterling and the Belgian franc is £1 = 100 francs, and that the price of a return flight between the two capitals has been fixed at £100 if the ticket is bought in the first country, and 10,000 francs if bought in the second. The price of the return ticket is the same in both countries.

Next, suppose that the value of the pound slips on international currency markets, so that it now buys 99 francs. However, the airlines maintain their pricing at £100 and 10,000 francs. Strictly speaking, the price is no longer the same in both capitals, but the difference is so minor that it will have little practical effect. If however the value of the pound slips to, say, 60 francs, and prices of the airline ticket remain unchanged at £100 and 10,000 francs respectively, then there is a significant price difference. Regular travellers will try to ensure that they always pay £100, and never 10,000 francs, for the privilege of flying between the two capitals (one of the authors indulged in such antics in the mid-1970s). In a world of perfect competition these price differences could not persist as the law of one price holds that arbitrage operations will ensue until the price difference does not exceed the transactions costs incurred in transferring the product between the two markets. The market for airline tickets was less than perfectly competitive (for example the cheaper ticket could only be bought in the UK, never in Belgium, and the return ticket had to start in London, never Brussels) and so price differences could persist. A common price level would have meant changing one or both of the sterling and Belgian franc prices every time there was a non-negligible movement in the sterling–Belgian franc exchange rate.

Similarly with electrical goods, motor cars, tractors, and agrochemicals, in the currency disturbances of the 1970s and 1980s with national currency prices moving at a slower rate than exchange rate change, buyers often found they could purchase more cheaply in another member state, and all sorts of schemes — and shams — were put in place to evade the barriers that manufacturers and retailers used to segment markets. Parallel imports — cross border movements of goods by traders not sanctioned by the manufacturer — flourished.

It is hardly surprising that the CAP encountered similar problems when member states emphasised stability in national support prices in the face of exchange rate change. The problem materialised in the late 1960s, soon after common pricing had first been introduced into the CAP. In particular, in 1969 there was a devaluation of the French franc and a revaluation of the German mark. Figure 6.1 is a simplified version of Fig.1.1, illustrating the typical CAP support system of the time. The simplification is to condense all the CAP administered prices into one, described as the EU price. This is achieved by import levies (ML) for a deficit commodity, and a mixture of intervention and export refunds (subsidies — XS) for surplus products. The single support price means that the import levies and export subsidies will be the same, bridging the gap between the international (world) price (WP) and the EU price (EUP).

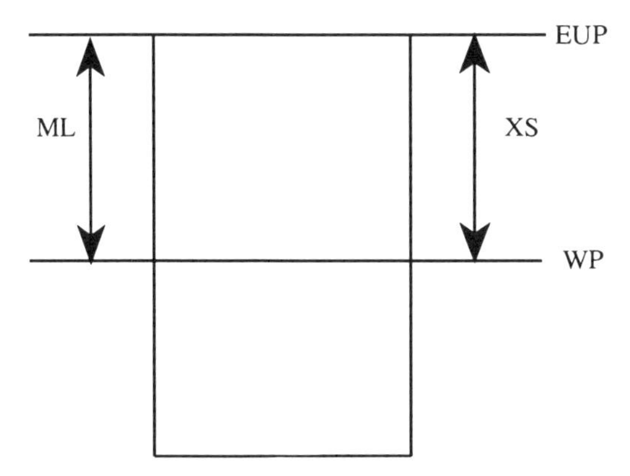

Fig.6.1: Simplified model of CAP price support system.

Figure 6.1 implies, however, that the EU is a single entity, with import levies (ML), intervention prices (EUP) and export refunds (XS) denominated in a single currency. Figure 6.2 introduces the currency dimension, *but with further simplifications*. We assume throughout this section of this chapter that the EU consists of only two member states. These are Germany, with a currency called the deutschmark (DM), and France with the franc (FF). The EU had to choose a currency in which to denominate administered prices under its common policy for agriculture. In terms of our example, it could have chosen the DM or the FF, or it could choose — as it did — to invent a new notional currency (or accounting unit) specific for the task. This initial accounting unit was called the unit of account (ua), and as it was only a notional accounting unit — that is, not traded with other currencies on foreign exchange markets, or at least not yet — it had to be given a particular value. This, in effect, meant tying it to another currency, or basket of currencies. Again the DM or FF, or a mixture of the two, could have been chosen, but in this simple model — and indeed in reality — the ua was tied to a third currency, the dollar ($), which also happens to be the currency in which international commodity prices are usually quoted. One unit of account was made equal to one dollar.

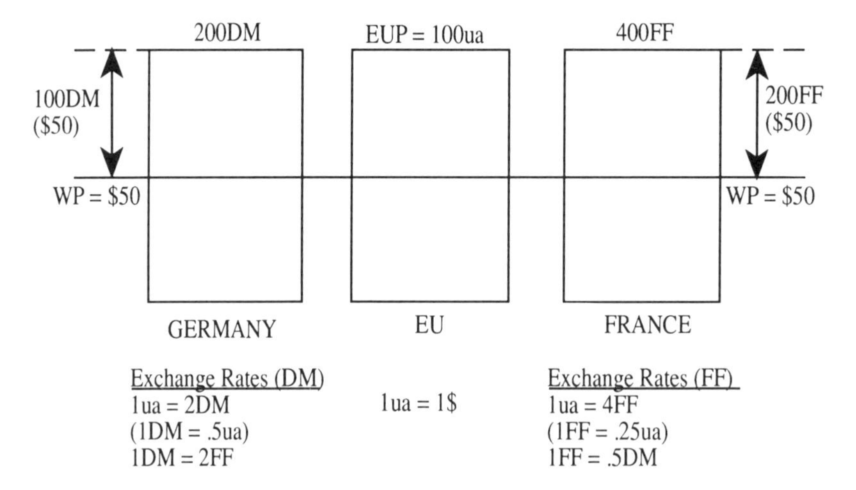

Fig.6.2: Two country CAP model with fixed exchange rates.

In the 1960s, under the International Monetary Fund's (IMF) old par value system, both the DM and FF had fixed exchange rates *vis-à-vis* the dollar. This meant that a central bank had to maintain the exchange rate of its currency with respect to the dollar within narrow margins either side of this fixed rate. As the ua had the same value as the US$, these fixed exchange rates under the IMF rules were directly translated into the conversion rates used in the CAP. A set of hypothetical, but broadly plausible, rates are introduced into the example in Fig.6.2. Thus 1ua = $1 = 2DM = 4FF. This means that a Council decision to introduce a support price at 100ua (for example, for wheat per tonne) will imply that the German intervention authorities will pay 200DM, and the French authorities 400FF. Produce will only flow between the two countries on account of normal commercial pressures, because with 1DM exchangeable for 2FF on foreign exchange markets the administered price levels are the same in the two national economies. Similarly, in terms of the example in Fig.6.2, any imports or exports will be subject to the same levy or subsidy of 100DM in Germany and 200FF in France — both equivalent to $50.

As with airline tickets, new cars and other goods, the complications arise when exchange rates change. First assume a rise[2] in the rate of the DM relative to other currencies. This means that it will require fewer DMs to purchase one unit of foreign currency, which is the same as saying it will require more foreign currency to buy 1DM. Specifically, let us assume that the new exchange rates are:

$1 [i.e. 1ua] = 1.5DM = 4FF

or, the same exchange rate expressed in a different format, that:

$0.667 [i.e. 0.667ua] = 1DM = 2.667FF

or, that: 1FF = 0.375DM.

The obvious, and administratively attractive, consequence for the operation of the CAP market policy would simply be to use the new exchange rate in fixing CAP administered prices in Germany. In terms of Fig.6.2, the number at the top of the German column would no longer read 200DM, but now only 150DM, and the German import levy/export subsidy would read 75 rather than 100DM. The exchange rates tabulated below Fig.6.2 would of course also change, but all the other details of Fig.6.2 — most notably the positions of the price lines — would remain unaltered. Although the DM price has fallen — and German farmers will not like this — the intervention price in France (400FF) equals that in Germany because 400FF at the new exchange rate of 1FF = 0.375DM equates with the new German intervention price of 150DM. What happens, though, if the German Government — as it did — refuses to reduce its DM support prices for agricultural products in response to the revaluation of the deutschmark?

Monetary Compensatory Amounts (MCAs)

We now need a new diagram (Fig.6.3), the most significant feature of which is the rise in the German price level when expressed in other currencies (to approximately $133 and 533FF). Without any other action, the consequences for the marketing of agricultural products would be disastrous. Since it is now possible to buy the product (say wheat) in France for 400FF, sell it in Germany for 200DM, and then convert back the 200DM into 533FF on foreign exchange markets, wheat would flow from France into Germany, with a shortage in France pulling up prices in the French market. German intervention stores would overflow. In turn, French merchants would find it profitable to import into France from international markets; buying at $50 (=200FF) on the international market, paying an import levy of 200FF, but then selling at more than 400FF into a French market otherwise denuded of its own domestically produced supplies. Indeed, the market disruptions would extend further than this, for merchants could then tranship the product to Germany to take advantage of the higher market prices there supported by German intervention buying at 200DM. This outcome is unsustainable. Pushed to its logical conclusion, wheat would flow into intervention in Germany until the international price was pulled up to the German price level (less the French import levy). The EU would be operating, and funding, its own *international* commodity agreement!

If the price gap between Germany and France, depicted in Fig.6.3, is to be sustained, without the adverse consequences outlined in the previous paragraph, it is necessary to increase the import levy on imports into Germany from the world market, and place a tax (called an MCA) on exports from France to Germany, as shown in Fig.6.3. Similarly, any exports from Germany to France would receive the MCA as a subsidy, and any exports from Germany to world markets would receive a higher export subsidy. In the example depicted in Fig.6.3, the MCA is equal to 50DM.

What has happened in Fig.6.3 is that the conversion rate used to convert CAP support prices, expressed in units of account, into deutschmarks, no longer corresponds to the exchange rates actually experienced on international currency markets. In fact, in the example used, the green rate — *the green mark* — is equal to the old exchange rate that was in force prior to the revaluation of the

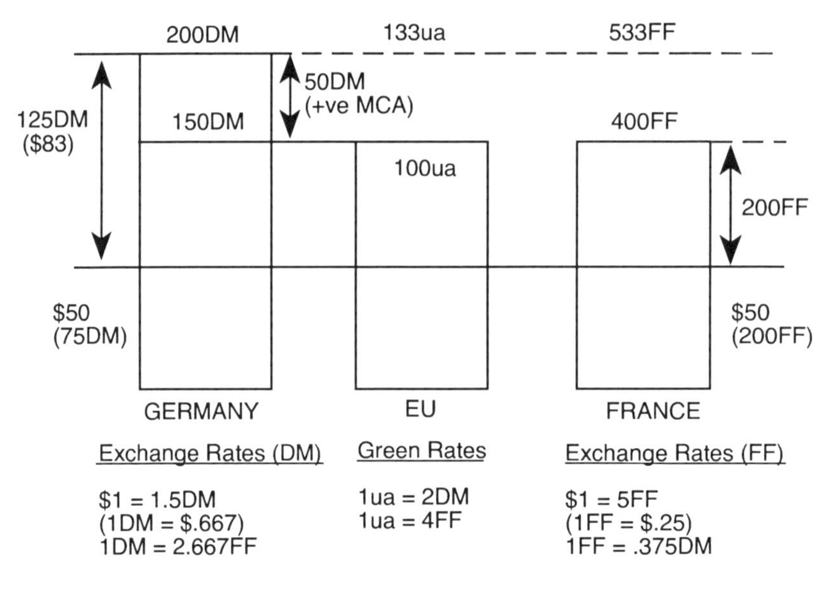

Fig.6.3: Two country CAP model: positive MCAs.

deutschmark. Germany has achieved the political imperative of maintaining stable CAP support prices in national currency terms, during a period of exchange rate turmoil, but the cost has been the fragmentation of the EU's common market. CAP support prices are no longer common between the member states: indeed, they are so far apart that border taxes and subsidies — MCAs — have to be applied on intra-Community trade (with similar adjustments to import levies and export subsidies on trade with third countries) if speculative trade flows are to be avoided.

The fall (depreciation, or devaluation) of a currency can lead to corresponding problems, and to complete the picture Fig.6.4 depicts a fall in the market value of the French franc (from 1FF = $0.25 to 1FF = $0.20), whilst maintaining the green conversion rate — the green franc — at 1ua = 4FF. CAP prices in France (at 400FF in the example in Fig.6.4) are now lower than they would have been if the new market rate for the French franc had been used in the CAP, and an MCA tax (of 100FF in Fig.6.4) is required on exports of produce from France to other member states, with a similar MCA subsidy on imports into France. On trade with third countries, the EU's import levy and export refund also have to be adjusted to take account of the lower support prices now applied in France. With French prices below the common price level, requiring MCA taxes on exports from France and subsidies on imports, France is said to have a negative MCA (labelled –ve MCA in Fig.6.4). Germany, with prices above the common price level, is said to have a positive MCA.

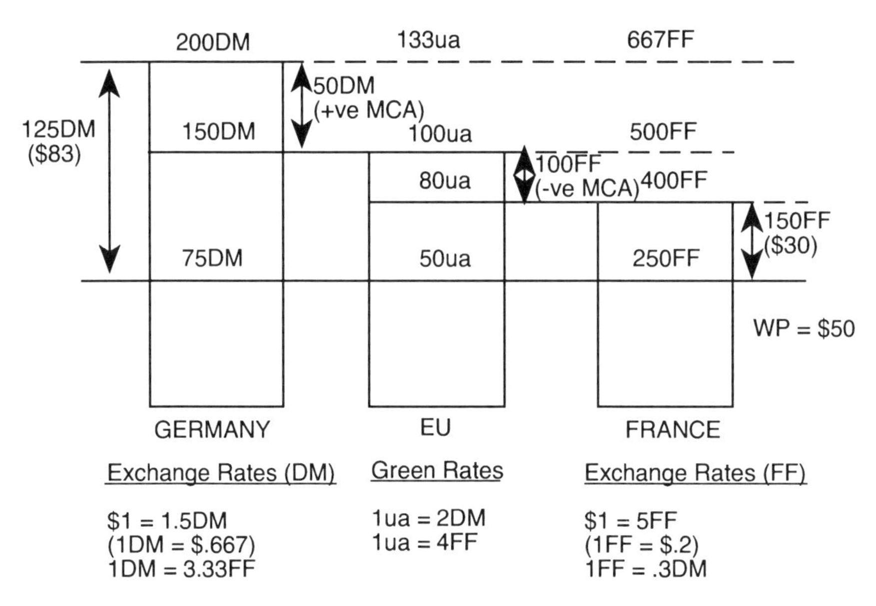

Fig.6.4: Two country CAP model: positive and negative MCAs.

Eliminating MCAs

In this section we consider a number of possible ways in which the CAP can return to common prices following the impact of currency instability, that is to return from Fig.6.4 to Fig.6.2; or alternatively to respond to currency movements in a way which does *not* involve the introduction of taxes and subsidies on trade between the member states. Where appropriate, we continue to use the simple model to illustrate particular points.

Two fundamental issues underlie all this. First, the presence of taxes and subsidies on agricultural trade between member states makes a nonsense of the concept of a common market, and thus the European Commission has always striven to remove MCAs. It is difficult to think of a policy mechanism which distorts competition more than one which fixes producer prices at a much higher level in one member state compared with another; and MCAs make the distortion so conspicuous that it leads to great perceptions of 'unfairness' within the agricultural sectors of some member states. Second, and following from this, it has proved much easier to reduce price gaps by raising the farm product price levels in member states with depreciating currencies than it has been by lowering prices in countries with strong (appreciating) currencies.

Administratively the task is straightforward: one simply adjusts the green rate (perhaps in a series of steps) to become equal to the market rate. For example, in the case of Fig.6.4, the French franc green rate is altered to 1ua = 5FF, leading to a rise in the intervention price in France to 500FF and the elimination of the negative MCA of 100FF. Indeed this adjustment to the green rate (known as devaluing the green rate) might occur immediately, without the introduction of a MCA. This might be possible for small changes in market rates of exchange, particularly for depreciating currency countries. It is much more difficult to take this course of action with appreciating currency countries, say by revaluing the green mark to 1ua = 1.5DM in Fig.6.4, as this would reduce the intervention price to 150DM. Farmers in countries with appreciating currencies will undoubtedly resist green rate revaluations, because of the impact this has on domestic support prices.

In contrast, green rate *devaluations* raise CAP support prices in local currency terms, and will be welcomed by local farm lobby groups. Food industry and consumer groups, however, will oppose green rate devaluations, and governments may worry about the impact on food prices and hence inflation. However, it should be remembered that one of the reasons for exchange rate change is different rates of inflation: as the currency loses internal value, it loses its international value. Thus, in the case of Fig.6.4, France has probably experienced higher rates of inflation than has Germany, generating the exchange rate change that led to the policy dilemma. French farmers, therefore, have some justification in arguing that, as their costs have risen during the inflationary period, their output prices should also be allowed to increase as the exchange rate falls.

A depreciation/devaluation would usually be reflected in an increase in the domestic price of traded goods; and the magnitude of that price change would be dependent upon domestic and international supply and demand price elasticities. In a world of perfect competition, if prices are in effect determined on world markets and the country is a price taker, the domestic price of traded goods would increase immediately by the full percentage effect of the depreciation. In the regulated world of the CAP, this is equivalent to when green rate devaluations exactly match currency depreciations. However, if prices are exclusively determined on domestic markets, at least in the immediate aftermath of the depreciation — as is the case for non-traded goods — then the depreciation will have no effect on local currency prices. In the CAP, this is when there is no green rate devaluation following the currency's depreciation. Neither of these extremes matches the outcome that would be expected for agricultural product prices in unregulated markets; but we should recognise that the CAP's internal logic suggests that — for an individual country — support prices are fixed externally, and hence that a change in the country's exchange rate should be reflected immediately by a similar percentage change in CAP prices in local currency terms.

More generally, inflation might provide a solution for the dilemma faced by Germany in Fig.6.4. If there is a more general inflation, with world market prices increasing, then at the EU's annual farm price review there will be a call to increase CAP support prices. Figure 6.5 illustrates a possible outcome. The world market price has risen to $80. In the annual farm price review, EU ministers have increased the EU's support price from 100 to 133.33ua (and the

French price level, following a devaluation of the green franc to 1ua = 5FF, has thus risen to 667FF), leaving the dollar value of import levies and export subsidies approximately the same as before. However, *at the same time*, the green mark has been revalued so that 1ua now equals 1.5DM. Consequently,

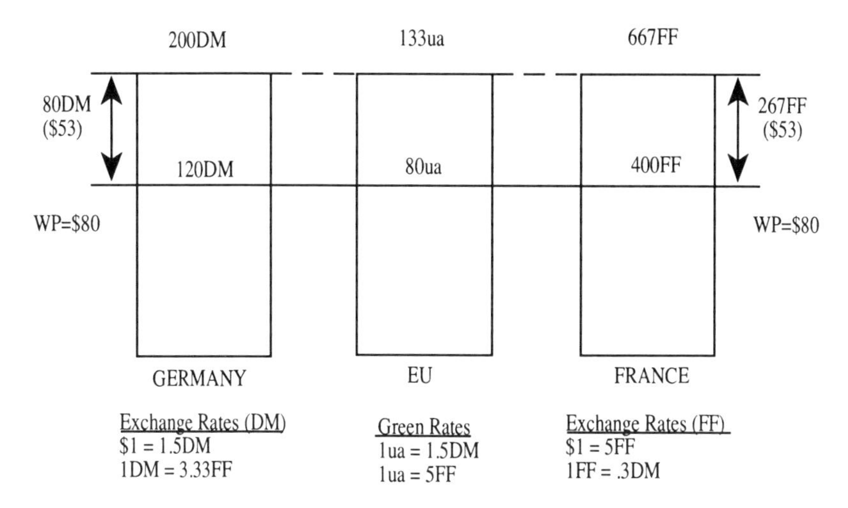

Fig.6.5: Eliminating MCAs under world price inflation.

CAP prices in Germany remain unchanged at 200DM. German farmers have suffered no fall in deutschmark prices, but MCAs have been eliminated.

In a very simplified manner, Fig.6.5 does illustrate one of the ways in which the CAP has moved back towards common prices following the distortion introduced by exchange rate movements. This particular solution has been made possible by the inflationary rise in international prices and thus the justification (or at least pressure) to increase 'common' prices in terms of units of account, together with an element of 'money illusion' — prices to German farmers may have fallen in real terms, but prices in deutschmarks have not been cut.

This kind of solution would be much more difficult to implement if international prices had remained at $50. It would have required a decision by the Council to increase agricultural support prices, in terms of units of account, without any associated cost pressure or rise in international commodity prices. Agricultural support prices throughout the EU would rise to the German price level, increasing the gap between EU and international prices, and increasing the levies on imports and subsidies on exports.

This leads on to the second way in which, in various guises, MCAs have been reduced or eliminated. Put crudely, this is to allow the unit of account (or at least the unit of account used by the CAP) to appreciate together with the strongest of the EU's currencies. The effect of this can be illustrated if we return to Fig.6.3. If we had originally fixed the value of the unit of account in terms of deutschmarks, rather than dollars, at 1ua = 2DM, then Fig.6.3 would instead have looked like Fig.6.6 following the appreciation of the deutschmark. There is no change in the 'common' support price of 100ua, but this is now equivalent to $133; and the appreciation of the deutschmark has created a negative MCA of 133FF for France, rather than a positive MCA for Germany. Common prices can now be restored by devaluing the green rate for France, thereby increasing prices in France to 533FF, which is politically a much more feasible course of action than reducing prices in Germany.

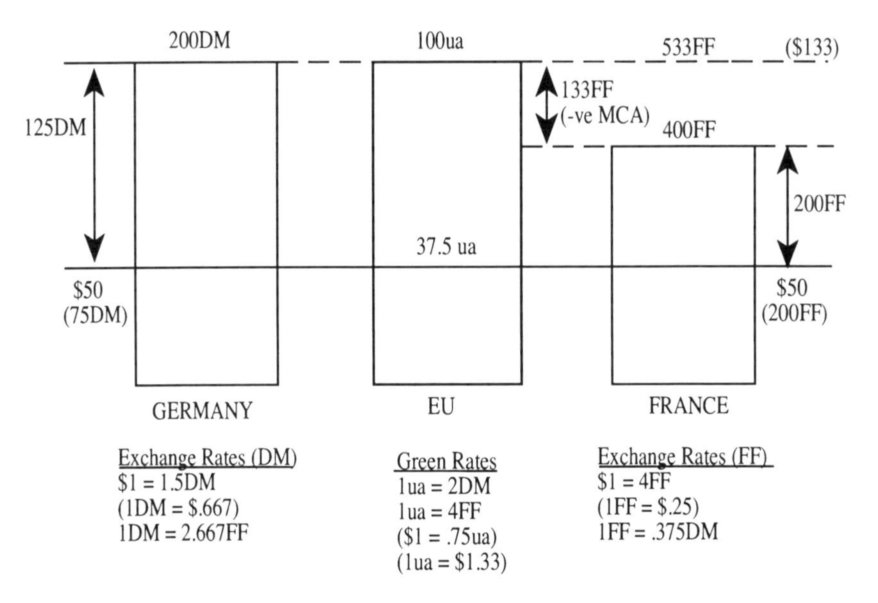

Fig.6.6: Eliminating MCAs by tying ua to strong currency.

A third solution, which has been adopted intermittently, is to revalue the green rate of an appreciating country (essentially Germany), thus reducing support prices in that country, but to permit the government of the country concerned to compensate farmers by some sort of special payment to offset the revenue loss associated with the fall in product prices. This might involve national financing, but equally the EU's budget might also be called upon to meet part of the cost of such payments.

Green money: A brief history - from the Treaty of Rome to EMS

Preceding pages have, in fact, given a potted, albeit stylised, history of green money: all the eventualities discussed above can be found in the contorted history of green money from the 1960s to the present day.[3] In the 1950s, the European Coal and Steel Community adopted a unit of account for budgetary and other purposes. With the formation of the European Economic Community (EEC), on the coming into force of the Treaty of Rome, the same unit of account was incorporated into its budgetary procedures, and — from 1962 — into the CAP. At the time, under the IMF's par value system, international currency markets were characterised by fixed exchanges rates denominated in gold, although in practice the US dollar played the pivotal role in the system. The European Community's (EC's) unit of account had the same gold content — and hence the same value — as one US dollar.

In August 1969 the French franc was devalued; and later the same year the German mark was revalued. In an attempt to maintain domestic price stability, green conversion rates that diverged from IMF par values came into play, and MCAs, in effect, were invented. France was given time to dismantle MCAs gradually (there was concern in France about the inflationary impact on food prices); and in Germany, where the pace of change was quicker, farmers received a measure of revenue compensation from the Government.

There then followed a complicated period with respect to the international monetary system which culminated, as far as green money is concerned, with the decision in 1973 to tie the unit of account to the 'joint float'. The joint float consisted of the currencies of Germany, Denmark, the Benelux countries, and briefly France. (Denmark, Ireland and the UK had just become members of the EC.) The governments concerned had agreed to restrict to a very narrow band any movement in exchange rates between their currencies (in effect fixing exchange rates within the joint float), but to allow their currencies to float collectively against other currencies.

The result was two sorts of MCAs: 'fixed' MCAs for countries participating in the joint float, in that the MCA gap between a country's green rate and its market rate against the unit of account remained fixed apart from periodic revaluations and devaluations within the 'joint float'. The other member states (UK, Ireland, France and Italy) had 'variable' MCAs: fixed green conversion rates kept CAP support prices constant in national currency terms, whilst MCAs expanded or contracted to compensate for their fluctuating market rate against the unit of account. In practice it was mainly 'expand', since all four currencies tended to depreciate against the 'joint float'. This created negative MCAs, which from time to time were reduced by devaluations of green rates pushing up support prices in terms of national currencies.

The deutschmark was revalued within the 'joint float', creating significant positive MCAs for Germany. However, eliminating positive MCAs is, as we have seen, much more difficult that the elimination of negative MCAs. In the inflationary 1970s, with the Council of farm ministers awarding generous price increases in the annual farm price reviews, it was possible to reduce positive MCAs in part by offsetting green rate revaluations against increases in the 'common' price level in units of account. Thus the sequence of actions surrounding Fig.6.5 did apply intermittently throughout the 1970s.

However, although not generally realised at the time, the factors described by Fig.6.6 were also present. The deutschmark was a 'strong' currency and tended to dominate the 'joint float'. Thus, as the deutschmark rose in value, it both pulled up the values of the currencies of its smaller 'joint float' partners and of the international value of the unit of account that was defined in terms of the 'joint float'. Consequently, a proportion of the magnitudes of the non-'joint float' countries' negative MCAs was being created by the rise in the value of the mark, and the effect was to increase agricultural support prices within the EC by a greater amount than implied by the Council decision to increase unit of account prices.

Figure 6.7 illustrates vividly the impact of MCAs during the 1970s, a period of unstable international commodity prices and generally high inflation. It shows wheat intervention prices in the big three original Common Market countries (France, Germany and Italy) and the 1973 newcomer, the UK, converted into US$ at annual average market rates of exchange. The 'world price' is derived from quarterly data for the Rotterdam landed price of comparable Canadian wheat. A number of interesting comments can be made on the price levels experienced during the 1960s. First, the international price is remarkably stable. Second, the EC's intervention prices, as originally fixed, were well in excess of the international price (about 60% above). Third, as the EC moved towards common prices in the late 1960s it was necessary for Germany to reduce its prices and France to increase. In fact, Germany delayed a price reduction, whereas France had already anticipated higher common prices earlier in the 1960s and, as pointed out by Tracy (1994), the common price chosen for grain exceeded the weighted average of pre-CAP prices (see also Chapter 2).

The first currency instability in 1969 can be seen, with the refusal of the French and German governments to alter intervention prices in response to the revaluation of the mark and devaluation of the franc, leading to a gap between French and German prices, and the first MCAs.

The British support price for grain was much lower than that in the EC and thus, when in 1973 Britain joined, it was necessary for there to be a series of upward adjustments in price. Therefore, during the 'transitional period', another sort of compensatory amount, called accession compensatory amounts (ACA), applied. However, this orderly transition was soon destroyed by the explosion in international commodity prices in 1973, with the world grain prices bursting through CAP levels. Levies (taxes) on exports were introduced to try to prevent internal EC prices rising with international prices.
The world price of wheat came down as fast as it went up, but by then world-wide inflationary pressure was developing and all prices rose in dollar terms. The inflationary currency movements put great pressure on the CAP and extremely wide differences in national support prices developed, requiring substantial MCAs on intra-community trade. There is little question that during this period the green money system prevented a collapse of the CAP. As argued by Ritson and Tangermann (1979) at the time, it allowed countries experiencing different rates of inflation, and with different degrees of pressure from their farm lobbies, to achieve price levels which matched national requirements.

The decade concluded with a complete reversal in the fortune of the pound sterling, which moved rapidly from being regarded as a weak to a strong currency, mainly on the back of North Sea oil. In 1978 all the European intervention prices rose substantially in dollar terms because of a fall in the value of the dollar. In the UK the Conservative Government elected in 1979 promised to remove negative MCAs, but it was the strong pound that caused the UK — albeit briefly — to become the country with the highest agricultural support price level in the EC.

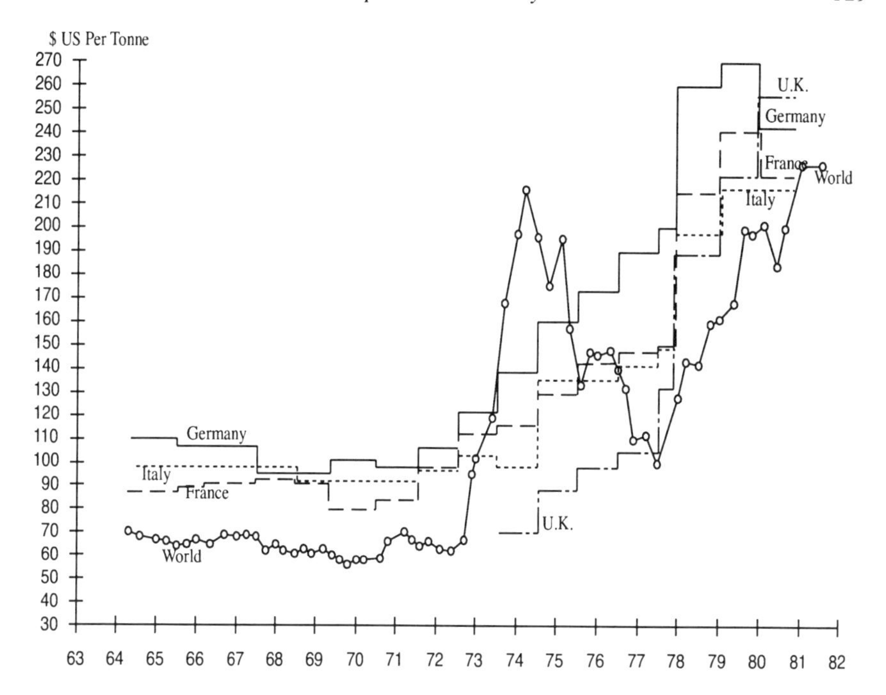

Fig.6.7: Intervention prices for wheat (US$). Source: Calculated from EC and International Wheat Council statistics.

Green money: A brief history - from EMS to the single market

In 1979 the member states decided to make further progress with their collaborative efforts in exchange rate management by establishing a European Monetary System (EMS). A central feature of EMS was the ERM which replaced the old 'joint float'. The Exchange Rate Mechanism (ERM) was based upon a European Currency Unit, today's ecu defined in Table 6.1. This meant that the unit of account was no more, and had to be replaced within the CAP by the ecu. But one ecu was worth less than one unit of account, and so in order to maintain all national currency prices unchanged, all CAP support prices expressed in units of account had to be multiplied by a coefficient. In short, 1.208953 ecu = 1 old unit of account (see Table 6.2).

Until August 1993 a key feature of ERM was that certain countries maintained fixed exchange rates between their currencies, with margins of fluctuation either side of fixed central rates of ± 2.25%, whereas other countries — for both political and economic reasons — preferred not to participate in ERM, or else did so with much larger margins of fluctuation around fixed central rates. Thus the distinction between fixed MCAs (for currencies within the narrow fluctuation margins), and variable MCAs (for the currencies with much wider bands, or engaging in independent floats) persisted.

As we saw in Table 6.1, the ecu is literally a 'basket' of currencies; and although its composition had changed on a number of occasions (notably to include the Mediterranean states following their accession to the Community) its composition goes back to the 1970s. Inevitably, with fixed currency components, the stronger currencies will over time tend to dominate the value of the ecu: weaker currencies lose market value and hence contribute proportionally less to the ecu's value, in contrast to stronger currencies that are now worth more. Thus the ecu too has tended increasingly to become a proxy for the deutschmark zone.

Table 6.2: Discontinuities in the units of account used in CAP pricing.

- From 1962 to 1979, CAP prices were fixed in units of account (ua). Initially 1ua = 1 US$, but from 1973 the ua was linked to the EU currencies involved in the 'joint float'.
- 9 April 1979. The ecu replaced the ua. As the ecu was worth less than the ua, all CAP prices had to be multiplied, and all green conversion rates divided, by the coefficient 1.208953 to maintain price levels in national currency terms.

From April 1984, with the introduction of the switchover, a green ecu was in effect created which increased in value in comparison with the commercial ecu.

- 1 February 1995. The commercial ecu replaced the green ecu. As the commercial ecu was worth less than the green ecu, all CAP prices had to be multiplied, and all green conversion rates divided, by the coefficient 1.207509 to maintain price levels in national currency terms.

The most significant development in the 1980s was the introduction of the switchover mechanism in 1984. Germany 'suffered' from the persistent strength of its currency, resulting in revaluations of the deutschmark in first the 'joint float', and then later in the ERM against the ecu. Thus large positive MCAs were built up, but as the inflation of the 1970s abated, and the CAP faced enhanced budgetary pressure, it became increasingly difficult to reduce positive MCAs by offsetting a green mark revaluation against an ecu price increase in the context of the annual farm price review. (Britain's positive MCAs of the early 1980s disappeared as the pound depreciated against the deutschmark and ecu.) The switchover, modelled on the outcome earlier depicted in Fig.6.6, was the ingenious mechanism for squaring the circle. Figure 6.8 summarises the impact of the switchover on the 1984 price fixing.

Germany had a MCA gap of +10.8%. The Ministers decided that this had to be reduced by 3 percentage points, to +7.8%. However, the green mark was not to be revalued; and ecu prices were not to be increased. Instead, the ecu was redefined so that the new ecu now used in the CAP (which we can label the *green ecu*) was worth 3.4% more than the ecu otherwise used by the Community (which we can call the *commercial ecu*).

German price levels remained unchanged, but the German MCA was reduced. The effect was to increase the 'common' price level by 3.4%, and everyone else's negative MCAs. Thus the French MCA gap increased from –5.9% to –9.5%. Countries wishing to demonstrate their *communautaire* spirit could now secure additional devaluations of their green conversion rates, delivering an extra 3.4% in national support prices, without the political embarrassment of having to vote for an additional price increase in ecu terms. The system was far from transparent — indeed it is not clear that all Ministers themselves realised that what they had in effect done was to raise prices — and hence was liable to be hijacked by the lobby groups that did understand what was going on.

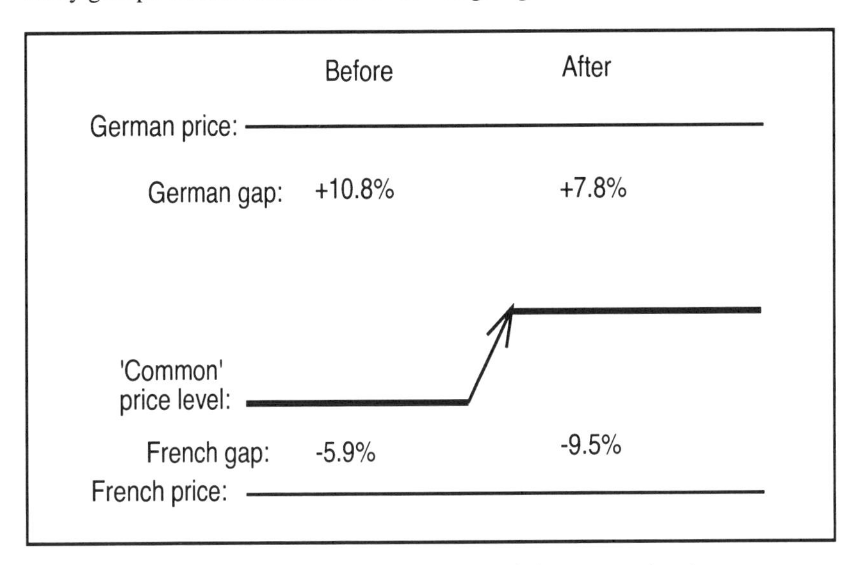

Fig.6.8: The 1984 price settlement and the switchover mechanism.

Germany's positive MCA was eliminated, in part by further increases in the switchover coefficient, and in part by green mark revaluations with offsetting compensation payments to German farmers. Furthermore, a new rule was introduced. This said that if any future exchange rate change in the ERM would create fixed positive MCAs (i.e. positive MCAs for currencies participating in the narrow band of ERM) then a new switchover coefficient would be determined, which would switch the positive MCA gap into negative MCAs for everyone else. Germany's position was secured. For a while there was a good deal of currency stability within the ERM, so much so that by the time the Maastricht Treaty was being negotiated some innocent observers tended to believe that irrevocably fixing those ERM central rates was an easy next step on the road to monetary union.

Although fixed positive MCAs had been abolished by the late 1980s, negative MCAs — and variable positive MCAs for currencies not participating in the ERM narrow band — were still applied on intra-Community trade; and as 1 January 1993 approached, this was clearly incompatible with the EU's quest for a single market, "an area without internal frontiers in which goods could pass freely from one member state to another". At the eleventh hour, in December 1992, the Council agreed rules which allowed the more frequent fixing of green conversion rates — outlined briefly in the next section — ensuring that market rates and green rates would not in future diverge significantly, and sweeping away MCAs. Traders were thus able to consign their MCA manuals to the waste paper bin. But the green money rules remained complex, and in particular — at German insistence — the switchover mechanism had been retained. This meant that as the deutschmark and other narrow band currencies appreciated, amendments to the switchover coefficient were triggered which protected farmers in those currencies from green rate revaluations, but otherwise green rate revaluations and devaluations would automatically occur as green rates were kept more or less in line with market rates.

The new arrangements were however short-lived, for in August 1993 in order to avoid a forced devaluation of the French franc within the ERM, all the ERM currencies widened significantly their margins of fluctuation. There were no longer any narrow band currencies within the meaning of the agri-monetary regulations. This meant that German farmers were no longer protected by the switchover, and — faced with currency movements which should have led to automatic revaluations of what were formerly narrow band currencies — an impasse arose. Politically it was deemed impossible for the Commission to impose a green currency revaluation on any of the strong currency countries. By mid-1995, however, after a number of false starts, a revised set of rules were in place which are outlined in the next section.

These alterations to the system enabled the abandonment of the green ecu which had increasingly become an anachronism. The world at large dealt with commercial ecu, and indeed in its commitments entered into in the GATT Agreement the EU had written all its new agricultural tariffs in terms of commercial ecu. By December 1994 however, on the eve of the implementation of the GATT Agreement, one green ecu in which CAP prices were still denominated was worth 21% more than the commercial ecu. Hence, when the green ecu was abandoned on 1 February 1995, all CAP prices had to be multiplied by the coefficient 1.207509 to ensure that price levels in national currency terms remained unchanged (see Table 6.2). As with the switch in 1979 in the unit of account used in the CAP, researchers need to exercise care in interpreting CAP prices: price increases in 1979 and 1995 might be more apparent than real.

Green money in 1995

As mentioned above, a new agri-monetary system came into force on 1 January 1993, along with the single market. MCAs were abolished as a result, and green conversion rates are now changed automatically so that they more or less match the real exchange rates between national currencies and the ecu. However, the rules governing the new agri-monetary system have been changed on a number of occasions since their adoption in December 1992, most recently — at the time of writing — in the context of the 1995-96 farm price settlement. Furthermore, although seemingly precise rules have been laid down in regulations, the rules have been suspended when they threatened to produce outcomes untenable to one or more member states. Thus, anyone liable to make or receive CAP payments expressed in ecu faces an exchange rate risk: will the Commission obey the rules laid down, and are the rules sufficiently transparent for the company's management to predict accurately a green conversion rate re- or devaluation? A further complication is that green conversion rates are not used for the new tariffs fixed under the GATT Agreement. Instead special conversion rates are fixed on a monthly basis, reflecting the actual market rate of the previous month. Again there are exceptions to the general rule. Thus when CAP policy mechanisms intervene, as for example in the cereals sector where import charges are fixed on a fortnightly basis to avoid infringement of the GATT constraint that says that the landed price of cereal imports may not exceed 155% of the intervention price, it is the agricultural conversion rate that applies.

There are three monitoring periods during the course of a month, and if the gap — between a green conversion rate and the currency's market rate with the ecu — recorded in any monitoring period exceeds specified thresholds, then the green conversion rate will be changed. In times of currency turmoil, shorter monitoring periods can be invoked. However, asymmetries exist, and devaluations of green conversion rates are still more easily secured than are revaluations. These asymmetries are illustrated by Fig.6.9 and the following (simplified) discussion.

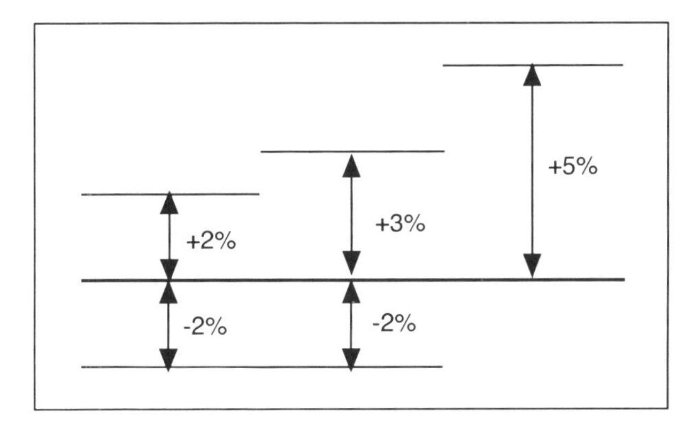

Fig.6.9: Thresholds for triggering green conversion rate changes.

The three monitoring periods are:

1st to the 10th,
11th to the 20th, and
21st to the end of the month.

During the course of each monitoring period, a *monetary gap* for each EU currency is determined. This is the percentage difference between the actual market rate of the currency concerned against the ecu, and the country's agricultural conversion rate. When the system was first introduced, the basic rule was that if the monetary gap in any monitoring period exceeded ±2%, then a re- or devaluation of that country's green rate would be triggered (see the first part of

Fig.6.9). What would happen was that the monetary gap would be halved (a gap of –2.8% becoming –1.4% for example), and halved again if necessary to come below the 2% threshold. However, from the outset the system was not symmetrical, because the switchover coefficient protected countries participating in the narrow band of ERM. Rather than a revaluation of a narrow band currency, an increase in the switchover coefficient would further increase the international value of the green ecu, thereby increasing the negative monetary gap of the other currencies, and forcing them into further devaluations of their green conversion rates so as to avoid breaching the –2% threshold depicted in Fig.6.9. From January to August 1993 the switchover coefficient increased from 1.195066 to 1.207509.

In August 1993, the ERM was revised, and consequently no currencies were participating in the narrow band. Thus an increase in the switchover coefficient could no longer be invoked to protect a green conversion rate from revaluation. The rules were suspended, and months of negotiations were required before new rules came into force on 1 January 1994. Figure 6.9 illustrates the changes.

Instead of the thresholds being set at ±2%, they were now fixed at +3% and –2%. Furthermore, it became a 'floating franchise'. That is to say, the maximum gap between any two currencies was to be 5 percentage points (e.g. +3% and –2%), within the range –2% to +5% but with devaluations of green conversion rates taking precedence over revaluations. Thus the effective range became –0% to +5%. Because devaluations take precedence over revaluations, CAP prices in international terms are rather higher than they otherwise would be.

Even this was not sufficient to assuage the farm lobby in countries facing green rate revaluations. A devaluation, to halve the monetary gap or reduce it to zero, would take place immediately. A revaluation would be delayed for one or more 'confirmation' periods — just to make sure that the green rate revaluation really is unavoidable![4] In the simplest case, a positive monetary gap of 5% or more would have to be confirmed over a second 10-day monitoring period before the revaluation is triggered.[5] However, if that revaluation were deemed to be 'appreciable' then the rules get really complicated. Some revaluations simply reverse devaluations of a month or so earlier. An 'appreciable' revaluation is one that does not simply overturn the effect of green rate devaluations experienced in the course of the previous three years.[6] Under these circumstances, five 10-day confirmation periods would be required before the revaluation is carried out, and in mid-1995 the Commission seemed to be making up the rules in response to changing circumstances (and, one suspects, to appease various lobbies). Once a revaluation had been confirmed, the normal rule would be to halve the gap. Thus a gap of +5.2% would be halved to +2.6% — quite a sizeable revaluation. Therefore the Council reserves the right to intervene at this juncture and decide upon a smaller revaluation! Little wonder that businesses with contracts denominated in ecu claim they would prefer to face the exchange rate risk, rather than these administrative and political uncertainties.[7]

Even so, farmers in countries facing green conversion rate revaluations worry about the future. In June 1995 it was decided that the Mac Sharry area and headage payments would be fully protected from appreciable revaluations.[8] When the switchover was abolished in February 1995, the price that the strong currency countries had exacted was that there would be an automatic increase *in ecu terms* in the level of the Mac Sharry payments to offset any appreciable revaluation.

Thus *all* farmers in both revaluing and devaluing countries stood to gain, payment levels and budgetary costs were bound to escalate, and the Peace Clause in the GATT Agreement — which only applies whilst decoupled payments respect levels of support agreed upon in 1992 — would, according to some commentators, have been violated. Understandably this had been dubbed the 'mini-switchover'.

The revised scheme allows countries that suffer an appreciable revaluation between 23 June 1995 and 31 December 1995 to continue to use the green rate of 23 June 1995, in the determination of Mac Sharry payments, until 1 January 1999. Whilst this revised scheme ends the farce of all the EU's farmers receiving compensation because of a revaluation of one EU green conversion rate, some commentators have suggested that this decision has begun the renationalisation of the CAP, and clearly, the stage is set for yet another crisis in the green money system once this text has gone to press. The new arrangement only protects against appreciable revaluations up until the end of 1995: what happens when a green conversion rate experiences an appreciable revaluation after 1 January 1996? In addition to all this, member states are also entitled to compensate farmers for reductions in CAP support prices stemming from appreciable revaluations.

Member states have proved remarkably adept at devising schemes for compensating the farm sector for any perceived loss of advantage stemming from the agri-monetary system. The latest wheeze — dreamt up by France, and approved in the context of the 1995-96 farm price settlement — allows a member state to grant compensation payments to its farmers if they suffer as a consequence of price competition from another member state which devalues its green conversion rate. French farmers have frequently felt threatened by cheap imports from other member states, be it wine from Italy, strawberries from Spain, or lamb from Britain. The European Commission, at the time of writing, is trying to contain the impact of this scheme. Failure to do so would signal to the farm lobby that on the flimsiest of excuses they should petition their own governments for additional national support, in the belief that it would be approved by the EU.

Concluding comments

We hope that there will be no need for a chapter on green money in the third edition of this book. The green money system is an anachronism which has been retained for political purposes, and which has been exploited by the farm lobby. Increasingly, however, trade groups are urging that all CAP transactions should be conducted in ecu, with the food manufacturer, trader or farmer directly incurring the exchange rate risk and the bank charges for exchanging ecu into national currency. We would be firm advocates of such a change. For large companies, the benefits of deregulation would outweigh the cost of the additional bank charges incurred. For small farmers receiving modest payments in ecu, the bank charges could be significant, and for this reason it may be necessary to permit national agencies to pay small sums in national currency. Nonetheless, a simple decision to do away with green money — as should have been decided with the abolition of MCAs and the advent of a single market on 1 January 1993 – would do much to simplify the system.

Even if green money is abolished, however, the legacy of the system will live on. Future generations of CAP analysts will have to grapple with its complexities, even though they make no pretence to being students of economic history, for any data series which makes use of ecu could easily become unintelligible unless the econometrician involved has at least a rudimentary understanding of green money.

References

Harris, S.A., Swinbank, A. and Wilkinson, G.A. (1983) *The Food and Farm Policies of the European Community*, John Wiley, Chichester.

Leenders, T. (1992) The Agrimonetary System Post 1992, paper presented at the conference *Post CAP Reform*, Agra Europe (London) Ltd, July 1992.

Ritson, C. and Tangermann, S. (1979) The Economics and Politics of Monetary Compensatory Amounts, *European Review of Agricultural Economics* 6(1), 119-164.

Swinbank, A. (1988) *Green 'Money', MCAs and the Green ECU: Policy Contortions in the 1980s*, Agra Europe Special Report No. 47, Agra Europe (London) Ltd, Tunbridge Wells.

Tracy, M. (1994) The Spirit of Stresa, *European Review of Agricultural Economics* 21(3/4), 357-374.

---

[1]Toon Leenders was at the time the Commission official in charge of agrimoney.

[2]There is a specialist vocabulary used by economists and financial experts to talk about exchange rate change. A rise in the value of a currency is referred to as an *appreciation* when the currency's value is freely determined by market forces (a so-called 'floating' currency) or as a *revaluation* when the Government or central bank concerned fixes the value of the currency. Similarly, for a fall in the value of a currency we talk about a *depreciation* or a *devaluation* of the currency.

[3]For further details see Chapter 8 of Harris, Swinbank and Wilkinson (1983), and Swinbank (1988).

[4]There is also an emergency three-day rule. If, over any three working days the total monetary gap between any two currencies exceeds 6 percentage points, then green conversion rate changes are triggered. Again, however, devaluations take precedence over revaluations, and revaluations are themselves subject to one or more confirmation periods.

[5]In certain circumstances, invoking one or more confirmation periods would also delay a devaluation.

[6]The threshold is a weighted average of the full devaluation of the last 12 months, one half that of the previous 12 months, and one third that of the first 12 months.

[7]When traders take out export licences they can 'advance fix' the green conversion rate. Within specified limits, the green conversion rate applied when the export licence is used will be the green conversion rate applicable on the date the licence

was issued. Similar arrangements apply for other subsidy schemes. Attempts by commercial organisations to mimic a futures market in green conversion rates have not, however, been successful.

[8]For all transactions, an 'operative event' has to be defined. For trade with third countries it is customs clearance, and for internal activities it is the date when "the economic objective of the operation is achieved". For the Mac Sharry area payments, it is 1 July — which explains why in May/June 1995 countries facing green conversion rate revaluations were pressurising the Commission to delay revaluations until after 1 July 1995.

Chapter 7

Some Microeconomic Analysis of CAP Market Regimes

Allan Buckwell

Introduction

The purpose of this chapter is to outline the principal economic effects of the commodity market regulations operated under the CAP. This is potentially a very large task because there is a great variety of regulations using a bewildering range of instruments affecting almost all commodities grown in the EU.[1] To compound the problem, these regulations are not static, they have been continually modified, often with significant economic effects: also they are not applied in exactly the same way in all member states. The classic example of the latter point is the operation of milk quotas; these are tightly implemented but freely tradable in the UK, yet in Italy they have been operated with less rigour and are non-tradable. Also, despite its name, the CAP has not produced identical prices in all member states. The complexities of the agri-monetary system (see Chapter 6) have enabled member states to pursue quite different price policies for significant periods of time. No attempt will be made to deal with all these details; the emphasis will be to develop a basic toolkit for explaining the effects of the principal regimes affecting the major products in the hope that the serious student of the CAP can then extend these analytical tools to explore the intricacies of other regimes.

Since its inception, it was always acknowledged that the commodity market organisations (CMOs) were one of two elements of the CAP. The second element is the structural policy. This has always been the poor relation in terms of the budgetary funds available, but it has increased in importance over time and may continue to do so. However, no attempt will be made here to analyse structural policy. Its mode of working does not operate directly through commodity markets and this makes it difficult to apply conventional microeconomic analysis to the effects of structural measures. Structural policy is referred to in Chapters 2, 10 and 13.

The form of economic analysis presented here will also be restricted. The most frequently encountered tool of analysis is still the comparative-static, partial equilibrium market model, and that is what will be developed here. This has the enormous strengths of simplicity and flexibility, but it cannot tell the whole story. The model is based on the supply and demand of individual commodities at the national or regional level. It is deceptively simple (especially when expressed in the classic market diagram) because underlying both the supply curve and demand curve is an astonishingly sophisticated theoretical framework.[2] The geometric form is of course limited to two dimensions (price and quantity) for a single product or group of products. In this version the interactions between commodities can be illustrated but not truly incorporated. As it will be shown, these interactions are depicted as shifts in supply or demand curves. Supply interactions occur as products compete for resources, and as outputs of some markets are inputs into others (e.g. seeds and feeds). Demand interactions occur as products 'compete' for the scarce budgets of consumers and substitute one for another in consumption. Of course two or more commodity markets can be illustrated side-by-side, but there is no simple geometric device to 'solve' or endogenize the effects of one market on the others. With many commodity markets interacting in both supply and demand there is no possibility to analyse them geometrically. An algebraic version then has to be used. This can handle any number of commodities, and can be manipulated rigorously to demonstrate certain propositions. Since the advent in the early 1980s of the micro-computer spreadsheet, and with the appropriate data, the algebraic version is now very easily expressed numerically. The spreadsheet model can be solved for the set of prices which creates equilibrium across all markets taking into account the myriad of instruments governments use in the name of agricultural policy. This is the basis of many simulation models of the agricultural sector.[3] In such models, individual supply and demand functions for all major commodities are specified, and the models used to explore the effects of changes in CMOs. The results (production, consumption, prices, incomes, expenditures and trade) can be aggregated over all commodities to get a picture of the sector wide effects.

There are ways to combine the (static, partial equilibrium) market effects of the CMOs to get a picture of the overall economic effects of policy or policy change on the whole agricultural sector. The most frequently used of these are the calculations of the welfare effects of farm policy using the concepts of producer and consumer surplus, and the calculation of producer and consumer subsidy equivalents, PSEs and CSEs. These will both be explained.

Despite the apparent complexity of the models and calculations mentioned so far, they do not usually embrace a great many important aspects of agriculture and agricultural policy. The effects of changing farm structures, farm income issues, risk and uncertainty, investment and the dynamics of supply are generally ignored or handled in a very crude way by the approaches listed. Neither do they take account of the interactions between agriculture and the rest of the economy. Some of these shortcomings and extensions will be briefly discussed at the end of the chapter, and are explored further in the next chapter.

The structure of what follows is first to explain the workings of the classic old-CAP triple-barrelled support system (illustrated first in this book by Fig 1.1) of the major commodities, based on domestic support through the intervention system, import protection using variable levies and the promotion of exports with variable export restitutions. It will be shown how this system contained the seeds of its own demise and attracted a series of modifications in the form of co-responsibility levies, consumption subsidies, quantitative restrictions and other forms of stabilisers, and led eventually to the 'new-CAP' following the so-called Mac Sharry reforms of 1992, which introduced compensation payments combined with restrictions on land use through the set-aside scheme (Chapter 5). It might be argued that there is no point in describing the classic old-CAP CMO for cereals as it is out of date. However, whilst the 1992 reforms combined with the Uruguay Round Agreement on Agriculture (URA) significantly modified the CMO for cereals, the three elements of intervention, import restriction and export subsidies remain. They also are still in place for dairy products, beef and sugar although in combination with other instruments, and they are still deeply embedded in the thinking of farmers and policy implementers around Europe. But in any case, understanding today's expression of any long-established policy is always easier when it is shown how it emerged from yesterday's policy. Given the annual, incremental, commodity-balanced packages of measures in the process of changing the CAP, and given the multi-national, multi-tiered bargaining procedures which characterise the political process in the EU, the best explanation for today's policy is always yesterday's policy.

The triple-barrelled support system for cereals

As was explained in earlier chapters, the essence of the market price policy for the principal commodities was to support prices above world market levels using mechanisms which would give greater internal price stability. It was predictable, and predicted (Mansholt, 1968), that this secure protected market environment would stimulate productivity and output. It was also clear that maintaining this secure regime would require strict controls to ensure that neither domestic overproduction nor foreign supplies would depress the local market prices. Three instruments were developed to achieve this, the domestic intervention[4] system, variable import levies and variable export refunds. The precise operation of these three instruments has a massive influence on market prices in the EU and of course the level of production and consumption and net trade. The instruments are not independent and their interactions are complex and subtle. Rather artificially, but for pedagogic reasons, each will be described separately before their effects are assembled, and as is usually the case, we shall use cereals by way of example.

Intervention

Politically determined intervention prices, agreed annually in ecu and translated into member state currency through the green rates were set for each cereal. The importance of these prices is that they represent a contract between the EU and farmers in which an appointed intervention agent will undertake to buy any volume of grain at this price provided it satisfies the conditions laid down and is delivered to the agency (the intervention boards). Although intervention appeared to be an open-ended commitment by the EC, in practice, the Community evolved many twists on the basic system to enable it to manipulate, usually restrict, the volume of grains purchased. The principal such variations are: the actual price offered was the buying-in price which was below the intervention price by a changeable margin; the quality of grain purchased (eg moisture content, purity, bread-making quality) could be manipulated from time to time; and the minimum quantity accepted for purchase, the window during which intervention was open, and the delay before sellers are paid for their grain could also be varied. By manipulating these variables the effective price at which intervention took place could be raised or lowered. This process is carried out by (usually, weekly) management committees held in Brussels, the function of which was described in detail in Chapter 3. These turned out to be sophisticated tools for market management.

To analyse the market effects of the intervention system it is first necessary to build up the market analysis framework. This is shown in the diptych in Fig.7.1. The left panel depicts the cereal market of the EC[5] in, say, 1967. It shows the supply and demand curves for grain in the range of prices experienced at that time. These curves represent the quantities of grain offered for sale and purchased over a single production period at different grain prices, *ceteris paribus*. As supply lies to the left of demand over most of this range this indicates the EC was a net grain importer. The net trade schedule is explicitly shown in the right panel, T_E. Geometrically, it is constructed as the horizontal difference between the domestic supply and demand curves, the negative portion indicates net imports, the positive segment net exports. In principle there exists corresponding supply and demand curves for grain in the rest of the world (denoted RoW), the horizontal difference between them is the RoW net trade curve, T_R. This is also shown on the right panel. In the absence of any policy interferences in the EC or the RoW, the world market price would be the intersection of the two net trade curves at price P_e and traded quantity M_e, which flows from the RoW to the EC. This is the free market partial (i.e. single commodity) equilibrium.

If, *ceteris paribus,* the EC now were to introduce an intervention system which had an open-ended commitment to buy all grain offered at price P_i, what would be the outcome? To answer this, the intervention system has to be bolted onto the EC's domestic demand. The simplest version is that the intervention boards, collectively, in the member states have a perfectly elastic demand at price P_i. Thus the new EC domestic demand schedule is the kinked line DCD'and the corresponding EC net trade curve is also now kinked, $T_E B T_E'$. The new world market equilibrium would be at the EC's intervention price P_i and the Community would find itself importing M_i grain, of which M_c was for domestic consumption and the rest being purchased into intervention at the initial outlay of the intervention price times the volume purchased.[6] Whilst this measure has

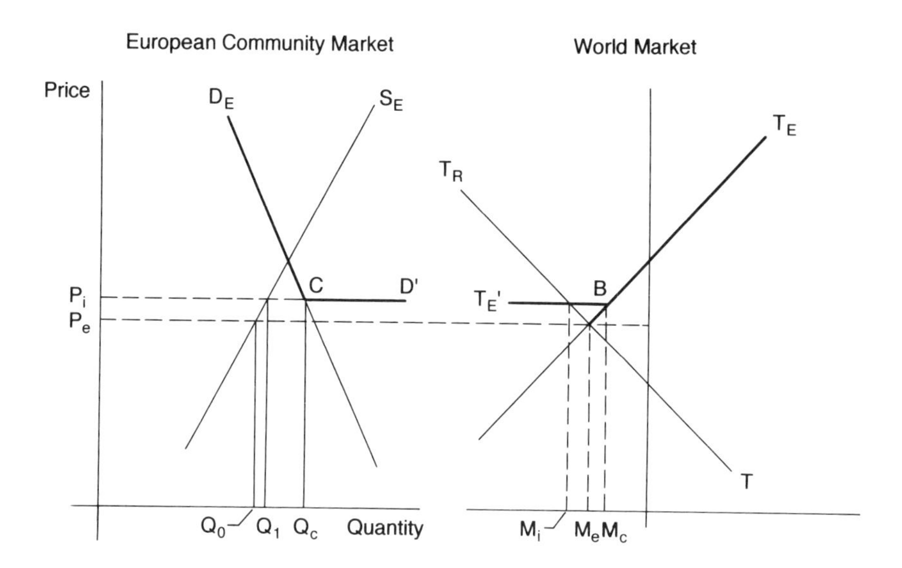

Fig.7.1: EC Market with intervention only.

the desired effect of raising the domestic price (P_e to P_i), thereby stimulating domestic production (Q_0 to Q_1), it is at the unacceptable cost of paying to buy up non-EC grain, and supporting the world grain price, to the benefit of all other producers in the world. These were certainly not the intended aims of intervention. This artificial analysis is sufficient to show why it was not possible to introduce intervention alone. It had to be accompanied by border measures to try and restrict the benefit of intervention to the domestic market.

Before introducing some import protection to the analysis, three qualifications should be mentioned. First, the diagram has been drawn with the EC shown as a net importer in the range of prices under discussion. The situation when the EC is a net exporter is shown later. Second, the world price must be assumed to refer to the same quality, time and location of grain as in the EC. These are simplifying assumptions which make the analysis tractable. In reality, grain is produced in a range of qualities, its price varies over the year (as storage costs add to the production costs) and over space (the further from the farm gate the higher the price as transport, handling and insurance costs are added). In most analyses it is usually assumed, even if not stated, that the grain market supply and demand are pictured at wholesale level and world price is the cif[7] price at a comparable location. Third, whether the intervention price actually determines the ultimate market price in the EC depends on the rigour with which intervention purchasing is pursued. If intervention is literally the obligatory purchase by the authorities of *all* suitable supplies offered to it at the intervention (or buying-in) price and assuming this is known to all market agents, then the above analysis holds. In most diagrammatic and numerical analyses, this is exactly what is assumed, intervention under the CAP is obligatory expenditure, and information about the intervention price is very widely disseminated by the authorities and the private sector. However, reality is more complex than the simple diagram allows, and because of quality, time and distance effects and imperfect information, intervention in practice is rather less automatic than this and the market price can fall below the intervention price. It may also be that as more produce is bought into intervention, the authorities tighten-up the conditions so that the demand curve for intervention is less than perfectly elastic. Thus in most real situations the market price will be below the intervention price, the amount below depending on a variety of details concerning policy implementation and practical market conditions (see Colman, 1995). For simplicity, rather than introduce yet another price variable, in all that follows, the line labelled P_i should be interpreted as the price at which intervention takes place and it is assumed that this is also the market wholesale price.

Variable import levies

It is fairly obvious that if EC authorities say they will guarantee to purchase grain offered to them at a price which is above international market levels, there would be lots of foreign grain suppliers who would be pleased to sell into EC intervention! Thus a domestic intervention system has to be protected against cheaper imports. Before the 1994 URA, the method chosen by the EC to do this was the use of variable import levies. The Community determined a threshold price, which was generally above the intervention price, below which imports were not allowed into the EC. [8] A border tax, or levy, was calculated as the difference between the threshold price and the world offer price. This gap varies from week to week as world market conditions fluctuate. The threshold prices themselves were generally changed annually at the spring 'price fixing' meeting of the Council of Ministers alongside changes in other institutional prices (Chapter 3).

In Fig.7.2, the threshold price P_t is shown at the same level as the intervention price (in reality it is higher). The corresponding variable levy is determined in the diagram as the vertical distance between the EC and RoW net trade curves at this price, V_l.[9] The effect of such a system can be depicted diagrammatically as 'swinging' the EC trade curve vertically downwards from point B, i.e. line T_EBT_E''. This illustrates the smallest levy which will prevent the 'flood' of imports into the Community which are trying to take advantage of intervention. The economic effects of this combination of the two price instruments (P_t very slightly above P_i), the variable levy V_l and the readiness to buy up all grain offered at the intervention price, can be deduced from Fig.7.2. The EC price will be at P_i, (actually slightly above), so EC production and consumption are Q_l and Q_c respectively. Imports are M_c (= Q_c - Q_l), world price settles at P_w and the Community raises revenue from the variable levies (worth $V_l x M_c$ - the shaded area). Compared to the 'intervention alone' situation,

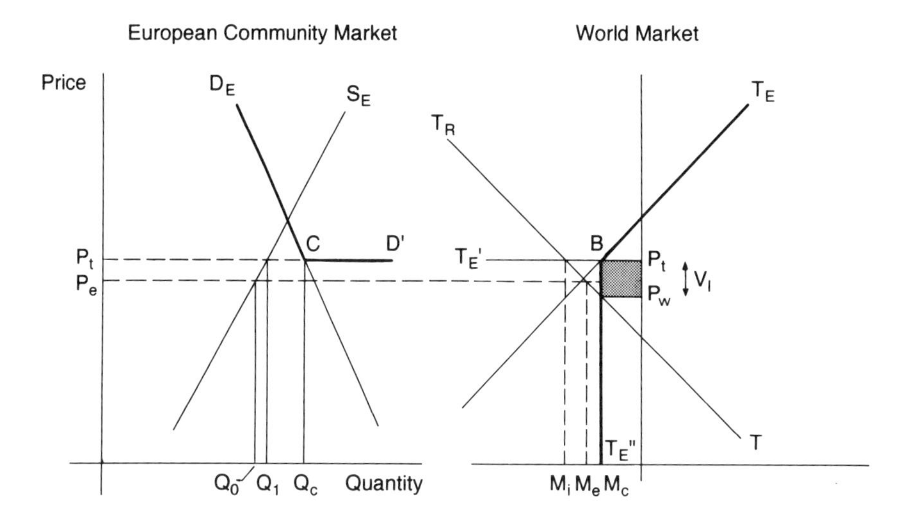

Fig.7.2: EC market with variable levy.

the domestic *market* is unchanged, though the budgetary situation is radically different. Now the budget collects the import levy revenues rather than incurring expenditure on intervention purchase and storage. In this situation all the costs of the support to farmers are borne by consumers, and taxpayers benefit from the revenues on grain imports. The world market is significantly affected by the introduction of the variable levies, as both volume traded and world price are pushed down, constituting a net loss to producers elsewhere and gain to their consumers.

The chosen instruments of intervention and threshold prices are very effective in insulating the EC market from supply instability at home or abroad. Domestic effects of any fluctuation in EC production, caused for example by weather and disease, will change the volume imported and therefore the levy revenue, but not domestic price or consumption. The main economic adjustments will be made on the international market where the world price will adjust (up or down) as the EC domestic availability changes (down or up). Similarly, if there are supply shortfalls or bumper harvests in the RoW, their impacts will be felt only on international prices and thus RoW producers and consumers. The EC net trade curve is unaffected and given the unchanging EC intervention and threshold prices the upheavals in the world market go unnoticed in the Community.[10]

The situation described so far illustrates the grain market reality of the late 1960s for the EC-6. During the next two decades several developments took place. In the EC, given the economic incentives provided by the high and relatively stable price regime, there was a rapid technical and structural development of European agriculture.[11] Farms enlarged to take advantage of the innovations in farm mechanisation, crop breeding, fertilisation and crop protection. The environment was favourable for new investment which embodied this new technology. In economic terms the overall impact of these changes in scale and technology was that real production costs per unit declined and thus the supply curve shifted down (or to the right — which is the same thing in the simple market diagram). Demand was not static either. The real income growth during this period encouraged greater meat consumption in the diet which boosted the demand for grain, especially as much of this growth was in the white or 'pinker' meats (poultry, pork and young beef) which utilise relatively more grain in the feed rather than forage based feeds. These trends occurred in all EC countries, and in addition the EC itself expanded from six to nine to ten to twelve in this period. Thus more and more farmers were embraced by the protective price regime. There is little doubt that the rate of productivity growth exceeded the growth in consumption brought about by the slowly growing population of richer people. This age-old tendency causes real food prices to fall, which they did in the EC. However, at the same time the development of price policy in the Community was dominated by a tendency to increase, rather than reduce the institutional prices. This was particularly caused by the way the EC reacted to the monetary instability of the 1970s and 1980s creating an agri-monetary system which consistently gave an upward twist to institutional prices each time it was modified, as outlined in Chapter 6.

Meanwhile there were technical developments in agriculture in the rest of the world too. The RoW supply curve was also moving steadily down and to the right. But, compared to Europe, there was much faster growth in population in many parts of the world, and also in some countries, especially south Asia, faster real income growth too. Both of these boosted the effective demand for grain. There was a tendency for more countries to import gain. This came both from the poorest developing countries whose grain production could not keep pace with population growth and the newly industrialising countries whose grain consumption per head was rising as more livestock products entered their diet. The net effect of all these factors in the RoW was a slight excess of supply shift over demand shift causing real prices to fall.

The combined effect of these market developments was that the growing EC steadily moved from being a net grain importer in the 1960s to a position of grain market balance at the end of the 1970s and then to a progressively larger trade surplus. Although this has been described for grains, it affected all the major agricultural products so that by the end of the 1980s the Community was more than 100% self sufficient in all the major indigenous products except maize, oilseeds, sheepmeat and fresh fruit. (Table 1.3).

The underlying grain market situation of, say, 1988 thus looked as depicted in Fig.7.3 (focus only on the supply and demand curves and trade curves). The net trade curve of the (enlarged) EC has shifted right or down compared to Fig.7.2. The net trade curve of the RoW has shifted up and to the right. In the relevant range of prices the EC has switched to become a net exporter, and correspondingly, of course, the RoW has switched to being a net importer. The intersection of these curves is now to the right of the vertical axis. This meant that the Community had increasingly to make use of its third instrument for defending its high domestic prices, the export subsidy.

Export subsidies

Throughout the 1970s and 1980s whilst the market developments described above were taking place, the political situation in the EC was such that it proved impossible to reduce the nominal support prices. Even though prices may have fallen in real terms it was quite clear that the community was operating at prices well above point E in Fig.7.3. The evidence for this was the build-up in public and private stocks[12] which occurred; this indicates that at the protected internal price, domestic consumption was systematically lower than production. This process cannot go on for ever. There is an opportunity cost of holding such stocks especially as during much of the period when high stocks were held interest rates were also high and real grain prices were falling and expected to continue to do so. Such costs are in addition to the physical costs of the treatment (drying and pesticide treatment), storage and deterioration of the grain, and the threat to the market of having a large over-hang of stocks. But what can be done with such stocks which have been purchased at prices above international market levels? They cannot be released onto the domestic market because that would undermine the price which was the explicit target of policy to support. Neither can they be sold commercially abroad at the EC price level. The only solutions available to dispose of the stocks without undermining the domestic market are either to destroy the grain or to subsidise its sale outside the EC. The former solution is rejected for grain; it would be potentially environmentally damaging and it would be morally and politically unacceptable systematically to destroy significant quantities of grain at the same time as we perceive periodic food shortages in other parts of the world.[13]

For these reasons, during the 1970s there was a rapid growth in FEOGA financed subsidisation of exports. The system employed for this was a market based process. Private grain traders would be invited to 'bid' for export licences stating the volume of various grains they wished to export, the destination and the unit export subsidy required (technically called refunds or restitutions). The subsidy they require is the difference between the price they paid for the grain from the EC farmer (plus any costs they incurred in moving and storing it) less the price they can sell it for (and their costs of transporting it to the destination). The cereals management committees open the bids and decide which to accept. This competitive tendering process is intended to ensure that there are incentives for traders to find the market opportunities requiring least subsidies (i.e. not paying too much to EC farmers and getting the maximum from the purchaser abroad). This process places very large responsibilities on the management committees and the Commission officials which service them to have good market intelligence about the state of the domestic and foreign markets and their likely developments. Through this process of export subsidies the EC became a major exporter on world grain markets, the second in importance after the USA.

The domestic economic effects of the subsidised exports of the grain policy with all three instruments in place are illustrated in Fig.7.3. The high internal price supported by the intervention and threshold prices means that consumers pay more (P_i) for a lower quantity of grain (Q_c) than in the absence of the support system. The size of this effect is conventionally measured as the change (loss) of consumer surplus with the support compared to the 'free market'

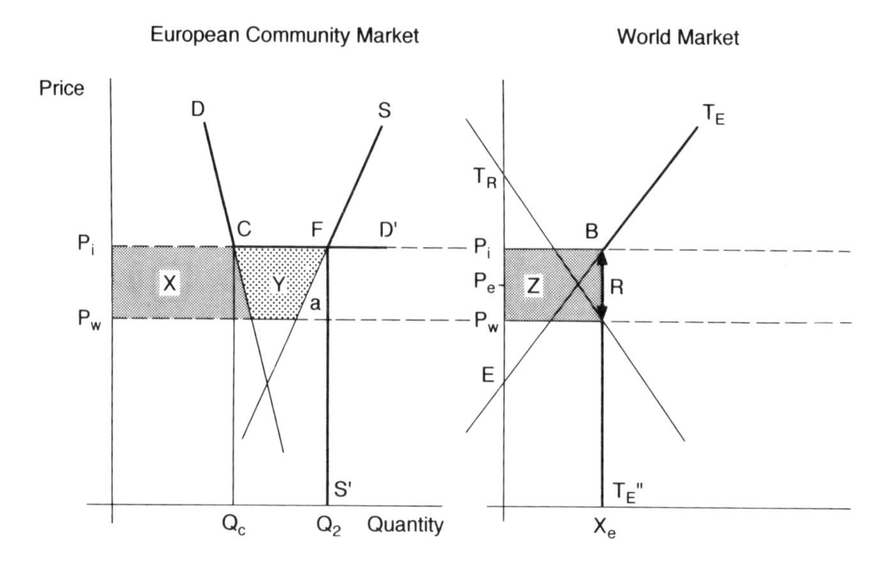

Fig.7.3: EC market with export subsidies.

situation. This is area X in Fig.7.3. The corresponding measure of the benefit to producers is their producer surplus gain resulting from selling more (Q_2) at the higher intervention price, less the marginal costs of the additional production. This is illustrated as area X+Y in Fig.7.3. These transfer payments are measured against the reference of the world market price P_W.[14]

In the simplistic world of these two-dimensional models, there are only exports and not imports. In reality, there may be both, some imports arise because of different qualities or special temporal or spatial conditions. To the extent that there are such imports, there will be tax revenues from the import duties on such trade. This is a source of budget revenue. However as a net exporter of cereals any such effect is outweighed in the EC by the large exports. The simple analysis summarised in the market diagram also does not incorporate the modelling of stocks so the simplest analytical assumption is therefore that all of the surplus production ((Q_2–Q_c) in Fig.7.3) is exported in the same period it is produced. Therefore whatever stocks existed at the beginning of the period are unchanged at the end of the period. The cost of disposal of the excess production is therefore the volume of exports multiplied by the size of the export refund per unit (P_i–P_w). This cost is borne by European taxpayers, and is denoted area Z in the figure. Provided we are willing to value an ecu gained by producers at the expense of consumers and taxpayers equally to all three parties, we can calculate the net domestic welfare effect as (X+Y)–X–Z = Y–Z, which with conventionally sloped supply and demand curves will always be negative, indicating the net domestic social cost of the policy. Compared to the situation of the 1960s when the Community was an importer, the switch to exporting means that both consumers and taxpayers are paying for the transfers to producers.

The analysis of the combined effect of the three measures on the world market now focuses on the effect of the export subsidy. This has the same effect of making the net trade curve 'vertical' as did the variable import levy described already, though the explanation is slightly different. Given the domestic signals that by the combination of the three support measures, prices will be maintained at (or around) P_i and any surplus production will be bought up and disposed onto the world market, the domestic demand schedule becomes the kinked line DCD' (as in Fig.7.1) and effectively the supply curve the line SFS'. In other words, the market support arrangements are such that an international price below P_i has no relevance for domestic consumption or production. The corresponding EC net trade curve is then again the kinked line T_EBT_E." The size of the refund per tonne of grain required to dispose of the EC surplus X_e is the vertical amount R. The total budget cost of these exports is the unit refund times the volume exported, the shaded area. From the diagram it is plain to see that the effect of the subsidised export has been to depress the world price to P_w below the equilibrium level P_e. This may be seen as a benefit by consumers in countries which import from the world market, but will damage the interests of producers in all countries outside the EC. As world market conditions change, whether for example through systematic growth in demand, or through once-off harvest short-falls, the main adjustment affecting the EC is the size of the export refund. This means there is no impact on the EC grain market, only on the total cost of refunds to EC taxpayers. If EC domestic supply or demand change then all the impact shows up as a change in the export surplus and thence into a change in the world market price, and again the taxpayer cost of refunds. In this way none of the world market instability is imported into the EC and domestic instability is exported to the world market.

The reader is invited now to consider what happened as the technical progress combined with slower growth in grain demand continue to push the net trade curve to the right. The answer is that, *ceteris paribus*, the additional exports push the world price down further causing an explosive growth in budget costs. Something had to be done, and it was. From the mid-1980s the various 'reforms' described in Chapter 4 (successively co-responsibility levies, stabilisers, guaranteed thresholds and voluntary set-aside) were introduced to the armoury of CAP instruments. These will now be analysed.

Attempts to save the classic CAP

Co-responsibility levies (CRL) were first introduced into the milk regime and then extended to cereals. The concept was a pragmatic, budget-driven idea to try and send the signal to farmers that they had to share in the responsibility of dealing with surplus production. It was a device in which the intervention price was untouched, so the consumer price was unchanged, but producers were 'taxed' by receiving a few percentage points less than the intervention price. Thus the producer price falls, and it was hoped production too, thereby curbing some of the costs of surplus disposal. In addition the levy raises tax revenue which could be used to contribute towards the cost of surplus disposal.[15] This is illustrated in Fig.7.4. The small levy or tax reduces the producer price, inducing an even smaller cut in production (given the inelasticity of supply) and raises some revenues to offset the budget costs of surplus disposal. Consumers are evidently unaffected by the levy. From a budget point of view it can be demonstrated that with production in excess of consumption, the levy will always raise more revenue that would be saved by a cut in intervention price by the same percentage as the CRL; this seems reasonable from the geometry of the diagram, but it can be proved more rigorously.

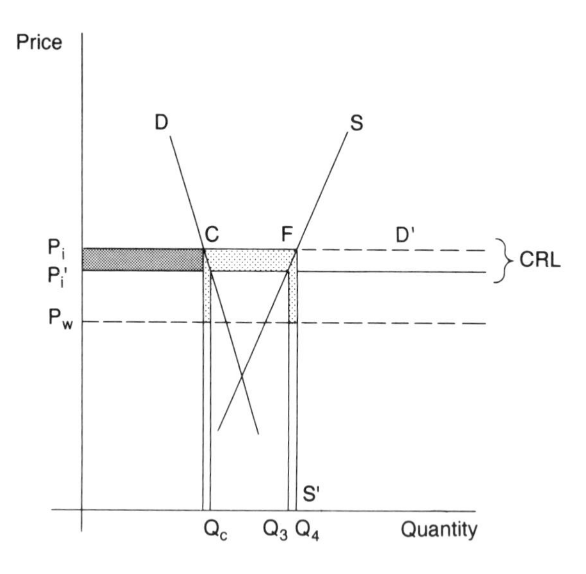

Fig.7.4: EC market with co-responsibility levy.

In the cereal sector, this instrument was never extended to make a serious contribution to funding export subsidies - the maximum cereal co-responsibility levy was 2%. Until 1992, it proved simply impossible, politically, to persuade the Council of Ministers to impose significant nominal price cuts on grain farmers.

To digress a little from cereals, in principle, such a levy or tax could be increased until the revenue from the levy more or less covers the cost of remaining subsidised exports. This is close to the logic of the sugar regime, and during 1996 was being actively promoted as the salvation of the milk regime. The sugar regime works by defining three quotas or physical volumes of sugar for each member state. The 'A' quota is, roughly speaking, the volume of domestic consumption, a 'B' quota is some export surplus which will receive a protected price, and the remainder, 'C' quota sugar, is production in excess of the domestic market requirements which has to be exported at world market prices.[16] For 'A' and 'B' sugar there is a small basic levy (also around 2%) which, diagrammatically looks exactly as the CRL shown in Fig.7.4. For 'B' sugar there is an additional levy (it is much bigger and can be up to 37%), which really does collect significant revenue towards the cost of export refunds. The regime is therefore held up by its supporters to be 'budget neutral' and thus a model CMO.

The view taken of this argument depends on the importance attached to the high consumer cost of such a regime and the particular necessity to protect sugar producers in this way. The accusation can be made that the unsubsidised exports of 'C' sugar are actually cross-subsidised by the higher payments (net of the levies) received for 'A+B' sugar.

Coming back to the development of the cereals CMO, in 1988, following a budget crisis in the Community, a new general scheme of budget 'stabilisers' was introduced. In practice, as we saw in Chapter 4, the scheme turned out to be non-functional because there was not the political will to implement it. However, formally speaking, the concept of stabilisers remains in the regulations, so it is worth summarising. A physical volume of grain, the maximum guaranteed quantity (MGQ) was defined for the Community — it was set at 160 million tonnes. Producers, collectively, were told that if production exceeded this quantity in any year, the following year the co-responsibility will be increased and the intervention price reduced. These changes were to be cumulative. The economic effects of such an arrangement may be analysed straight forwardly as a succession of either increases in the CRL or cuts in the intervention price, the former offering no benefit to consumers, the latter offering lower priced grain.

Given the absence of will in the late 1980s to reduce producer grain prices, it was inevitable that sooner or later the dairy model of supply controls would be introduced. They were in a small way in 1988 with a voluntary set-aside scheme, and then with much greater effect in the 1992 reform to which attention is now turned.

The 1992 Mac Sharry — Uruguay Round reforms

These are bracketed together because, although for domestic political reasons they were deemed to be separate, they were, as explained in Chapter 5, quite clearly inextricably linked. The package of measures adopted had to satisfy domestic ambitions for the CAP and the outcome of the Uruguay Round. From the point of view of analysis of policy change, by 1995 there were five significant changes to the old-CAP. These mostly affected grains, oilseeds, beef and sheep but only those for grains will be analysed. The changes were: (a) 30% cuts in institutional prices; (b) full, average revenue compensation for the price cut which, for farmers producing more than 92 tonnes of grain, was conditional on; (c) set-aside of a portion of the farmers (1991) base arable area and, in addition; (d) variable import levies were converted to fixed tariffs; and (e) export subsidies were limited by volume and expenditure of individual products. The price cuts were phased in over the three marketing years[17] 1993/94, 1994/95 and 1995/96 and the compensation payments increased correspondingly. Set-aside became operational in the 1992/93 production year and, because of EU and international market developments, the required proportion of land to be set aside decreased from 15% to 12% to 10% in the first three years of operation. The switch to tariff equivalents and the constraints on export subsidies both became operational for the first time in the 1995/96 marketing year.

The nature and magnitude of these changes is very significant. As will be pointed out in the final section, the balance of support in the cereals sector has changed away from market price support towards a much greater reliance on the compensation payments. These 'direct' payments and the introduction of individual farm supply management for most of EU grain production has dramatically changed the basis of the regulation of the grain market, from essentially manipulation of the market at the wholesale level to direct support at the individual farm level. It has instantly made the process of farm support very much more visible. The payments are absolutely transparent, and whilst most non-farmers cannot tell the difference between a crop of wheat and a field of set-aside, the public has a very clear impression of what is going on through the much-repeated phrase 'they are being paid to do nothing'.

In analysing the economic effects of the changes, the price cuts pose no difficulty. This is precisely what market models are designed to do. Although the size of the cut was large, introducing it over three years moderated its effects and also made it more reasonable to analyse using conventional elasticities. In practice because there were large changes in the European monetary system in the autumn just as the Mac Sharry reforms were being introduced, the changes in producer prices were not felt uniformly throughout the EU. Farmers in countries which left the exchange rate mechanism or devalued against the ecu enjoyed increases in prices (and compensatory payments) in their own currencies. However, to the extent that price cuts did take place, the economic effects are to stimulate grain utilisation and to curb the growth in production, if not actually reduce it. This would be expected to reduce the export surplus of grains which should help towards respecting the URA commitment on subsidised exports. There is tangible evidence of the consumption effect as grain has substituted for some non-grain feed ingredients. The production effect of the price cut is much more difficult to discern because of the confounding influences of the agri-monetary disturbances, set-aside and the arable payments, drought, and the changes in other regimes.

The compensation payments do pose problems for market analysis. The payments are calculated as the cut in institutional price (in ecu per tonne) multiplied by a regional average yield per hectare and are paid in two instalments for each hectare of cereals, oilseeds and proteins registered on the farm for the base year. The size of the payments is a significant part of the gross margin,[18] thus there is a strong incentive for farmers to ensure that they plant the whole of their base area. The only situation where it would be rational not to plant some of the base area is if the product price is so low that the losses would be made even on the minimal variable inputs to establish a crop, and these losses exceed the compensation payment per hectare. This is very unlikely, so it is reasonable to claim that the arable area payments are not decoupled from the decision on the area of grain to plant. Turning to the intensity of production on this base area, logically, the decision about the use of variable inputs should be made solely on the basis of the marginal costs of production and the marginal revenues from producing grain. The existence of the payments should have no bearing on this decision at all. The producer will get the payment whatever his input use and yield. Unfortunately there are many, from farm management advisors to market analysts and policy administrators who add back the payment per tonne to the price. This can only encourage irrational decision making on the part of farmers and it makes it difficult to argue that the payments are decoupled from production decisions at all.

From a theoretical point of view there are at least two other routes through which the arable payments can have an effect on production decisions. The first operates through the capital market. Phimister (1995) has shown that in situations of credit scarcity and rural capital market imperfections, a dependable source of liquidity such as direct payments to farmers can influence production decisions. The second involves the trade-off found for those working on family farms between allocating their labour to farm work, off-farm work and leisure. In this context, direct payments can be shown to have systematic impacts on the labour allocated to farm production and therefore on agricultural output (Kjeldahl, 1995). There has been insufficient experience with the EU arable payments to be able yet to measure empirically the importance of these effects and to know what impact they have had on grain production. However, these ideas do illustrate that as policy moves away from instruments which have their effect through influencing prices or quantities, the conventional tools of sectoral-level market analysis will become less helpful. Other models may have to be developed which are based more explicitly at the level of the farm decision process. This in turn will not entirely solve the problem because the results of such models are notoriously difficult to aggregate to the market level.

In principle, a simple compulsory or quasi-compulsory set-aside scheme would not be difficult to incorporate into conventional market models. If farmers are required or induced to remove a proportion of their land from production, this will reduce the quantity which can be supplied at each price and therefore shift the supply curve to the left. The size of this shift is not too difficult to calculate. The effect of this on production, net trade, world price and thus budget costs are straightforward to trace through the market models as described above. However, as always in the EU, the set-aside scheme is not so straightforward. First, farmers producing under 92 tonnes may participate in what is called the simplified scheme in which they are exempt from any set-aside at all. Thus the leftward shift of the aggregate national or EU supply curve is moderated depending on the relative importance of such small producers. Second, several variants on the basic scheme have been introduced to try and capture some environmental benefits through set-aside. The basic scheme, whose primary purpose is to reduce production, and thus the exportable surplus, is the rotational set-aside. The intention is that the farmer must put a different part of his farm out of production every year; this avoids the possibility that he always retires the least productive land. From an environmental perspective it might be preferable if the least productive land *was* removed from production. This is the basis of the permanent set-aside scheme. Analytically, these variants on the set-aside scheme make it necessary to have detailed knowledge of the distribution of farm size and even of land quality within farms if careful analysis is to be done to estimate the leftward supply shift for a given array of set-aside programmes.

The changes in the border regime agreed in the URA should also have a significant economic impact in the medium to longer run. Switching from variable import levies to a tariff — whether fixed (x ecu/tonne) or *ad valorem* (y%) — instantly reconnects the EU market to the international market. In principle, instead of the horizontal (Fig.7.1) or vertical (Fig.7.3) net trade functions for the EU, the switch to a tariff and the curtailment of the export subsidies should mean that there is a wedge between EU and world prices, and whenever fundamentals either inside or outside the EU shift the net trade curves, the impact is transmitted fully between the markets. In practice this may not happen for some time. The political negotiation which characterised the Uruguay Round resulted in bound maximum tariffs for the EU which are a long way above the applied tariffs. Even when the ceiling bindings are reduced under the terms of the agreement, they will not mean a relaxation of the terms of import access compared to the 1995 situation. Quite simply the tariffs are mostly prohibitively high and have the effect of shutting out imports (except for maize and rice). The situation for export subsidies is a little clearer although there was no experience gained in the 1995/96 year of how the commitments on export subsidy volumes and values will operate in practice. The reason for this is that world grain prices rose so high during the autumn of 1995 that the EU switched from export subsidies to export taxes. This was done to prevent the domestic market price rising as high as the world market price in order to protect domestic grain users, the target of this protection was the intensive livestock industry.

In summary, by 1996, the EU grain market had changed in appearance quite significantly from that pictured in Fig.7.3. The gap between EU and world market prices had closed, and even reversed because EU prices had fallen and because the RoW net trade curve had shifted up,[19] the EU domestic supply curve had shifted left because of the set-aside programme — probably for the first time in 30 years, the protection to the domestic market was provided by a fixed tariff and not a variable levy,[20] and exports were taxed, not subsidised. However in case the reader who has struggled this far is wondering why so much trouble has been taken to describe a redundant model, the conventional wisdom is that the first and last of these differences are generally expected to be temporary and so a modified version of Fig.7.3 should continue to be useful for some time.

Combining the economic effects

The Uruguay Round negotiations on agriculture took nearly eight years to complete because not only was agricultural protectionism so deeply embedded in most countries of the world, but it involved many commodities each protected to a different extent employing a multitude of different economic instruments in each country. The key mechanisms which made it possible eventually to find a solution were the agreement to focus on three negotiating areas: import access, export subsidies and domestic supports together with the development of the Producer Subsidy Equivalent/Aggregate Measure of Support (PSE/AMS) calculations as a way of summarising domestic support. This section will briefly review the economic logic (and difficulties) of these measures. The interested reader is referred to the OECD annual outlook and monitoring reports for full descriptions of the methodology and results of the PSE calculations; these are now routinely calculated for all OECD member countries and sometimes for other countries too (e.g. Poland, Hungary and the Czech Republic).

The PSE is intended to represent the cash subsidy which would have to be given to farmers to make them no worse off if all the current supports they are receiving through all public policy measures were to be removed. It is calculated commodity by commodity because that is the way most agricultural support has conventionally been made (though this is now changing in the EU) and the commodity PSEs may be aggregated to find the total PSE for the whole sector. The PSE is calculated as an absolute amount of money, but to enable comparisons to be made between commodities and between countries it is expressed as a percentage of the domestic value of the product, the percentage PSE, or expressed per tonne of the product to calculate the unit PSE.

The practical calculation of PSEs involves adding four components, the market element, direct payments, reduction in input costs and general services. The market element is a static, simplified version of the economic concept of producer surplus. It is the price gap between domestic and international price of the commodity multiplied by the volume of domestic production. In Fig.7.3 this can be seen to be bigger than producer surplus by area 'a'. If supply is inelastic then area 'a' is a pretty small part of the PSE so the magnitude of 'error' is small. The beauty of the calculation is twofold. First, it does not need elasticities which are notoriously difficult to estimate — and to agree in an international negotiation. Second, this measure embraces all the domestic and border measures which create the price gap. Thus no matter whether the gap is caused by border measures employing: import quotas, import bans, voluntary export restraints, health and hygiene regulations, export taxes, *ad valorem* tariffs, variable levies, specific duties, minimum import prices or export subsidies and irrespective of whether the country uses for domestic market support: intervention purchasing, quotas, set-aside, deficiency payments, or loan rates, the net effect is that domestic price is different than the international price for the commodity in question. The market element measures this price wedge or gap to indicate the extent of the support, enabling international and commodity comparisons.

There are many problems in doing the calculation. What commodities are included? The ideal answer is all commodities for which there are market support regimes. However this poses very difficult problems for fruit, vegetables, wine and tobacco, not least because there are so many different varieties of these products. The practical result of this issue is that in international comparisons, the aggregate PSE measured by the OECD for seventeen products relates to a bigger proportion of support in some countries (e.g. 94% for Finland in 1994) than others (e.g. 40% for Turkey). What is the domestic producer price? This should ideally be an average taking account of quality, season and location, weighting carefully the relative quantities sold in each of these categories.[21] Even more difficult is to measure the international price. It should relate to the same quality of product, at the same location and time period as the domestic price. Furthermore, if it is to have relevance it should be possible to buy or sell significant quantities at that price. The first requirement is conceptually straightforward, if practically complex; the second requirement is conceptually difficult for large countries like the EU and USA. Plainly, for some products, e.g. sugar or dairy products for the EU, the international price is very sensitive to EU internal policy. It means that the PSE measure of policy support is using as a reference point a price which itself depends on the policy; it has been likened to measuring distance with a stretchy ruler! This problem together with the volatility of world prices explains why the EU has always been hesitant about using PSEs as the measure of domestic protection in the Uruguay Round. Part of the compromise they suggested was to define an Aggregate Measure of Support (AMS) based on the concept of a world reference price which is an average of several years' world prices (1986–1989 for the URA). The other problem in calculating the PSE (or AMS) is that international prices have to be converted to domestic currency. In times of unstable exchange rates this poses another practical difficulty of choosing the appropriate exchange rate. It also means that the measure of support and how it changes is partly dependent on the currency in which it is measured.

For livestock products a further refinement of the PSE calculations is that the market support of livestock products calculated exactly as described above is adjusted for any support or taxation of feed costs. This is particularly important in the EU where grain prices are high; if this adjustment was not done it would suggest that livestock production is supported more than it actually is. In fact, the market regimes for pigs and poultry are mostly in place to offset the high grain prices so the measured PSE per unit should be small; there is no intervention, or supply control for these products and practically no export refunds. The feed adjustments are calculated by computing typical grain inputs per tonne of livestock product and the additional domestic costs per tonne for livestock feeds compared to international prices.

The three other components of the PSE calculations involve fewer conceptual difficulties, the main problems are political — which programmes will be included — and practical — how to allocate general non-commodity related programmes to each product. The direct payments are measures which transfer money directly from taxpayers to producers without raising prices to consumers. For the CAP they include the arable area compensatory payments, set-aside payments, the suckler cow and male bovine premiums, the sheep premiums, the hill livestock compensatory allowances and other Less Favoured

Area payments. In general these also include any farmer tax exemptions or favourable fiscal treatment. Reduction in input costs embraces any schemes to subsidise credit and capital, favourable tax treatment of fuels or energy, transport subsidies, insurance assistance and subsidised irrigation water. General services include: research, development, extension, training, product inspection, pest and disease controls, infrastructural development and assistance to adapt farming structures, assistance with marketing and promotion and disaster assistance. In moving from the PSE which includes all these measures to the AMS there was a strong political argument about which should be included. This story is related elsewhere (see Ingersent *et al.*, 1993), but it suffices to mention that this debate concluded with the colour coding of direct and indirect payments. Payments which are decoupled from production and consumption decisions may be in the 'Green' box and are exempt from reduction commitments and are thus excluded from AMS calculations. Payments which were deemed to be sufficiently decoupled to be permitted without reduction commitments in the Uruguay Round but which may come under review after the so-called Peace Clause expires in 2003 are classified as 'Blue' box, and also excluded from the AMS calculation. The latter includes the EU's arable and livestock compensation payments.

In parallel to the concept of the producer subsidy equivalent is the Consumer Subsidy Equivalent (CSE). This measures in an analogous way the economic effect of various support (and taxation) policies on consumers. The logic of its calculation is that if through domestic subsidies consumers pay a lower price for foodstuffs than the international price, then the subsidy equivalent is the amount of cash which would have to be given to consumers to make them no worse off if the price support was removed.

Tables 7.1 and 7.2 illustrate these concepts for the EU. Table 7.1 summarises the development of the EU's support over the last decade (from 1986–1989 to 1995) as measured by the PSE and CSE. The total nominal support has risen from 62 b.ecu to 74 b.ecu (20%) over this period. Part of

Table 7.1: Producer and consumer subsidy equivalents in the European Union, 1986 to 1995.

	1986–89	1994e	1995p	% change 1986–89 to 1995
Producer Subsidy Equivalent (PSE) b. ecu	61.7	68.6	74.1	20.1
Percentage PSE	48	49	49	
Consumer Subsidy Equivalent (CSE) b.ecu	51.2	46.1	43.4	-16.0
Percentage CSE	44	38	34	

Source: OECD (1996).

Notes: e-estimated p-provisional, EU-12 for 1986–89, EU-15 for 1995, includes ex-GDR for 1994 and 1995.

this increase is an inflationary phenomenon, and part is due to the expansion of the EU from 12 to 15 member states in 1995 and the reunification of Germany. The percentage PSE has hardly changed (48% to 49%) over the period. A breakdown of the PSE into the market element and the indirect and direct payments would also show that the proportion of support being given through direct payments is also increasing as a result of the switch in the cereals and oilseed regimes from price support to direct payments. This latter effect also shows up in the consumer subsidy equivalents, which have fallen significantly both in absolute terms (despite the enlargement of the Union) from 52 b.ecu to 43 b.ecu, and in percentage terms from 44% to 34%.

Table 7.2 illustrates for 1995 how the producer support is made up for each of the 16 commodities. For the cereals and oilseeds over 60% of the 1995 support came from the direct payments. This is a big change from the situation pre-Mac Sharry. In the sheepmeat regime, similarly, a majority (55%) of support is based on the direct payments. For beef the proportion is lower (14%) and for milk it is negligible. It can be seen that the support to wheat as expressed by the percentage PSE is the lowest of the cereals and oilseeds. The highest support is

Table 7.2: Producer subsidy equivalents in the European Union, 1995 by commodity and by component.

	Market Price Support	Direct Pay-ments	Cost Red-uction	Feed Adjust	Total PSE	Unit PSE ecu/t	% PSE
Common Wh.	1407	4511	785		6703	84	43
Durum Wh.	52	1881	78		2012	296	63
Maize	1431	1106	253		2790	98	50
Barley	1990	3640	293		5923	135	61
Oats	151	643	11		804	134	61
Rice	505		38		543	248	61
Soya		133	9		142	152	44
Rape		1231	67		1298	159	48
Sunflower		1087	39		1126	309	60
Sugar	2762		43		2682	166	59
Milk	20465	9	3011	-702	22782	189	63
Beef/veal	15539	2935	2058	-259	20273	2478	65
Pigmeat	2393		1912	-2393	1912	121	9
Poultry	2152		655	-795	2011	261	26
Eggs	450		391	-560	281	52	5
Sheepmeat	1117	1589	209	-57	2858	2453	59

Notes: EU-15, Total, unit and percentage PSEs are gross for crop products and net of the feed adjustment for livestock products. First five columns are m.ecu.

Source: OECD (1996).

for milk and beef. The intensive livestock sectors are much less protected, indeed the poultry figure at 26% is surprisingly high as it has been a sector in which the product price support was intended solely to offset the disadvantage of the high cereal price.

These calculations attract a lot of attention, and are politically very sensitive. They are often expressed per full time farmer equivalent or per hectare. When this is done the very high levels of protection in the EU appear lower than in, for example, the USA. Thus the EU total PSE for 1995 is much higher than in the USA and likewise the percentage PSE (49% for the EU against 18% for the US). When expressed per farmer, because there are very many more farmers in the EU, who are of course the objects of the support, the total transfers per farmer in the EU at just under $20,000 is two-thirds of the level per US farmer. On the other hand the support per hectare in the US is considerably lower than in the EU. These figures illustrate why apparently simple questions, like who supports and protects their farmers most, turn out not to have simple answers.

Concluding remarks

The simple analysis summarised in this chapter contains the essence of the way policy questions are considered at government level in most countries. Whilst the academic literature takes these ideas to considerably greater sophistication, the public debate rarely involves anything more complicated than the ideas reviewed here. The simple comparative static partial equilibrium model centres on the competitive market model of profit maximising farmers and utility maximising consumers, because most agricultural support has been based on market interventions domestically and at the border, the interconnections in supply and demand are the basics of the practical economists' stock in trade. Whenever a policy change is contemplated the first things asked concern the production, consumption, trade, expenditure, revenue, price and budget effects of the changes. This is *par excellence* what the models described here offer.

However this analysis also misses a great deal. In particular, it omits explicit treatment of farmer behaviour and managerial skills, structures and structural change, technology, dynamics and risk, the food industry and input industry and their relationships with farmers, the factor market adjustments, pluriactivity, the environmental impacts of different policies, and social and rural development. All of these are important and some of which are growing in importance. Also the model, focused on economics, omits the political dimension of policy. Some of these themes are taken up in subsequent chapters.

References

Begg, D., Fishers S. and Dornbusch, R. (1994) *Economics,* McGraw-Hill, London.

Buckwell, A., Harvey, D.R., Thomson, K.J. and Parton, A. (1984) *The Costs of the Common Agricultural Policy,* Croom Helm, London.

Colman, D. (1995) Problems of measuring price distortion and price transmission: a framework for analysis. *Oxford Agrarian Studies,* 23(1), 3–13.

Ingersent, K.A., Rayner, A.J. and Hirne, R.C. (1993), *Agriculture in the Uruguay Round,* Macmillan, London.

Kjeldahl, R. (1995) *Direct income payments to farmers. Uses implications and an empirical investigation of labour supply response in a sample of Danish farm households,* pp171. Report No. 85. Danish Institute of Agricultural and Fisheries Economics, Copenhagen.

Mansholt (1968) *Memorandum on the reform of agriculture in the European Economic Community,* Commission of the European Communities, Brussels.

Nunez-Ferrer, J. and Buckwell, A. (1996) *Using ESIM to model economic impacts of enlargement of the European Union to the Central and Eastern European Countries,* Wye College, University of London.

OECD (1996) *Agricultural Policies Markets and Trade in OECD Countries: Monitoring and Outlook Report 1996,* Paris.

Phimister, E. (1995) Farm consumption behaviour in the presence of uncertainty and restrictions on credit. *American Journal of Agricultural Economics,* 77(4), 952–959.

Varian, H.R. (1992) *Intermediate Microeconomics a Modern Approach,* 3rd ed, Norton, New York.

Varian, H.R. (1993) *Microeconomic Analysis* 3rd ed, Norton, New York, London.

---

[1]The only significant crop for which there are no CAP regulations is potatoes.

[2]This can be studied in any textbook of microeconomics; three which are recommended in increasing order of sophistication and difficulty are Begg, Fisher and Dornbusch (1993), Varian (1992) and Varian (1993).

[3]Two examples of these partial equilibrium models are the so-called Newcastle CAP model which existed in the early 1980s but was initially based in Fortran code on a mainframe computer, Buckwell *et al.* (1982); and a more recent model, ESIM, constructed by Josling and Tangermann and others to examine the effects of eastern enlargement of the European Union, (Nunez-Ferrer and Buckwell, 1995).

[4]The word intervention is a good example of how general words acquire a special meaning in the context of the CAP. The phrase 'intervention in the market' refers in principle to *any* collective or public action which influences or disturbs the result which the individual market agents (buyers and sellers) would have created on their own. In this sense, all the measures under the CMOs of the CAP are interventions, the border measures, the producer or consumer subsidies, quotas, set-aside, the lot. However in all discussion of the CAP the word has a much more restrictive meaning. Intervention refers to the procedure whereby farmers – actually it is usually not farmers directly but traders or first stage processors – can choose to sell produce to a government-backed intervention agent at a politically determined intervention price (or more usually some fraction of that price).

[5]It is a fairly inconsequential point to use EC to denote the European Communities back in the 1960s, but it does serve to remind that we are discussing the past system, and is the convention adopted in this book.
[6]This is described as the *initial* outlay because it is hoped that the grain purchased can be released onto either the domestic or world market at some later date when there is a shortage of supply. This instantly alerts the analyst to another gap in this simple model – that it contains no provision for stocks to be built up or run down.
[7] This denotes that the prices are inclusive of costs of insurance and freight.
[8]There is a strict relationship between intervention and threshold prices. The intervention price was defined for the lowest price region in the Community (Ormes in France), handling and transport costs per tonne from here to the most deficit point in the EC (Duisburg, Germany) were added to determine the target price. The threshold price is then derived from the target price by deducting the unit transport and handling costs for grain between Duisburg and an appropriate port, e.g. Rotterdam.
[9]In practice the levies are determined by the market management committees of the EC who receive offer prices from potential exporters to the EC, and choose the lowest such price as the basis from which to calculate the levies for the week. The figure shows the smallest levy required to prevent foreign supplies being bought into intervention. There was little to stop larger protection being applied which raises domestic EC prices and reduces trade and world price still further.
[10]Student readers are urged to draw versions of Fig. 7.2 to illustrate these features for themselves. For example, sketch the situation when EC short-run supplies are abnormally small or large (i.e. show a vertical 'supply' curve to the left or right of Q_1), show how this affects the net trade curve and the levy required to protect the EC market at the threshold price. Then do the same thing for both supply and demand shifts in the rest of the world market by adding a third panel to the diagram for the RoW.
[11]The degree to which the high rate of technical progress and farm structural change in Europe was 'caused' by the high and stable price regime is unknown. Europe is a sufficiently large market for it to be reasonable to assume that there was some degree of induced innovation in the crop protection, nutrition and mechanisation industries. However these industries operate worldwide and the same process of 'intensification' of crop production was taking place in markets much less protected than Europe. On structural change it can be argued that the support system protected the inefficient and thus the observed changes were slower than they would have been under a more liberal regime. These are complex and not well understood areas, but Chapter 9 analyses the relationship between the CAP and technical change.

---

[12]Not all of intervention stocks are physically held in publicly owned intervention stores; there are also schemes whereby private merchants can receive support for storing grain to keep market prices up.
[13]This chapter is not the place to discuss the merits and demerits of food aid, and it should also be noted that a further option for utilising more EC grain is to stimulate its use as a source of energy. The economics of both of these options leads to their rejection as anything more than minor, occasionally interesting avenues to pursue.
[14]It can be seen in the diagram that in the 'big-country' case such as the EC, this is a dubious thing to do. If the counterfactual policy was no intervention, no import protection or export subsidies, then the world price would be significantly higher at P_e. The consumer loss and producer gains are therefore exaggerated if these effects on the world market price are not endogenized in the model.
[15]In one of the greatest perversities of the CAP, in the milk sector the co-responsibility levy revenues had to be used to promote milk consumption. Thus intervention prices (and after 1984, supply controls) were used to discourage consumption, and then co-responsibility levies (essentially a further tax on consumption) were used to stimulate consumption!
[16]A further complication of the sugar regime is that 1.3 million tonnes of sugar is allowed into the EU from the ACP countries under the Lomé agreement.
[17]For grains the production year 1992/93 refers to the period in which the crop is sown and then harvested, e.g. sowings in autumn 1992 and spring 1993 leading to harvest in summer 1993, and the marketing year 1993/94 for this crop is the period until the next harvest when the crop is sold, e.g. autumn 1993 to summer 1994.
[18]The gross margin is the revenue per hectare less the variable production costs - seeds, fertilisers, crop protection chemicals, and machinery variable costs.
[19]The cause of the latter is far from fully understood. There were multiple factors at work: historically low stocks combined with a poor year and thus low yields in some producing regions (e.g. USA and Russia), a drop in area cultivated (e.g. EU and parts of eastern Europe) and a rise in consumption following continued high economic growth (e.g. China).
[20]Though, as explained in Chapter 5, for cereals something similar to a variable import levy still applies.
[21]It is extremely easy to state that producer price should *take account* of quality, location and season, but the data requirements to do so are immense, requiring storage costs, transport and handling charges and detailed market information.

Chapter 8

Extensions and Political Analysis of the CAP

David Harvey

Introduction

The previous chapter explained the comparative static economic analysis of the CAP. This kind of analysis leads to the conclusion that the 'old' CAP was economically inefficient. Furthermore, the policy redistributes income from the less well-off to the better-off, since much of the support cost is borne by the consumer who pays according to the proportion of income spent on food, while the benefits accrue to farmers on the basis of their shares of production. Such observations are not new. Wilson's conclusion: "clearly agricultural support has been neither in the national interest nor justified by widely held perceptions of social justice" (Wilson, 1977) has been echoed by most other analysts and commentators and, at least until 1992, was an accurate, if academic, conclusion throughout the CAP's existence (Chapter 4). In addition, as explained in the previous chapter and further explored in Part IV, the CAP's costs extended beyond the boundaries of the European Union (EU) through its effects on world markets and thus on other developed and developing economies.

This raises two important questions. First, if the analysis is both correct and widely accepted, how did the CAP persist so long? Second, what additional analysis is needed to explain the robustness and persistence of the CAP? It is to these questions that this chapter turns. A third question — what has to change in order for the policy to change? — is the subject of the concluding chapter (18). Throughout, the analysis proceeds from an economist's perspective, betraying the background of the author. Other explanations of the policy development from different perspectives can be found in, for example, Lowe *et al.* (1994) especially Chapter 1. A final introductory caveat is in order. The arguments presented here are in outline form only. Footnotes are used extensively to provide some elaboration of points in the argument and to indicate relevant parts of the literature. Readers who simply want the general thrust are encouraged to leave reference to footnotes until later.

Interpretation of the comparative static analyses of the CAP

Comparative static analysis of the last chapter produces the kind of results illustrated in Table 8.1.[1] The economic inefficiency of the CAP is illustrated by the net welfare cost figures, showing a loss to the EC-12 of just over 13.7 billion ecu, some 0.25% of EC GDP, with three of the twelve member states (Denmark, Greece, Ireland) actually showing a gain from the policy compared with the alternative of unilateral free trade. These figures are simply the sum of the preceding rows in the table, showing the producers' gains, the users' losses and the taxpayers' cost respectively. What do these figures actually mean?

Producers' gain is a measure of the increase in *economic rent*[2] earned by factors engaged in agriculture over and above that which would be earned in the absence of the policy intervention (illustrated by area X + Y in Fig 7.3). In the case where all factors and inputs except land are available to agriculture in perfectly elastic supply (that is, the prices and returns of these factors and inputs do not change whatever the agricultural output and use levels), theory suggests that all of the policy benefits will accumulate in rents and agricultural values of land, so the figure is a measure of the annual gain to landowners (hence the expression of this gain on a per hectare basis in the second half of Table 8.1, though simple average figures can only be treated as illustrative). In practice, the assumption of perfectly elastic supply of all factors and inputs is extreme. Some fraction of the gain would be expected to accrue to the owners of other factors specifically associated with the industry, including those in industries 'up-stream' from the farm-gate.

Bearing this in mind, perhaps the simplest way of thinking of the producers' surplus gain is that it is an approximation of the increase in agriculture's gross product (gross value added) due to the policy, relative to the assumed counter-factual or base policy, in this case no intervention at all in the EU (shown in Table 8.1 as Gain: % GVA).

This gain can be expected to accrue to the owners of factors of production (land, labour management and capital) which are less than perfectly elastically supplied to the industry (or its up-stream suppliers). In turn, the market value of these factors can be expected to reflect the value of the support. Entrants to the industry after the introduction of the support system will have to pay for the privilege of obtaining the support, and thus transfer their gain to the previous owners. The gains identified in the simple welfare arithmetic, therefore, are likely to be extensively dispersed throughout the economy.

Notwithstanding the avalanche of empirical studies of the effects of the CAP during the 1980s (Demekas *et al.*, 1988), the literature on the dispersion of the 'gains' from the policy is extremely thin. However, Harvey (1991) does attempt a preliminary exploration of this issue for the UK. This analysis starts with an estimate that agricultural gross product in the UK was greater by 22% because of the CAP than it would have been under conditions of multilateral free trade in

Table 8.1: Benefits and costs of CAP versus unilateral (EC) free trade (1990).

	BLeu	DK	FRG	GR	SP	FR	IRL	It	NL	P	UK	EU 12
(Billion ecu)												
Producers' Gain	0.86	1.07	4.70	1.11	2.89	7.15	0.94	3.92	1.76	0.40	3.82	28.62
Users' Cost	0.82	0.42	4.86	0.76	2.51	5.13	0.33	4.15	1.17	0.56	4.31	24.98
Taxpayers' Cost	0.80	0.42	4,54	0.33	1.21	3.59	0.17	2.46	1.16	0.14	2.51	17.33
Net Welfare Cost	0.76	-0.23	4.69	-0.02	0.83	1.57	-0.43	2.69	0.57	0.26	3.01	13.79
Budget & Trade Effect	-0.27	1.05	-2.59	1.05	-0.51	1.24	1.62	-2.04	1.93	-0.17	-1.31	0
OECD 'PSE' (1)	2.1	2.55	14.8	2.05	6.71	17.4	2.48	9.0	4.76	1.06	8.69	71.58
(ecu or %)												
Gain/agric.worker	8669	11134	6164	1420	2063	4851	3943	1818	7486	461	8370	3340
Gain/"ft" farm (2)	14309	12663	9920	5151	4021	9587	5165	6184	19977	3788	18182	8135
Gain/ha. UAU (3)	578	382	396	193	107	234	165	228	872	88	207	224
Gain: % GVA (4)	29.8	32.9	36.4	15.6	19.6	28.2	41.2	13.5	21.0	20.7	44.6	24.6
Illustrative Costs:												
Cost/worker (5)	419	322	336	293	296	402	448	312	372	149	256	320
Cost: % GDP (6)	0.45	-0.27	0.32	-0.03	0.18	0.16	-1.15	0.29	0.24	0.30	0.35	0.26
Cost/benefit ratio	1.88	0.79	2.00	0.98	1.29	1.22	0.53	1.69	1.32	1.75	1.79	1.48

Table 8.1 Notes:

1. The PSE (Producers' Subsidy Equivalent) is a gross measure of the total consumer and taxpayer cost of providing farm support, here disaggregated from the EU total as estimated by the OECD on the basis of national shares of EU production. It was explained in detail in the previous chapter.
2. 'Full time' farms are here represented as farms of more than 5 ha of utilisable agricultural area.
3. UAU is Utilisable Agricultural Area.
4. Producers' gain expressed as a percentage of agricultural Gross Value Added (GVA).
5. The consumer and taxpayer cost expressed per head of the total civilian employment.
6. Net Welfare Cost expressed as a percentage of total GDP.

Source: European Commission, 1994, Table 30, and author's calculations.

1986, based on Buckwell *et al.*, 1982. On the basis of a simple land price model, Harvey estimated that land prices were inflated by 46% because of CAP support policies, on average.[3] At 1986 land prices, this increase amounted to £655/ha. This inflation in land prices can be associated with a policy-induced rent increase of £34/ha on average (based on an estimated relationship between rents and land prices provided by Lloyd, 1989). Applied to the UK agricultural land area, this amounted to £631m, which in turn was 55% of the estimated farming gain. Thus, the implication was that just over half the support provided to the agricultural sector through CAP price support policies was capitalised in land values and rents, while the remaining 45% was implicitly distributed through the factor and input markets to other resources (factors of production) used directly or indirectly by the industry.

Conversely, removal of support can be expected to reduce the value of assets (factors) employed in the industry. In this sense, the estimate of producers' surplus gain is an estimate of the extent to which the annual stream of agricultural returns to factors of production would be reduced if the support were to be removed. In turn, this reduction would be reflected in the capital value of these factors (assets), which, in the case of many farmers and their families, represents a fall in the value of their pension funds. Thus, farmers can be expected to resist support reductions strongly, or at least demand considerable compensation (redundancy payments) to be persuaded to forego further support. An illustration of the extent of these effects can be seen from the expression of the gains per head of the agricultural labour force and per 'full time' farm in Table 8.1 (though it should be noted that the definition of full time farm used here is particularly cavalier as simply those farms occupying more than 5 hectares).

Users' cost is, in principle, a measure of consumers' *willingness to pay* for the support price reduction. (It is illustrated by area X in Fig.7.3.) In other words, it is an estimate of the answer to the question: how much would you (the consumer) be willing to pay to forego the policy change? In principle, the alternative question could be asked: how much would you be willing to accept in order to put up with the policy rather than the alternative?[4] In essence, this is a measure of the additional spending on food caused by the support regime, including an allowance for the consumption foregone as a result of the higher prices, since consumption typically declines as prices increase, other things being equal. The measure can be taken as indicating the extent to which consumption of other goods and services or saving is reduced as a consequence of the support system.

Here, the measure is applied at the farm gate, and hence is labelled 'user' rather than consumer. However, this figure typically assumes that the downstream processing distribution and retailing (PDR) sectors are perfectly competitive and that price reductions at the farm gate are transmitted completely to the final consumer. If this assumption is not met, then there is scope for accumulation of part of this cost in the returns to the 'fixed' factors[5] employed in the downstream PDR sectors.

Taxpayers' cost is usually a measure of the direct exchequer spending on the programmes, and typically does not include either the administrative costs of the programmes or the indirect effects of taxation. Administrative costs are not insignificant, especially for those policies which require individual farmer application (such as 'direct' subsidies or deficiency payments) rather than market intervention (through intervention buying, import taxes and export subsidies). The indirect costs of taxation arise as a consequence of the effects that general taxation has on the efficiency of the economy, through the distortion of incentives and market signals throughout the economy.[6] Ignoring these costs thus tends to bias conclusions in favour of those policies which depend on taxpayer costs rather than on consumer costs. An illustration of the size of the taxpayer and consumer burden of the policy is provided by the expression of the sum of these two components per head of the civilian working population (who have to bear these costs) in Table 8.1.

The 'net welfare cost' of policies, as illustrated in Table 8.1 above, is simply the sum of these three components. What does it actually measure? The answer depends on the separation of *efficiency* issues from *distributional* or equity issues (see below). Efficiency refers to the capacity of the economy to produce (and thus consume) as much as possible from as little as possible (Just *et al.*, 1985), and the so-called welfare analysis in this sense simply identifies the extent to which the economy could produce more than it currently does with the existing level of resources. The expression of the net welfare cost as a percentage of total GDP (Cost: % GDP) in Table 8.1 illustrates the relative inefficiency of the CAP in the total economy, measured here at the partial level — which ignores the multiplier effects that such inefficiency can be expected to have elsewhere in the economy. The figure shown here for the EU of 0.25% of GDP is towards the lower end of the range of estimates from other studies (see Winters (1987b) and Demekas *et al.* (1988) for surveys), which tend to report efficiency losses of the order of 0.5%.

General equilibrium analysis

More sophisticated *general equilibrium* analysis of policy, explores the ramifications of policy change throughout the whole economy, as opposed to the more typical partial analysis (as illustrated in Table 8.1) which restricts attention to one (the agricultural) sector. The advantages of the general equilibrium approach (see, e.g. Winters, 1987b, 29–38) are that the interactions between agriculture and the rest of the economy are included, thus typically magnifying (multiplying) the costs of the policy, while the disadvantages are that the analysis becomes more complex and reliant on assumptions made about the behaviour and mechanisms of the rest of the economy as well as of the agricultural sector. While general equilibrium analyses are less common, those that have explored the EU's agricultural policy tend to report rather larger efficiency losses for the policy, of the order of 1% of national GDP, and as high as 2.7% (Burniaux and Waelbroeck, 1985), implying a multiplier effect of the order of 2 to 6.[7]

There is often some confusion about the precise meaning of general equilibrium model results. In essence, these models seek to capture the economy-wide effects of policy change on the productive performance potential of the economy. In the case of the policy alternative of eliminating farm protection and support, the models seek to measure the full 'deadweight loss' of existing policies for the whole economy. Thus the results answer the question: how much more real income *could* the economy generate without this policy than it does with it? These models do *not* answer the question of how much more income the economy *would* generate, which depends critically on the economic conditions and management of the economy. Since the models deal with general equilibrium, they produce results on changes in balances of trade and employment levels as well as changes in national income, and it is tempting to report results as indicating that current policies increase unemployment — and removal would reduce unemployment, as a mirror image of the potential increase in national income which could be achieved without the policies.

However, the effects on employment and balance of trade depend very heavily on the structure of the model, in particular, the 'closure' of the general equilibrium system, the nature of the factor supply conditions and substitution possibilities between factors and sectors (Winters, 1987b). If the model is closed (as is typical) by forcing trade accounts to balance and factor markets to clear at full employment, then much of the effect of policy change occurs in changing prices and wage rates, depending on the factor supply assumptions and on substitution between factors and sectors. If, on the other hand, the model is closed through rigid real wage rates and flexible exchange rates, the consequences of policy change can be rather differently portrayed, in this case resulting in different levels of employment. Thus, under flexible wages and full employment, Stoekel (1985) estimates that employment in manufacturing was reduced under the old CAP by about 1.5% (about 400,000) in Europe, while employment in the agricultural and food sectors was correspondingly increased. Output and prices increase as a consequence of the CAP in the agricultural and food sectors and decline in other sectors (in response to relative price changes). Trade balances in the agri-food sector are improved by the CAP, but are reduced in the manufacturing sector. Under rigid wages, the same basic model produces a result of increased aggregate unemployment by 1%, spread across the non-agricultural sectors, since under this formulation, most of the adjustment is forced into a quantity adjustment in the labour market.

As Winters (1987b) concludes, "by promoting agriculture, EEC countries waste resources, cut wages and/or employment and curtail manufacturing" (p37). In seeking to identify the detail of these economy-wide interactions, CGE models are a clear advance on the more typical partial equilibrium approaches in the literature. However, such extensions come at the cost of pre-conditioning results by the structure and parameters of the models, and it is difficult to be sure of the extent to which actual results are a reflection of reality rather than of the structure of the model. Nevertheless, the logic underlying such an extension would seem inescapable. It must surely be true that partial equilibrium models cannot possibly capture the full effects of farm policy intervention, which must necessarily be an order of magnitude greater than these first round partial effects.

Distributional issues

Policy analysis typically requires that something be said about the *distribution* of gains and losses amongst the electorate. Here, the analytical foundation becomes more difficult. The fundamental problem is that conventional economic theory has nothing to say about how to compare the welfare of different people — the problem of interpersonal comparisons. The only criterion accepted by this theory is that a policy change which makes at least one person better off and makes no one else worse off can be regarded as an unambiguous improvement — the Pareto criterion. Unfortunately, practically no policy change falls into this convenient box. Certainly, a change in agricultural policy would make some people worse off. Economic theory escapes from this dilemma, as might be expected, by making a series of heroic assumptions. Suppose that everyone attaches the same (identical) value to a change in their income, regardless of the amount of income they presently receive (in economic terminology, the *marginal utility of income* is both constant and identical for all people). In this case, it is possible to compare (rank) people's gains and losses. If it is also assumed that this constant marginal utility of income is equal to one (in other words, that a £ is a £ regardless of from whom it comes or to whom it goes), then it is also possible to add gains and losses and treat the result as if it were a measure of economic welfare. As such, the Pareto criterion can be rescued, in the sense that those made better off by the policy change could compensate the losers and still remain better off — the compensation principle.[8]

Such assumptions clearly are nonsensical as reflections of the political judgements most often averred (if not always followed) by politicians. The fact that most political systems ascribe to (generate) progressive income tax schedules demonstrates that the political system is likely to judge the marginal utility of income of the well-off as being lower than those of the poor. But, having laid out the analysis in this way, a solution becomes apparent — determine (assume) an appropriate reflection of the political judgement about marginal utilities of income (from a social as opposed to a private point of view) and use these 'weights' to adjust the elements of the welfare sum. Unfortunately, determination of the political values turns out to be a seriously difficult problem, not least because appeal to typical rules of democracy about voting practice appear to deny a generally useful *social value function* [9] with which to compare different policy outcomes. There is no such thing as a well-behaved social value function.

It is tempting to ignore this theoretical result as an intellectual curiosity and either a) assume some political preference weights, or b) estimate political preferences on the basis of the history of government decisions. The former is not commonly accepted practice in the literature, while the latter proves extremely difficult given the wide variety of factors and influences on political decisions. Practical use of welfare analysis therefore requires an intelligent interpretation of what it says and, more importantly, what it does not say. The most concrete conclusions that can be drawn from welfare analysis of the CAP are that a) there are more efficient ways of organising the economy; and b) that it should be possible, at least in principle, for the gainers from a change in the policy to compensate (fully) the losers and for everyone to be no worse off, while at least some would be better off. The framework of the analysis is simply not capable of allowing an answer to the question of why present policies are pursued and what might cause them to change and in what direction. Before turning to these questions, however, some further remarks about the results of comparative static analysis, illustrated in Table 8.1, are helpful.

The *budget and trade effect,* in contrast to the previous figures, is somewhat less contentious. These figures measure the effect of each member state 'opting out' of the *common* policy and simply pursuing exactly the same policy instruments at the same level of prices, but as independent national states. In this case, since prices and other policy instruments are not changed, quantities produced and consumed do not change and there is no change in producer or consumer welfare. The results of this hypothetical experiment are, of course, no change at the EU level, since there is no difference between a common policy and the equivalent set of independent national policies at the EU level. However, abandoning both free trade at common prices and the common financing provisions of the CAP does affect the national balances (flows of income between the member states). In essence, nationalisation of the CAP implies that trade would take place at world prices rather than the preferential common prices. Thus, in each country, imports from elsewhere in the Community (as well as from the rest of the world) would raise import levy revenues while exports to other member states (as well as to the rest of the world) would incur export refund spending. The difference between these balances at the national level, and those incurred as a result of the CAP is expressed as the 'budget and trade effect.[10] The results illustrate the familiar proposition that the common features of the policy generate inter-community transfers between member states — some countries would win and others would lose from a nationalisation of the policy. The particular case of the UK under this alternative was the basis of the UK government's argument in the early 1980s about the inequity of the policy, and led to the Fontainbleu agreement under which the UK gets a budgetary rebate.[11]

As noted above, expressing welfare gains and losses on a per head basis has the advantage of indicating the relative importance of the policy to the interest groups, notwithstanding the caveats already outlined about the underlying figures.[12] These relativities perhaps provide a clue to the persistence of the policy. The cost/benefit ratio indicates, *inter alia*, the relative political weights (in favour of producers) which would be necessary to make the 1990 policy more acceptable than the alternative of free trade. However, when the opening sentences of this chapter are remembered, the results seem amply to bear out Wilson's conclusion.

Public choice analysis and the CAP

The public choice approach to policy development and policy choice stems from the obvious unsatisfactory nature of neoclassical economic analysis of policy, but is borne of a deep-seated reluctance to discard the general presumptions of this analytical framework. As Just (1988) remarks (p448): "In spite of widespread advice by welfare economists, many major policies that are apparently contrary to accepted principles of welfare economics continue to be employed. As a result, a common view among economists is that the policy process is often misguided. On the other hand, those closer to the policy process argue that (welfare) economists are misguided".[13]

The basic proposition of virtually all strands of the now large and rapidly growing public choice literature is that policy decisions are made in a political 'market place' in which the essential currency is votes, backed up with political contributions and support and coloured by power motives, including those deriving from the size and scope of politically-dependent bureaucracies, typically treated as analogous to the profit motive underlying the supply-side of neoclassical economics.

Indeed, the general logic of the need to regenerate the political side of 'political economy' is readily apparent from a consideration of the fundamental picture of the world as perceived in the neoclassical story. Consider the implications of profit-seeking firms and utility-(income)-seeking consumers combined (as the theory admits it must be) with a government whose major function is the redistribution of income and wealth. The workings of the competitive market mean that this redistribution, even if entirely resource neutral,[14] will need to be continuous, not once and for all, since the pursuit of profit will (in the short-term) lead to accumulation of wealth. Even in the absence of market imperfections and failures, then, the market model includes a government continually engaged in economic activity, taking and redistributing income.

The existence of a government, responsive to public wishes about distribution, provides entrepreneurs, consumers and taxpayers with the means to influence their economic environment, which now includes the government, to their own ends. Add to this model the evident gains to be made from collective action (especially but not only in the labour market) as opposed to individual action, and the advantages of collective and impersonal power associated with the limited company. The implicit pressures in favour of the maintenance of, at least, workable competition in the traditional theory, namely the continual pursuit of personal satisfaction and profit, are now turned against that maintenance and in favour of winning control over the government, as well as the market place. This is the essence of much of the public choice literature, epitomised by Rausser (1982) in the classification of PERTs (legitimate engineering and maintenance transfers designed to correct market failures, provide for public goods and foster/improve economic efficiency) and PESTs — the rent-seeking transfers which arise as self-seeking *Homus economicus* struggles to enhance their positions and improve their lot through winning control over government.[15]

The public choice literature solves the impossibility problem which denies the existence of a well-behaved social welfare function, by defining the problem of public choice as a two stage process — first, the formulation of constitutional rules and prescribed procedures for determining public choices; second, the operation of this system of rules in making "political exchanges" between interested parties (Buchanan and Tollison, 1984, p16). The arguments about appropriate and realistic constitutions make fascinating reading, developed on the basis of such pathbreaking works as Rawls (1971), in which rational choice of equity rules are most unambiguously determined behind the 'veil of ignorance' — what rules would you choose to live with presuming that you do not know the circumstances in which you will find yourself when the rules are operated.[16]

The results of political exchanges, determined through the rules and practices embodied in the constitution and associated institutions are seen as implying a 'political preference function' (PPF). Given an assumption that observed outcomes reflect equilibria on such a PPF, it is possible to estimate the parameters (weights) of the underlying function (see Swinnen and van der Zee, 1993, 264ff. for a survey). Given such weights, they can then be applied to the conventional welfare arithmetic to produce trade-offs and outcomes which might more closely resemble the real world. However, the restrictive assumptions necessary for such an approach and difficulties of estimation seem bound to limit severely its practicality and policy utility. As Swinnen and van der Zee conclude: "PPF studies include the underlying structure of government preferences, but do not provide any explanation of the preference structure itself. The government is still regarded as a single entity that acts according to some behavioural function. The abstract policy-maker is an artificial concept to circumvent the modelling of the political market" (p. 266). In effect, the PPF can be seen as being a sort of 'reduced form' of the political market models, in which the interesting and potentially useful features of the policy process are suppressed.

These difficulties with the PPF (or "Self-willed Government" — MacLaren) approach led to the development of more explicit models of the political market place (Swinnen and Van der Zee, 1993, 266ff) or the "Clearinghouse Government" model (MacLaren, 1992). Although MacLaren characterises the latter as representing the outcome of pressure group interaction and the government's response, political market models encompass frameworks which describe the process of government as the interaction between interest groups, voters, politicians and bureaucracies rather than separating government from this process. There are a number of different strands to this literature, stemming from the Downs' (1957) model of political parties as vote-maximising firms, through

Tullock (1965) modelling bureaucrats maximising power and influence approximated by size, to Olson (1965) considering the power of interest groups through the costs and benefits of lobbying.

Voting models, in which the electorate is seen as trading votes for the delivery of their preferred policies (and prevention of those most damaging), typically include costs of voting and the rational ignorance hypothesis (Downs, 1957) under which the costs of gathering and assimilating information on fringe issues are not outweighed by the potential benefits of information. There is a distinction between the discrete voting model tradition (either for or against a policy proposition), in which the average and median voter issues predominate, and the 'proportional' voting models which concentrate on political support as opposed single issue votes. The latter are closely associated with (possibly more realistic) pressure group models, in which the linkage between pressure group behaviour and policy choice can be either explicit or implicit. The latter, following the Olson tradition, have been particularly influential in explaining farm policies (Swinnen and van der Zee, 1993), stylised as showing an inverse relationship between protection (support) and relative size of the farm population, explained as a result of the costs of lobbying when the lobby is large but the individual benefits are small and vice versa. Gardner (1983), for instance, finds considerable empirical support for this proposition for US farm programmes.

There are, however, some reasons for doubting that the lobbying cost argument is the whole story as far as farm policies are concerned. Indeed, de Gorter and Tsur (1991) present a model which does not include any lobbying effects but simply relies on differences between a) *relative* income of the (farming) group compared with the income of the other (non-farming) group; and b) the *redistributed* income of each group, i.e. how much more they get with the transfer than without it. These differences drive a political support function, which in turn delivers farm support. On the basis of conventional assumptions about the form of these functions, de Gorter and Tsur conclude that many of the stylised facts of agricultural support can be explained by such a model.[17] They argue that the Olson-type model is not able to explain, for example, the observed preference of West Germany for high farm support versus the UK preference for lower levels of support. Such a result is intriguing for those interested in minimising the complexity of public choice models, since the avoidance of appeal to either lobbying costs or imperfect information considerably reduces room for manoeuvre in the face of contradictory evidence.

de Gorter *et al.* (1991) employ the voter-politician model to the CAP, using MCAs as an indication of national preferences. They find some empirical support for their hypothesis, supporting the proposition that, given the budgetary situation, the closer are farmers' endowment incomes (those without policy intervention) to parity with the rest of the economy, the lower the political support for intervention, while the greater the effects of price intervention in increasing incomes (producer surplus gains), the greater the political support. As an illustration of the relative power of a parsimonious model, this is modestly impressive, yet to the seasoned CAP-watcher, the explanation appears rather too parsimonious and ignorant of other potentially key variables.

Munk (1994) presents one of the few and recent attempts to integrate the public choice and public finance approaches to explaining the farm policy of the EU, complete with a brief but comprehensive account of many of the theoretical issues associated with such explanations. Munk also includes discussion of the important but often neglected indirect costs of policy intervention associated with administrative and transactions costs and the secondary market distortions resulting from general taxation necessary to support farm policy. He develops a quantitative model of farm support (measured as the OECD's Producer Subsidy Equivalent — PSE[18]) where the level of support across countries and over time is hypothesised to depend on the difference between average agricultural incomes (per head values added in agriculture relative to per head GDP in the whole economy) and on the 'net export intensity'. The latter variable is a reflection of the fact that the larger the export share of domestic production, the more costly (to taxpayers) it is to support domestic agriculture and typically the lower is the level of support. This model is estimated over 11 OECD countries (treating the EU as a single country) over the period 1979–90. Munk's EU results (Munk, 1994) are shown in Fig.8.1, but they have been extended to include both 1979 and 1991–1993.[19] Munk's model incorporates a 'partial adjustment' mechanism whereby the actual (current) PSE is hypothesised to adjust slowly towards its equilibrium value given the major determinants of the PSE level (relative agricultural incomes and net export intensity). When allowing for gradual adjustment of support to 'equilibrium' levels and also for country specific relationship differences, this model accounts for 95% of the total variations in PSE over time and between countries, with 74% of the variation being accounted for by differences in export intensity and agricultural relative income (the remainder being accounted for by the country specific term — possibly reflecting omitted variables).

Two major features of this political economy model are apparent. First, from 1986 onwards the model suggests that the general level of protection for the EC is adjusting downwards towards a gradually declining equilibrium level of protection. This reflects, particularly, export intensity tending to increase throughout this period, since relative agricultural incomes as measured in this model are not generally improving — though see below for further discussion of this point. Second, as Munk admits, the "complicated adjustment process" of PSEs over time is not well captured by this model, which appears to be out of step with reality by one year.

Despite this growing literature, however, the contribution of formal public choice analysis to an explanation of the CAP remains distinctly marginal, as noted *inter alia* by Swinnen and van der Zee (1993), Maclaren (1992) and Munk (1994). This is illustrated by Munk's formal model of CAP development above. In particular, characterisation of the CAP simply as changes in PSE is fraught with dangers and simplifications (some of which are highlighted in Chapter 17), while attempts to reduce the policy process to a simple linear equation driven largely by two highly simplified variables is not likely to prove very convincing or enlightening. It says much for the Munk model that it performs as 'well' as it does. Notice, however, that this model holds out little prospect for an early reduction in the level of agricultural protection, and certainly offers very little clue about the need for or fact of the 1992 CAP reform package.

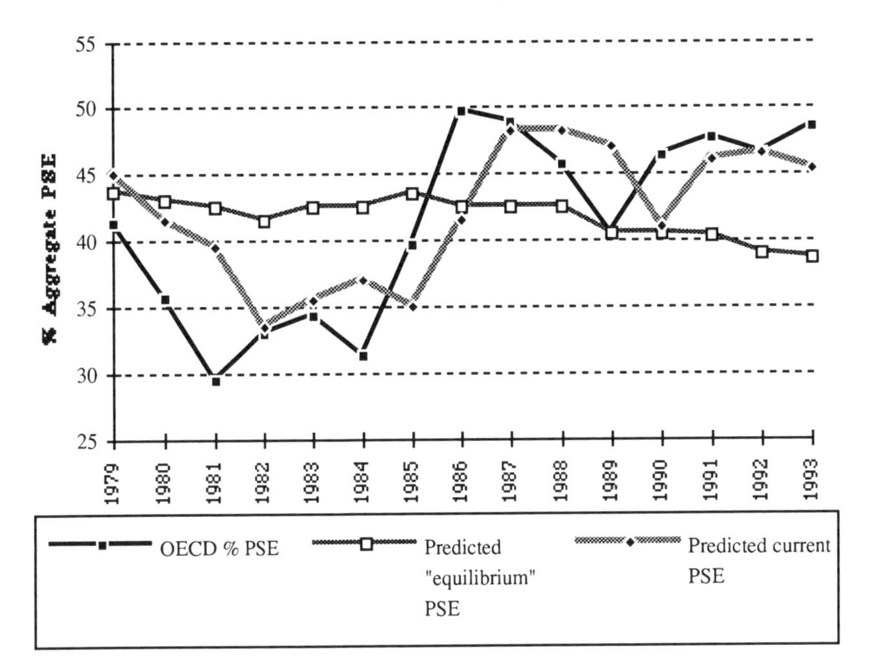

Fig. 8.1: Actual and predicted EU % PSE, 1979–1993.

Source: OECD PSE tables, 1993, Munk (1994) and author's estimates.

Among the most common reasons advanced for this gap are: that a supranational policy is intrinsically more difficult to deal with in formal choice models, especially since the policy is not 'purely' supranational, there being some scope for national discretion in application and instrument setting; that such models do not produce the erratic or chaotic policy decisions observed in practice; and that the complex additional motives apparently important in policy decisions, especially connected with the environment (Chapter 13) and rural development, are not easily catered for in such models. In any case, the present state of public choice theory and derived explanations of farm policy do not seem capable of providing clear indications of the circumstances or conditions under which we would expect policy to change. There seems to be little in the literature which would point to the 1992 reform package for the CAP.

Fearne (1989) adopts a more direct approach, though reaches similar conclusions. He considers the decision making body (the European Commission and Council of Ministers) as a 'satisficing' organism rather than as optimising a PPF. In addition, the body is seen as reacting to disparities between observed current and 'target' levels of the key variables (farm incomes and budgetary expenditure, identified on the basis of previous literature) only as these disparities reach trigger levels, hence producing discontinuous policy response. Although the underlying framework admits of the possibility that both targets may reach crisis simultaneously, hence prompting a search for alternative instruments (for example, the introduction of dairy quotas in 1984), these decision points are classified as 'non-routine' and are not included in the subsequent empirical analysis. Using the weighted average real national support price change as the dependent variable, Fearne finds considerable support for the satisficing model, albeit that budgetary expenditure appeared not to be a significant determinant of real price changes. However, he is forced to conclude that "there are too many judgmental inputs for the model to be used for forecasting beyond one year ahead", while the key policy decision — that of a reform or change in instruments — is explicitly excluded from consideration in this version of the model, as it is (though implicitly) in virtually all other formal versions of public choice models.

Other analysts tend to concentrate on possible variations to the decision-making process itself to make the CAP more responsive to a perceived conventional rationality (see Swinnen and van der Zee, 1993, 273ff), though seldom offer explanations of either why these alternatives are not followed or what circumstances might encourage their adoption. At a more pragmatic level, there are two pieces of analysis, now both somewhat historical, which do seek to explain the condition and development of the CAP. In both cases, they can now be seen as attempts to bridge the considerable gap between formal public choice modelling and the traditional discursive policy analysis. Building on the welfare analysis of the CAP, Harvey (1982) considers the question of the apparent conflict between the demonstrated economic inefficiency and inequity of the CAP with its persistence and robustness. His basic approach is to compare simple measures of national interests in agricultural policy in terms of: (a) maintaining and improving farm incomes; (b) minimising consumer costs; and (c) minimising taxpayer costs with associated performance of the CAP in achieving these objectives. The structure of the comparison is shown in Table 8.2.

The evidence strongly suggests that the 'old' CAP was a unifying policy, in the sense that the differential national effects of the CAP did not correspond to expected national priorities (correlations between national preferences and CAP effects across countries are insignificant or perverse), but that the CAP is broadly acceptable (correlations between CAP effects and national interests within countries between objectives are more significant and positive).

This illustrates the familiar proposition that the old CAP was inefficient (costly) in meeting national objectives, and leads to the conclusions:

- that attempts to satisfy member state objectives for improving farm incomes will typically lead to higher support prices since those with the greatest interest will demand such higher prices (FRG) — echoing both the Fearne and de Gorter *et al.* results;
- that reform of the CAP will entail an element of 're-nationalisation' of the policy to provide greater flexibility for member states to achieve their own objectives;
- that inbuilt conflicts within the policy (escalating budgetary costs associated with continued attempts to maintain farm incomes, let alone improve distribution of income) will eventually force such a reform.

Table 8.2: CAP effects and national interests: comparison of rank orders.

Objectives	FRG	Fr	It	Ne	B/L	UK	Ir	Dk	Corr. Coeff.
1. Farm Income Levels:									
a. CAP effects	1	7	8	5	3	6	3	2	+0.04
b. National Interests	1	3	2	6	7	8	4	4	
2. Farm Income improvement									
a. CAP effects.	5	7	8	2	1	4	6	3	-0.64*
b. National Interests	1	3	2	6	7	8	5	4	
3. Consumer Costs									
a. CAP effects	6	5	8	3	4	7	2	1	-0.17
b. National Interests	4	6	7	3	2	1	8	5	
4. Taxpayer Costs									
a. CAP Effects	2	3	5	6	4	1	8	7	+0.12
b. National Interests	2	5	7	4	3	6	8	1	
Correlation Coeff:	+.76*	+.56	+.08	+.75*	+.32	+.01	+.51	+.32	

Notes: * signifies significantly different from zero with 95% confidence
1a = Producer Surplus Gain (PSG) w.r.t. free trade at world prices as % of ag. GDP
1b = Agriculture's share of GDP/Agriculture's share of labour force
2a = Change in PSG/head of ag. labour force from 10% change in support prices
2b = 1b
3a = Change in Cons. Surplus per unit increase in PSG following a 10% price change
3b = Non agricultural labour force as % of total working population
4a = Change in tax contributions per unit increase in PSG following a 10% price change
4b = GDP/head

Source: Table 9, Harvey, 1982.

With the considerable benefit of hindsight, these 'predictions' appear modestly robust, though still somewhat inexplicit about the sort of reforms one might expect and the precise conditions under which one would expect them to occur. In common with all other published analyses of the CAP, this approach also failed to anticipate the introduction of dairy quotas in 1984, notwithstanding that policy discussions at the time recognised their near-inevitability. It was recognised that the Council of Ministers would be forced to take the quantitative control route eventually, to square the circle of rising budgetary costs associated with open-ended support with the apparent political imperative of preserving farm incomes (asset values), see, e.g. Harvey (1985).

Harvey and Thomson (1985), extending discursive analysis of the future prospects of the CAP based on traditional welfare analysis, concluded that:

- "although the Council has only put (dairy) quotas in place for five years, there is no prospect of the policy being repealed at the end of that time...Once [quotas] become tradable, the only acceptable way of getting rid of them is for the authorities to buy them in";

- "Council can be expected to quarrel more or less continuously about the national distribution of quota, with each country making more or less plausible cases as to why they should be granted more";
- The impending cereal regime crisis would be likely to include "incentives for farmers to retire land from agriculture, or some form of cereal area quotas. These suggestions will meet with considerable opposition from both farmers and bureaucrats, but, if linked to entitlement to other forms of support, could prove feasible and effective";
- "International repercussions of the CAP add further pressure...Resolution of these conflicts is difficult, but cannot be separated from the domestic development and evolution of the policy".

They also predicted, however, that: "it is most unlikely that adequate savings can be made in the market support policy in the medium term to provide enough funds to compensate for realistic and effective price reductions"; and suggested that the necessary raising of additional (budgetary) funds to allow maintenance of the existing policy set might usefully be linked to national proportions of farm production, thus providing an incentive to member states to moderate their own national policies to limit production increases.

Nevertheless, again with the benefit of hindsight, there is little evidence that the intelligent use of traditional policy analysis is yet markedly outperformed by more formal public choice analysis, fascinating though the latter proves to be. MacLaren reaches a similar conclusion. "The process of agricultural policy reform in the European Community is not well described by any single model of political economy. The institutions of the EC and their inter-relationships, the structure of the budget, the initiatives taken by key individuals, the role of interest groups and the response of the political parties, are all essential to an explanation of the pressures for reform and the sources of inertia which have prevented fundamental reforms of the CAP. What emerges from this web of interactions is perhaps best described by Lindblom's phrase 'the science of muddling through' rather than the systematic choice of the most efficient policy instrument". Having said that, however, the capacity of intelligent heads to use the tools of economic analysis to make some plausible and realistic assessments of policy development does suggest that there may exist some systematic framework. That it has not yet been found should not discourage the search.

Reflections on the dominance of farm income

The conclusions to be drawn from this brief survey of the caveats and extensions of the conventional neoclassical economic model of the farm sector is that public concern over farm incomes appears to be dominant in the determination of policy. Interest and pressure group theory is advanced by some, following Olson, as being heavily instrumental in 'perverting' social welfare in favour of the farm sector at everyone else's expense. However, as noted above, it is possible to explain (following de Gorter *et al.* arguments) the general political preference for farmers without such pressure group elaboration.[20] Neither Gardner (1987) nor Lindert (1989) find much empirical support for the pressure group theories in the US, though work still remains to be done for the EU. Whatever the outcome of such work, it would seem that a particularly blind, if not perverse, political system is required in the longer term to allow such pressures to dominate policy indefinitely. The fact that the 'farm income problem' appears to remain a dominant justification for socially inefficient and apparently inequitable farm policies seems to imply a continued acceptance of the farm income arguments by the electorates. If and when this legitimisation is withdrawn, then it seems likely that the domination of farm policy by sectional interests will also decline, pressure groups or no. Thus the 'farm income problem' is worth more careful consideration.

Gardner (1992) provides a systematic and comprehensive review of the farm income problem for the US. Although no such comprehensive treatment yet exists for the EU, the arguments apply here also, though may yet have some way to go before reaching the US position. The starting point of the logic is the particular circumstances of farm product markets. Both farm supply and demand for food and other farm products are typically held to be inelastic. Supply is so since increases in farm output can only be achieved through application of more inputs to a more or less fixed land base, or the use of marginal and less productive land. In either case, successive increases in output will be substantially harder and more expensive to win. Demand is inelastic since the capacity of well fed stomachs is clearly limited.[21] These circumstances, coupled with the fact that farm production is variable and unpredictable, lead to farm prices being volatile and, as a result, farm incomes being even more volatile (typically termed unstable). Variability of food supplies and farm incomes is likely to generate considerable public sympathy and support for government intervention to reduce both. Furthermore, demand grows slowly, if at all, as real incomes grow in the rest of the economy (the well-known Engel's Law), while technological change in agriculture generates continual productivity improvements and increases supply at rates of 1–3% per year. Farm product prices, therefore, trend downwards through time, reducing the real value of farm revenues, given the inelasticity of demand. Although technological change also reduces costs (otherwise it would not be successfully adopted), the general presumption is that falling revenues also lead to falling incomes.

Such arguments demonstrably hold in closed developed economies, the home of protective farm policies. However, their force is considerably reduced in the presence of international trade. In the world market, variations in farm product supplies tend to be reduced. Global crop failures and universal disease outbreaks are fortunately rare indeed, and in their event it is safe to assume that even the most protective farm policies will be practically impotent. The world market will be less volatile (more stable) than local domestic markets in the absence of government intervention.[22] Furthermore, declining terms of trade for the agricultural sector (falling product prices relative to input and other prices in the rest of the economy) is rather less reliable than within a developed and closed economy. In the global economy, the terms of trade for agriculture depend on the continual race between farm productivity improvements, population pressures and income growth. Engel's law has little relevance to the 80% of the world's population plagued (often literally) by actual or near starvation. However, post-war history shows a trend decrease in world agricultural prices of some 2% per year in real terms, though with some sharp increases in the late 40s and early 50s, and particularly in the 70s and mid-90s. Notwithstanding the adverse effects that industrial country protective policies have had on these prices (see Chapter 17), it has been generally accepted that such a downward trend is an established phenomenon (see, e.g. European Commission, 1994).

In this sense, then, chronically declining farm revenues are a condition selected by developed countries through their apparently deliberate isolation from world markets. The reasons why nation states should choose to ignore world markets are not hard to understand, especially among those with a recent national history of food shortage occasioned by war.[23] Yet the prolonged period of 'peace' since World War II substantially dims such memories, while the prospect of severe local food shortages is among the less frightening of the potential consequences of a third world war, whose prospects in turn appear less probable with the collapse of the former Soviet Union and the end of the Cold War. The reasons for isolation now appear much weaker, though the 'natural' inclination to prefer dependence on home rather than foreign supplies is still strong, in spite of the serious consequences of such policies for others in the world (see Chapter 17).

But, even given a remaining political preference for self-sufficiency in food, the logic of falling farm revenues has still to be associated with a consequent fall in farm incomes for protectionist policies to make sense. The disturbingly simple economic logic for a sector suffering declining revenues is that it should and, other things being equal, will contract in size. People will leave farming and seek employment and incomes elsewhere in the still growing economy. If we observe declining farm incomes, according to this story, it is because people are not leaving the industry fast enough. While it is acknowledged that there may be considerable adjustment difficulties (costs) in such a restructuring of the economy, and also that macroeconomic conditions of unemployment may make such adjustment more difficult in the short run (which may be distressingly long for those involved), nevertheless the workings of the general economy ought to result in a decline in farm labour forces (including farmers themselves) and maintenance of farm incomes in comparison with those which could be earned elsewhere.

In fact, farm labour forces do contract, in spite of farm policies. Since the UK's entry to the CAP in 1973, the percentage of the population engaged in agriculture in the EC has fallen from over 10% to barely more than 5%. Across the EU, the agricultural labour force has been declining at about 3% per year for the last two decades, tending to decline rather faster in those countries with substantial agricultural sectors.[24] In most member states this decline has been sufficient for individual or household incomes to keep pace with those in the rest of the economy, notwithstanding the decline in total agricultural sector revenues in real terms (see Table 8.3, below).

This raises a further question about the measurement of farm incomes. Superficially simple, this question rapidly becomes extremely complicated. Most obviously, simple measurement of farm incomes does not necessarily measure perceptions of well-being. If, as is at least possible, people get more satisfaction from working in farming (and living in the countryside) than factories, offices, shops or universities, then it is to be expected that they will be entirely willing to forego some monetary reward for the 'privilege' of remaining in agriculture and some divergence of farm to non-farm incomes would be expected. It may also be cheaper to live close to the work place in rural areas, making lower observed incomes worth more than their urban counterparts: housing and work related travel costs are likely to be lower; actual incomes may include considerable income-in-kind, in the form of housing, food and fuel; though it should also be recognised that other costs (especially transport to major shops, entertainments and services, prices of consumer goods) may well be higher in rural areas.

Most importantly, measurements of farm incomes are typically measures of income from farming rather than incomes of farming connected households. Small farms generate extremely small incomes, with or without government product related support. Not surprisingly, further investigation reveals that small farm households do not typically rely solely on farm incomes for their survival. They have and look for supplementary sources, or rely on other jobs and sources for the majority of their income. This observation is not restricted to small farms. The proportion of 'part time' farmers (with substantial fractions of income earned off-farm) is seemingly rising throughout Europe, or perhaps has always been large but previously imperfectly or unmeasured.

Mobility of labour off farm as revenues decline, therefore, need not mean people leaving their homes and backgrounds for distant urban environments. It may simply mean earning more incomes from other areas of economic activity, often within the same rural environment. Agriculture is accounting for increasingly smaller shares of total economic activity, even within rural areas, throughout much of Europe. Meanwhile, the attractions of working, living and playing in rural as opposed to urban areas is increasing, creating additional demands for non-farm but still rurally based goods and services. As this happens, so mobility of labour between farming and non-farm activity will become increasingly easy and less costly.[25]

Even where this is not the case, mobility of labour does not necessarily mean current employees and farmers leaving the countryside. It can be, and often is, accomplished through sons and daughters choosing other occupations (and often other locations). Both the trauma and the cost of such mobility is substantially smaller than for the existing workforce. It is the consequence of people taking rational, economically literate and wholly voluntary decisions that there are easier and more profitable occupations than farming in which to earn a living. Income disparities between farming and other sectors of the economy will, under these circumstances, be expected to reflect disparities in age, transferable skills, education and training. Those with better opportunities elsewhere than agriculture will 'vote with their feet', while those with lower 'opportunity costs' of staying in farming, because of age etc., and observed income disparities would be expected to reflect these differences. Furthermore, the more active and viable non-agricultural sectors of the local rural economy, the closer one would expect farm and non-farm incomes to be.

The 'farm income problem' is thus a symptom of economic growth and development, not a persistent and malignant feature of the market place. To the extent that it exists, it is a reflection of necessary (demand driven) changes in the economy. To the extent that it persists, it is a signal that necessary adjustments are not occurring. The 'cure' is then more closely related to assisting adjustments and encouraging local non-farm economies than to farm-related policies. In fact, the evidence is that the 'farm income problem' no longer exists to any substantive extent (at least on average) in most parts of the EU except Portugal and possibly Spain (Table 8.3) in spite of the fact that agricultural incomes (as measured by value added per head) continue to lag behind those elsewhere in the economy.[26]

This is not to say that farm or rural policies can totally ignore farm incomes. There are strong arguments that more people engaged in farming might well be in the public interest because the effects of farming (especially the more traditional types of small scale and/or more labour intensive systems) on wildlife, landscape and natural resources are beneficial to the public but will not be provided by a free enterprise economy (see, for example, Harvey and Whitby (1988) and Chapter 13 below). If farm incomes are insufficient to provide such conservation, amenity, recreation and environmental goods and services, then public support is required for their provision. However, the reasons for such support imply a rather different set of support policies than has been typical of the history of the CAP.

The effects of CAP product-related support on farm incomes are far from unambiguous. In general, the arguments above suggest that farm incomes are largely determined by off-farm opportunities. Farm policies do not change these. Product-related support merely increases economic activity in farming, encouraging farmers to produce more and hence increasing costs of production as more inputs are used on increasingly unproductive land. Farm incomes, as the difference between revenues and costs, remain unchanged. Furthermore, product-related support tends to increase income disparities within farming. Larger farms with more output get more support, and those farmers adopting new technologies improve their incomes while those who do not get left behind. It is apparent that the divergence between large, better resourced farm incomes and those of their smaller and less favoured neighbours has tended to increase rather than diminish during the period of high farm support under the CAP.

As the House of Lords (1991) commented: "in its 'Reflections' paper,[27] the [European] Commission succinctly outlines the deficiencies leading to the present crisis in the CAP. These include… the failure of agricultural support to maintain the incomes of farmers, particularly those with small and medium sized holdings. As it points out, the analysis is not new… The Commission concludes that 'the contrast between on the one hand such a rapidly growing budget and on the other agricultural income growing very slowly, as well as agricultural population in decline, shows clearly that the mechanisms of the CAP as currently applied are no longer in a position to attain certain objectives prescribed for the agricultural policy under article 39 of the Treaty of Rome".[28]

Table 8.3: Farm household income relative to all households, EU.

Country	Relative Income for Farm Households (%) [1]	Year	Relative Agricultural Income(%) [2]
Belgium	n.a.		78
Denmark	115	1988	67
Germany	110	1988	45
Greece	108	1985	64
France	108	1989	57
Ireland	112	1982	69
Italy	145	1988	42
Luxembourg	143	1985	78
Netherlands	228	1989	71
Portugal	81	1986	30
Spain	n.a.		43
UK	n.a.		64
EU (12)	n.a.		47

Notes: 1. Total disposable income per farm household relative to that of all households (for year shown in column 2).
2. Agricultural gross value added per person employed in agriculture relative to GDP per head in the whole economy, 1992 (from European Commission, Agricultural Situation in the Community).

Source: Adapted from European Economy, 1994, Table 1., (Relative Household income from SOEC. Total income of agricultural households, 1992 Report, Luxembourg).

The gradual realisation of these economic 'truths' both within and outside the farm sector have undermined the political case for continued product related support, despite sophisticated attempts to explain their persistence with political market place theories and in spite of more or less rigorous measurement of economic costs and benefits of farm policies. Perhaps, however, these can be seen as myopic gropings towards a more rational future. The story is taken up in the concluding chapter.

References

Alston, J.M. and Hurd, B.H. (1990) Some Neglected Social Costs of Government Spending in Farm Programmes. *American Journal of Agricultural Economics*, 72, 149–156.

Ballard, C.L. and Fullerton, D. (1992) Distortionary Taxes and the Provision of Public Goods. *Economic Perspectives*, 6, 117–131.

Brown, C.G. (1990) Distributional aspects of the CAP price support. *European Review of Agricultural Economics*, 17, 289–301.

Browning, E.K. (1987) On the Marginal Welfare Cost of Taxation. *American Economic Review*, 77, 11–23.

Buchanan, J.M. and Tullock, G. (1962) *The Calculus of Consent*, University of Michigan, Ann Arbour.
Buchanan, J.M. and Tollison, R.D. (1984) *The Theory of Public Choice – II*, University of Michigan, Ann Arbor.
Buckwell, A.B., Harvey, D.R., Parton, K.A. and Thomson, K.J. (1982) *The Costs of the Common Agricultural Policy*, Croom-Helm, London.
Burniaux, J.M. and Waelbroeck, J. (1985) The impact of the CAP on developing countries: a general equilibrium analysis. In: Stephens, C. and Veloren van Themat, J. (eds) *Pressure Groups, Policies and Development*, Hodder and Stoughton, London.
Chipman, J.S. and Moore, J.C. (1978) The New Welfare Economics 1939–1974. *International Economic Review*, 16, 547–581.
Commission of the European Communities (1991) *The Development and Future of the CAP*, Reflections paper of the Commission, Communication to Council, COM(91) 100 final, February.
de Gorter, H. and Swinnen, J.F.M. (1994) The Economic Polity of Farm Policy. *Journal of Agricultural Economics*, 45(3), 312–326.
de Gorter, H. and Tsur, Y. (1991) Explaining Price Bias in World Agriculture: the Calculus of Support-Maximising Politicians. *American Journal of Agricultural Economics*, 73, 1244–1254.
de Gorter, H., Swinnen, J. and Tsur, Y. (1991) The Political Economy of Price Policy Preferences in European Agriculture. Paper to the *Sixth Congress of the European Association of Agricultural Economists*, The Hague, Netherlands, September.
Demekas, D.C., Bartholdy, K., Gupta, S., Lipschitz, L. and Mayer, T. (1988) The Effects of the CAP of the European Community: a Survey of the Literature. *Journal of Common Market Studies*, 27(2), December, 113–145.
Downs, A. (1957) *An Economic Theory of Democracy*, Harper, New York.
Dunleavy, P. (1991) *Democracy, Bureaucracy and Public Choice*, Harvester Wheatsheaf, London.
European Commission (1994) EC Agricultural Policy for the 21st Century. *European Economy, Reports and Studies*, 4.
FAO (1975) Agric. Protection and Stabilisation Policies: A Framework for Measurement. *FAO, C75/LIM/2*, October.
Fearne, A.P. (1989) A 'Satisficing' Model of CAP Decision Making. *Journal of Agricultural Economics*, 40(1), 71–81.
Gardner, B.L. (1983) Efficient Redistribution through Commodity Markets. *American Journal of Agricultural Economics*, 65, 225–334.
Gardner, B.L. (1987) Causes of U.S. farm commodity programmes. *Journal of Political Economy*, 95(2), 290–310.
Gardner, B.L. (1992) Changing Economic Perspectives on the Farm Problem. *Journal of Economic Literature*, 30(March), 62–101.
Gasson, R., Crow, G., Errington, A., Hutson, J., Marsden, T. and Winter, M. (1988) The farm as a family business. *Journal of Agricultural Economics*, 39(1), 1–42.
Hallam, D., Machado, F. and Rapsomanikis, G. (1992) Co-integration analysis and the determinants of land prices. *Journal of Agricultural Economics*, 43(1), 28–37.

Harvey, D.R. (1982) National Interests and the CAP. *Food Policy*, 7(3), 174–190.

Harvey, D.R. (1985) *Quotas: Freedom or Serfdom?*, Centre for Agricultural Strategy, Reading.

Harvey, D.R. (1989a) The Economics of the Farmland Market. Paper to *Agricultural Economics Society One day Conference*, December.

Harvey, D.R. (1989b) Alternatives to Present Price Policies for the CAP. *European Review of Agricultural Economics,* 16, 83–111.

Harvey, D.R. (1991) Agriculture and the Environment: The Way Ahead?, Chapter 15 in Hanley, N. (ed) *Farming and the Countryside: An Economic Analysis of External Costs and Benefits*, CAB International, Wallingford.

Harvey, D.R. and Thomson, K.J. (1985) Costs, Benefits and the Future of the Common Agricultural Policy. *Journal of Common Market Studies*, 24(1), 1–20.

Harvey, D.R. and Whitby, M.C. (1988) Issues and Policies. Chapter 13 in: Whitby, M.C. and Ollerenshaw, J.H. (eds) *Land Use and the European Environment*, Belhaven, London.

Heap, S.H., Hollis, M., Lyons, B., Sugden, R. and Weale, A. (1992) *The Theory of Choice*, Blackwell, Oxford.

House of Lords (1991) *Development and Future of the CAP*, Select Committee on the European Communities, Session 1990–91, 16th Report, July.

Hubbard, L.J.(1995) General Equilibrium Analysis of the CAP using the GTAP Model. *Oxford Agrarian Studies*, 23 (2), 163–176.

Josling, T.E. and Hamway, D.M. (1972) Distribution of Costs and Benefits of Farm Policy. Chapter 4 in: Josling, T.E. (ed.), *Burdens and Benefits of Farm Support Policies*, Trade Policy Research Centre, London.

Just, R.E. (1988) Making Welfare Analysis Useful in the Policy Process: Implications of the Public Choice Literature. *American Journal of Agricultural Economics*, 70(3), 448–454.

Just, R.E., Hueth, D.L. and Schmitz, A. (1985) *Applied Welfare Economics and Public Policy*, Prentice Hall.

Lindert, P.H. (1989) Economic influences on the history of agricultural policy. *Working paper no. 58*, Agricultural History Centre, University of California, Davis.

Lloyd, T.A. (1989) A reconsideration of an agricultural land price model for the UK. *Discussion Paper No. 69*, Department of Economics, University of Nottingham.

Lowe, P.D., Marsden, T. and Whatmore, S. (1994) *Regulating Agriculture*, Volume V, Critical Perspectives on Rural Change, David Fulton, London.

MacLaren, D. (1992) The Political economy of agricultural policy reform in the EU and Australia. *Journal of Agricultural Economics*, 43(3), September, 424–439.

McLean, I. (1989) *Public Choice, an Introduction*, Blackwell, Oxford.

Munk, K.J. (1994) Explaining Agricultural Policy. Chapter C, European Commission (1994) EC Agricultural Policy for the 21st Century, *European Economy, Reports and Studies*, 4.

Munk, K.J. and Thomson, K.T. (1994) The Economic Costs of Agricultural Policy. Chapter B, European Commission (1994) EC Agricultural Policy for the 21st Century, *European Economy, Reports and Studies*, 4.

OECD (1994) *Agricultural Policies, Markets and Trade: Monitoring and Outlook*, Paris.

Olson, M. (1965) *The Logic of Collective Action*, Harvard University Press, Cambridge, Mass.

Phelps, E.S. (1985) *Political Economy, an Introductory Text*, Norton.

Rawls, J. (1971) *A Theory of Justice*, Harvard University Press, Cambridge, Mass.

Rausser, G. (1982) Political Economic Markets: PERTS and PESTS in food and agriculture. *American Journal of Agricultural Economics*, 64(5), December, 821–833.

Ritson, C. (1980) Self-sufficiency and Food Security. *CAS Paper 8*, Centre for Agricultural Strategy, University of Reading.

Shucksmith, D.M., Bryden, J., Rosenthall, P., Short, C. and Winter, D.M. (1989) Pluriactivity, Farm Structures and Rural Change, *Journal of Agricultural Economics*, 40(3), 345–360.

Stevens, J.B. (1993) *The Economics of Collective Choice*, Westview, Colorado.

Stoeckel, A. (1985) *Intersectoral Effects of the CAP: Growth, Trade and Unemployment*. Occasional paper 95, Bureau of Agricultural Economics, Canberra.

Sturgess, I. (1992) Self-sufficiency and food security in the UK and EC. *Journal of Agricultural Economics*, 43(3), 311–326.

Swinnen, J. and Van der Zee, F.A. (1993) The Political Economy of Agricultural Policies: A Survey. *European Review of Agicultural Economics,* 20(3), 261–290.

Thomson, K.T. (1994) EC Agriculture Past and Present. Chapter A in: European Commission. EC Agricultural Policy for the 21st Century, *European Economy, Reports and Studies*, 4.

Tullock, G. (1965) *The Politics of Bureaucracy*, Public Affairs Press, Washington.

Tyers, R. and Anderson, K. (1992) *Disarray in World Food Markets*, Cambridge University Press, Cambridge.

van den Doel, H. and van Velthoven, B. (1993) *Democracy and Welfare Economics*, Cambridge.

Wilson, G.K. (1977) *Special Interests and Policy Making*, Wiley, London.

Winters, L.A. (1987a) The Political Economy of the agricultural policy of developed countries, *European Review of Agicultural Economics,* 14, 285–304.

Winters, L.A. (1987b) The Economic Consequences of Agricultural Support: a Survey. *OECD Economic Studies*, 9, 7–54.

---

[1]The reader is advised not to pay too much attention to the precise numbers in this table. These are highly dependent on the assumptions made in the underlying model and the accuracy/relevance of input data. The general orders of magnitude, however, are broadly consistent with other estimates of the costs of the CAP, of which there were a substantial number made during the early to mid-1980s especially, reviewed and discussed in Demekas *et al.* (1988) and Winters (1978b). In this case, the alternative policy considered is that the EU abandons the CAP and leaves the agricultural sector to compete at world market prices (assuming other countries continue to follow their existing policies) and measures the direct effects, ignoring administrative and tax-related economic costs and also any offsetting policies which might be used in the absence of the CAP (European Commission, 1994, Chapter B, Munk, K.J. and Thomson, K.T., 1994).

[2]The definition of economic rent is that return earned by a factor of production (broadly, land, labour management and capital) over and above that which would be earned in the next best alternative use or occupation of the factor.

[3]Where land prices were estimated to be a function of, *inter alia*, gross values added (gross product) of the industry. See, however, Hallam *et al.* (1992) for a commentary on this model, in which it is noted that the time-series properties of this simple model do not appear to support the simple associations demonstrated by the regression results. In defence of the model, it might be noted that there are no other models in the literature which provide empirical evidence for the theoretical propositions either.

[4]Irritatingly, for the policy analyst, the answers to these two questions are not, in general, the same. It is only in the case where income levels have no effect on consumers demand decisions for the products in question that the two answers will be identical. In practice, for developed countries and for analyses focused on the markets for food at the farm-gate rather than the retail level, the differences can be expected to be very minor, and well within the errors associated with the input data and other aspects of the analysis.

[5]Here meaning those factors in less than perfectly elastic supply to the downstream sectors, in an analogous fashion to the discussion of the agricultural and upstream factors.

[6]See, for example, Alston and Hurd (1990); Browning (1987); Ballard and Fullerton (1992).

[7]A recent examination of the CAP using a general equilibrium approach can be found in Hubbard (1995), who finds an efficiency loss for the EU as a whole of some 0.8% of GDP, though more substantial reductions in manufacturing and services output (5 and 2% respectively) and exports (17 and 10%).

[8]In fact, even this compensation principle is not enough to ensure satisfaction of the Pareto criterion. There are circumstances in which the compensation actually needs to be made, rather than simply being possible, in order to be sure of an unambiguous improvement (see Just *et al.*, 1985, also Chipman and Moore, 1978).

[9]Often referred to in the literature as a social utility function or a political welfare function. The non-existence of a social value function arises because of Arrow's Impossibility Theorem (see, e.g. Heap *et al.*, 1992, p205–216 and 289–

291), which in essence states that it is not possible to respect the diversity of individual preferences while at the same time ensuring that the collective (public) choice will be coherent (i.e. be transitive — preferring outcome a to b to c without simultaneously preferring c to a).

[10]The terminology reflects the twin principles of the CAP: (a) Common financing requires a budgetary transfer between member states (the budget effect) since, for trade with the rest of the world, import levies are treated as Community income ('resources') rather than as national income to the member state actually importing, while export refunds are a Community, rather than national, expense. In addition, any necessary additional Community funds are contributed by member states in approximate proportion to their share of Community GDP. Under the nationalised alternative, levy income and subsidy expenses on trade with the rest of the world are counted in the member state in which they arise, with no cross-financing of Community level deficits; and (b) Community preference and free trade at common prices amounts to a common agreement to pay common (high) prices for products traded between member states rather than the prevailing world price (the trade effect). Under the nationalised alternative, these trade flows would also generate levy income and subsidy spending in the countries of origin.

[11]This agreement is currently set according to 66% of the difference between Britain's percentage contribution to revenue (excluding customs duties and agricultural levies) and its percentage share of EC allocated expenditure. If this rebate were to be included in these calculations, the disparity between the UK and other member states would, of course, be reduced. There was some pressure to re-open negotiations on (and thus reduce or eliminate) this budgetary rebate prior to the December 1992 European Council in Edinburgh. Six member states had formally registered their opposition to the rebate. It was this Council which fixed the outline of the EC budget over the period 1993–1999, (the so-called 'Delors II Budget Package'). However, a combination of skilful Presidency by the UK and the sensitivity of other member states to the British problems of ratification of Maastricht led to a shelving of the complaints and thus the rebate continues.

[12]For an early exploration of the distributional aspects of the CAP, see Josling and Hamway (1972). For a more recent exploration of the farm level distribution of CAP gains, see Brown (1990).

[13]Just goes on to point out that the perfectly competitive market model, on which most welfare economics is based, is subject to serious failure in the (common) event that real markets are inherently 'distorted', especially through the "presence of risk, risk aversion, incomplete risk markets and asymmetric information" (449), and that correction of these problems may well indicate (through the theory of second best) off-setting distortions elsewhere in the system. Just notes that "these problems break the identity between competitive equilibria and Pareto efficiency and raise serious questions about the applicability of the traditional competitive norm of efficiency" (449), and further that "transactions costs, imperfect information, moral hazard and externalities...block the separation of efficiency and equity" (451). See, also, Munk (1994) for discussion of these same points.

[14]That is, do not have any effects on supply and demand decisions taken by 'economically rational' people in the economy, and therefore have no efficiency costs associated with them. Such 'resource-neutral' transfers are extremely difficult, if not actually impossible, to design as practical policy alternatives.
[15]A useful survey of this literature as related to agricultural policy can be found in, *inter alia*, Swinnen and Van der Zee (1993); Winters (1987a); MacLaren (1992), de Gorter and Swinnen (1994) and Just (1988), from all of which this outline borrows heavily. More general treatments can be found in, for example, Phelps (1985); McLean (1989); Stevens (1993); Buchanan and Tollison (1984), van den Doel and van Velthoven (1993), Dunleavy, (1991).
[16]See, also, as the classic public choice articulation of these ideas, Buchanan and Tullock (1962).
[17]See, also, de Gorter and Swinnen (1993). It is of passing interest here that this voter-politician model relies purely on self-interested behaviour of both parties, where support depends on voters increasing their relative and redistributed incomes. In common with the vast majority of public choice models there is no room for altruism. However, it seems possible that the de Gorter, Swinnen, Tsur results can also be obtained from an alternative formulation of the model which would substitute 'income of relatives' for relative income, though this possibility needs further work for its firm establishment.
[18]Initially proposed and used by Josling for the FAO (FAO, 1975), this approach has been adopted by the OECD who now regularly release estimates of PSEs and CSEs for the member states (e.g. OECD, 1994). It is described in detail in the previous chapter.
[19]This has been done here by employing the estimated coefficients in Equation (9) of Munk (1994, p 115): $PSE(t) = 92.1 - 55.6RI(t) - 22.5EXP(t) - 0.76(PSE^*(t) - PSE(t-1)) + E(c) + E(t)$
where PSE is taken as being the OECD aggregate % PSE as reported in the OECD PSE tables (available in diskette form from OECD); RI(t) is the current value of the ratio of the share of agricultural Gross Value Added in the EU economy to the share of civilian employment in agriculture in the EU, as reported in the European Commission: Agricultural Situation in the Community (annual); EXP is farm production value less consumption value expressed as a ratio of production value (data from OECD PSE tables); E(c) is a country-specific term, here estimated from a calibration of the above equation to the specified data; and E(t) is a time-specific error term, here taken to average zero. Given this calibrated equation, it can be applied to data outside the estimation period as shown in Fig.8.1 for the years 1979 and 1991–1993.
[20]Though these authors admit that the existence and behaviour of active interest and pressure groups cannot be entirely accidental, they argue this to be a symptom rather than a cause of preferential treatment of the farm sector.
[21]Tyers and Anderson (1992), for instance report long run elasticities substantially less than one for price changes in all commodities (Table A3) and reported demand elasticities (Table A4) are also low.
[22]See Chapter 17 below for the effects of domestic policies on this world market, and Tyers and Anderson (1992, chapter 6, section 3), for a systematic empirical analysis of world market stability.

[23]See Ritson (1980), for a thorough analysis of these issues, also Sturgess (1992).
[24]See e.g. Thomson (1994).
[25]See, for example, Gasson *et al.* (1988) and Shucksmith *et al.* (1989) for discussion of the roles and behaviours of family farms and of the appropriate focus for policy. Thus, Shucksmith *et al.* "European policy is likely to focus increasingly on regional measures and developing off-farm rural labour markets, with less support for 'agricultural' measures" (358) and Gasson *et al.* "The changing objectives of agricultural policy require support to be more carefully targeted to particular types of farm" (35).
[26]It should be noted, however, that value added has to provide income for all resource or factor owners with a stake in the industry, not only labour, so this measure is always liable to conceal more than it reveals without further analysis.
[27]Commission of the European Communities, 1991(a).
[28]Commission of the European Communities, 1991(a), p. 65.

Chapter 9

The CAP and Technological Change

Arie Oskam and Spiro Stefanou

Introduction

The importance of technological change

Most attention in agricultural policy is usually directed towards prices and to market and price policy. The main driving force behind changing conditions in farming, however, is often technological change (sometimes referred to as just 'technical' change). This observation can be analysed intensively, but it is quite easy to see that technological change continues under various conditions of price levels and price developments. It is even very difficult to conclude whether relatively high or low prices most stimulates technological development (Van der Meer and Yamada, 1990; Kalaitzandonakes, 1994).

The basic characteristics of agricultural products have not changed very much. Wheat, potatoes, beef and milk produced in, let us say, the year 1900 would still be characterised today as similar products. That does not hold, however, for many inputs used in agriculture and for some outputs in, for example, horticulture. The use of totally different inputs to produce agricultural outputs is recognised easily as a change of technology. It starts to be less straightforward if nearly the same inputs are used in a more efficient way. Here it is a matter of convention whether one characterises such a development as 'technological change'. Economists are often inclined to use a broad definition of technological change, where: (a) different inputs and/or outputs; (b) different qualities of inputs and/or outputs; or (c) different processes, are all incorporated within the qualification 'change of technology'.

Whether a change of technology leads to higher productivity depends often on prices of inputs or outputs. Here productivity is defined as the output/input relation, where inputs and outputs receive weights according to their value share. It is easy to provide examples where under all price conditions a particular change of technology always leads to productivity increase. A new breed of cattle, which uses less feed of the same composition per kilogram of beef (of the same quality) is a clear example. In practice, however, price conditions play a crucial role in determining whether a technology is more or less productive. Only in the situation where a technology is more productive for present or expected future prices is a change characterised as technological improvement. Many technological changes that are unproductive for a low wage level might lead to a higher productivity for a high wage level and will then be characterised as technological improvements.

The basic nature of technological change

Here we define technological change as a development where the same input level allows a different output level, or where the same output level requires different inputs. In reality it is often a combination of both developments, along with changes in the quality of inputs and outputs. This real situation, however, is stylised in the form of quantity changes, where quality changes are incorporated either in the prices or in the quantities.

We start from the idea that there are two main reasons for output changes within a sector or within a total economy:

a) changes in input quantities;
b) changes in the technology.

A clever way of deducting input changes from output changes reveals the change in technology. It might be difficult, however, to make a sharp distinction between these two. Considering more inputs, let us say human capital and/or governmental infrastructure, reduces the contribution of technical change. Moreover, a substantial part of technological changes might be directly linked to inputs, with capital goods, seeds and plant material, etc. as clear examples. Therefore, it is often only by definition that these two elements can be distinguished.

Economic literature often considers the incentive of increased profitability at firm level to be the main driving force behind technological change. Firms apply new technologies because they expect higher profitability; and because individual firms in agriculture have no (or nearly no) market power, they consider prices as given. Their individual decisions do not influence input or output price levels. According to the rational expectation hypothesis (see Newbery and Stiglitz (1981) and e.g. Pindyck and Rotemberg (1983) for an application of this hypothesis) they incorporate the application of new technologies by the farming sector in their price expectations. Here, we identify one of the clear links between agricultural policy and technological change: prices not only influence input levels and therefore output levels, but also technology (and therefore again input levels and output levels).

Often, tax regulations, investment subsidies or interest subsidies form another important component of decisions on input use, especially with respect to investments in capital goods, land and production quota. In this chapter, we will not distinguish these elements separately. They are assumed to be a part of the price of the input. But it will be clear that regulations and subsidies can be important for decisions. Here we identify another link between the CAP (or national policies) and technology.

Besides straight economic motives, many other aspects play a role in the decision to apply new technologies. The availability of information might play a crucial role. Farmers cannot apply new technologies when they are unknown, or when the input/output relations of these technologies are highly uncertain. The availability of information includes also social aspects, because the behaviour of the reference group might play a role. A reason to postpone the application of new technologies lies in the expectation that better technologies are available quite soon. This, however, indicates the maximisation of long term profitability.

Main flows/approaches within the literature

A useful way to identify some of the main issues in the literature is to articulate the sequencing of activities and forces contributing to productivity gains and economic growth via technical change. A science-based technical change (after Sato and Suzawa, 1983) is illustrated in Fig.9.1. In the initial stages, specialised labour and capital resources such as scientists and laboratories are used to generate basic knowledge that is fundamental to understanding production relationships. This new knowledge contributes to the stock of knowledge generated from past efforts. It is this stock of knowledge that serves as a key input in the efforts to produce practical methods and inventions. However, the stock of knowledge is an intermediate output in the sense that it becomes an input to the development stage where innovative products and methods are readied for commercial use. The new products and methods will be implemented if they prove to provide an economic improvement (measured as either enhanced profits or welfare) over the current products and production practices. Productivity gains should be realised with the adoption and diffusion of new technologies.

The literature on the analysis of technical change and economic growth focus on different stages of this process. It is important to appreciate that there is an opportunity for government policy and economic forces to influence the process of technical change at each stage along the process.

Neoclassical versus modern models characterising technical change

In the neoclassical model, knowledge is a public good. Technology is a *non-rival* good implying one operator's use of the technology to produce a good or service does not preclude other operators from using the same technology simultaneously. This distinguishes the technology from a physical capital good which can be used only in one place at a time. Technology is typically a *non-excludable* good as well. Creators or owners of technological information often have difficulty in preventing others from making unauthorised use of it. This non-excludable character of technology in the neoclassical growth models is, again, unlike capital equipment which is readily excludable.

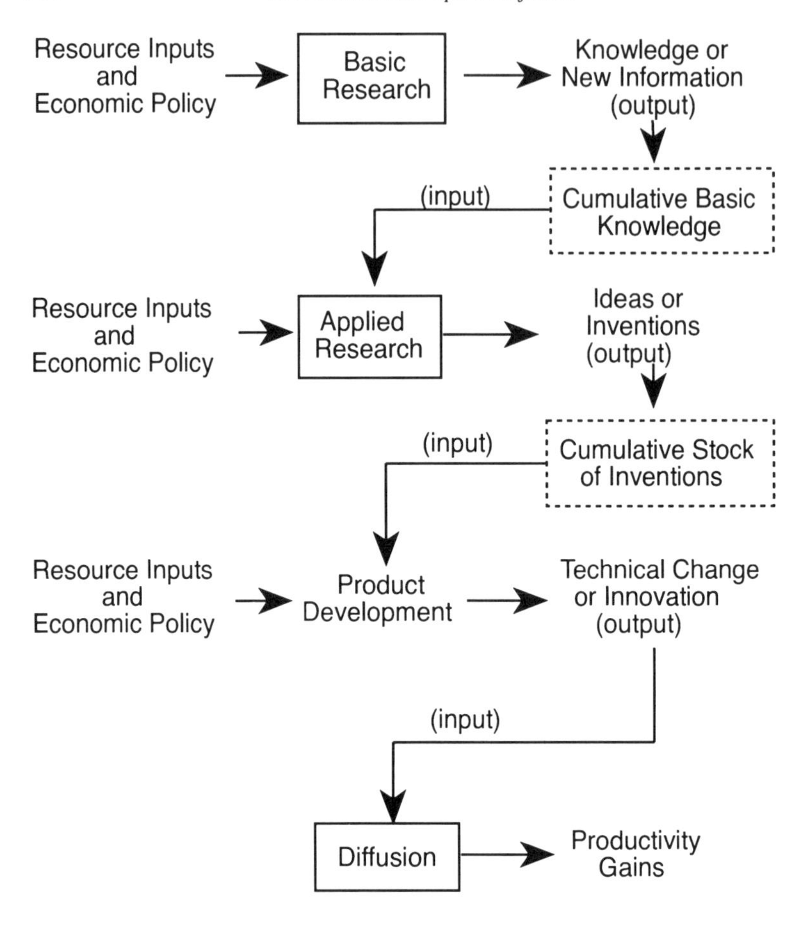

Fig.9.1: Organisation of science based technological change.

Modern growth models began with the learning-by-doing model introduced by Arrow (1962). In this model the manager must learn a number of things in order to thrive over the long run: market behaviour and demand for the products and services the operation produces, the market for inputs, and the production process. The production experience acquired is possessed by the individual leading to excludability. These new models of economic growth focus on the case where knowledge is a private good and it is knowledge serving as an engine of growth. One approach is Lucas (1988) specifying human capital in the form of knowledge which is the output of the research and education sector. This knowledge is acquired by the individual through intentional learning processes. Consequently, the knowledge acquired in this way is both rival and excludable. Labour decides how much time to spend on producing goods and how much time to divert to learning and education activities. Romer (1987) focuses on a research and education sector generating technical knowledge taking the form of designs for a new capital good used in the production of all commodities. In essence, the research and education sector produces blueprints which are excludable (via patents). Consequently, the results of the research and education sector can be sold and revenue per unit of skilled labour must be large enough relative to the alternative use of skilled labour.

The role of research and development in promoting technical change and growth

Incentives and allocation of resources to research and development (R&D) is a common instrument the public sector uses to promote growth. Most recognise the contributions of R&D expenditures on productivity growth into the future. A number of studies attempt to assess the lag length of R&D expenditures on productivity growth (e.g. Chavas and Cox (1992), Huffman and Evenson (1993)). However, most attempts at assessing the determinants of productivity growth take a linear view of this process by suggesting investment in R&D leads to productivity growth for a particular sector of the economy. Clearly, R&D in non-agricultural sectors has led to increased productivity in agriculture (e.g. machinery, computer hardware and software). In fact, R&D gains in one sector may lead to redirecting R&D directions in another sector (e.g. genetic engineering leading to biotechnologies).

Even when we restrict analysis to one sector's R&D and productivity growth, it is clear that initiatives to engage in R&D may arise from a need to enhance productivity growth. However, productivity growth implies resource use decisions affecting the quantity of resources available for investment in R&D, in particular, and activities, in general. Thus, it is reasonable to consider the prospect that there is a simultaneous relationship between productivity growth and R&D expenditures.

We can think of private sector R&D as an economically induced variety of R&D while the public sector provides an autonomous, a non-economically induced variety of R&D. Baumol and Wolfe (1983) discuss a model where R&D succeeds in increasing productivity growth. But in doing so it automatically increases its own relative costs in comparison with production cost leading to a reduction in the financial incentive of the R&D investment. Thus, the success of R&D activity serves to undermine its own demand. Unfortunately, the more impressive the record of past success of R&D activity, the more strongly it tends to constrain private demand for R&D. In addition, publicly provided R&D expenditures maintained at significant levels that are unresponsive to budgetary pressures facing the private providers of R&D tend to discourage a strong presence of private sector R&D. The private sector can free ride off the public sector R&D activity in times of budgetary pressures, implying that the private sector will not reinitiate its R&D efforts until well after the financial pressure has lifted. Periods of sluggish productivity growth may be a foreseeable consequence of the incentive mechanism for R&D.

Role of the market and infrastructure investment

Normally the exogenous growth school attributes a fairly high social return to technology investment (e.g. to produce the large residual in the aggregate production function) and therefore technology investment is likely to warrant ample public support. The modern growth school tends to argue that relative price levels will determine the rate and change of technological diffusion (and sometimes investment) and therefore one should essentially 'leave it to the markets' to signal how much and what sort of technology should be provided.

In addition to the purposeful accumulation of additional knowledge and expertise, there are other reasons for non-decreasing returns to private capital that relate to the type of capital present. Barro (1990) suggests that productive government spending, such as on public infrastructure, may be an input offsetting the diminishing private returns to capital. Rebalo (1991) suggests that, if there is a core of capital goods that can be produced without the use of labour and land, growth of sectors where capital as well as fixed factors are needed can be fuelled without bound. In such cases, the process of economic growth can proceed in an unbounded manner.

Concluding remarks

Once we have a more accurate empirical representation of technical change and the sources encouraging it we can attempt to attribute the influence of infrastructure investment, human capital and distribution of wealth on the pace of economic growth. By identifying the factors stimulating growth within the agricultural sector, the role of government support may be more accurately addressed. For example, which activities have significant impact on economic growth: autonomous government funding of R&D, strategic government R&D funding (e.g. funding of biotechnology research), government incentives to encourage private R&D investment, physical infrastructure investment, human capital infrastructure investment, price levels and price stability, etc?

Given the sequencing of activities and forces contributing to the realisation of technical change, it can be difficult to identify specifically the contribution of additional resources at a particular stage. For example, will an increase in government sponsored basic research funding have a greater payoff than more funds allocated to encourage the development of commercially viable products; or, should policy makers place greater emphasis on educational activities that can accelerate the diffusion of new technologies? Clearly, many difficulties can arise in fully identifying and measuring the contribution of each stage described in Fig.9.1. Often we observe technologies that have been implemented and not the wide range of available technologies that may be on the shelf and available when circumstances (economic and social) dictate.

Technological change within the CAP

Referring to the main approaches in the literature, technological change within the European Union contains at least elements of both approaches. A large part of technological change is exogenous to the agricultural sector: it has been developed more in general, is freely available, does not discriminate substantially between farms, etc. Another part, however, can be endogenous. Given the ruling prices and conditions, typical 'learning by doing' situations can arise (practical training for specialised groups, local/regional product development, study groups, producers organisations, etc.), which illustrate also the possible endogenous character of technical change.

Two elements have been very stimulating for technical change in agriculture under the CAP:

a) The relatively high price level of outputs attracts more inputs into agriculture and makes new technologies more profitable.
b) The CAP has contributed to a relatively stable price level of agricultural products within the EU. Price stability contributes to a higher input and output level (Newbery and Stiglitz, 1981). Although one might observe declining price stability within the CAP, this is still an important aspect.

Besides these stimulating aspects of the CAP, there are, however, also some other aspects that play a role. The high protection level kept many farms in operation which would have discontinued under a price regime with lower or more fluctuating prices. This holds also, or even more so, for a quota regime and for income policy. If land remains with these farms, that may influence the application of new technologies.

At present, there are many indications that small farmers and/or older farmers use their land and other inputs less intensively and efficiently (see Mansholt, 1986; Mulder and Poppe, 1993). This implies that policies where smaller and less efficient farms remain in business decrease agricultural output, compared with a situation where large farms or contractors operate the same area. Including all the labour of these small and less intensive farms in the productivity analysis leads to a relatively low level of productivity, unless a good estimation of opportunity costs is included.

The organisation of this chapter

After this extensive introduction, the next section will focus on the analysis of technological change and measurement issues. The third section returns to the characteristics of the CAP and tries to develop some hypotheses and results on the relation between the CAP and technological change. We then provide relevant information on this issue, using the results of empirical analyses. The fifth section takes a step outside into a more general approach of measuring technological change: external effects for the environment are incorporated. The last section provides conclusions.

The basic principles and analysis of technological change

The analysis of technological change

The effects of technical change are characterised by:[1]

a) more output can be produced given the same quantities of inputs (or, equivalently, the same output can be generated by smaller quantities of one or more inputs); or
b) existing outputs undergo qualitative improvement; or
c) totally new products are produced.

Three central questions emerge:

a) How important is technical change in the process of growth in the agricultural sector (or in the economy in general)?
b) What is the cause of technical progress? Is it exogenous or endogenous to the economic system? Although it is convenient to assume technical progress is exogenous, with the emphasis placed on R&D expenditures by government, industry and firms, endogenous technical change may be more appropriate. But with respect to individual farms, technical change is likely to be exogenous.
c) If technical change can be classified as labour saving, neutral or capital saving, is there any systematic bias in an industry towards any particular kind of technical change? If so, why?

Classification of technical change

Interest in bias of technical change focuses on classifying technical change as factor saving, neutral or using. There are three measures in the literature that are frequently used and are attributed to Hicks, Harrod and Solow. They differ in their basic assumptions. The most common one is the Hicks measure, which assumes a constant capital labour ratio in its measurement procedure.[2]

The change of output is measured along the line through the origin in Fig. 9.2. This figure provides two isoquants (points with equal production) in a two input space (K and L). The technology denoted by the Y'' isoquant is more capital intensive than the technology denoted by the Y' isoquant. Thus, for K/L fixed, the steeper the rate of technical substitution the more capital intensive the technology. Or in other words: to reduce one unit of capital with the same production level requires more labour. This makes it attractive to use more capital in a capital intensive technology. Now, assume that Y' and Y'' reflect the same production level, then the technological change can be measured either on the horizontal axis (K'' for Y'' relative to K' for Y'); the vertical axis; or along the fixed K/L ratio. Therefore the measures of technological change $F''F'/OF'$, $K''K'/OK'$ or $L''L'/OL'$ are identical. (A more formal analysis of

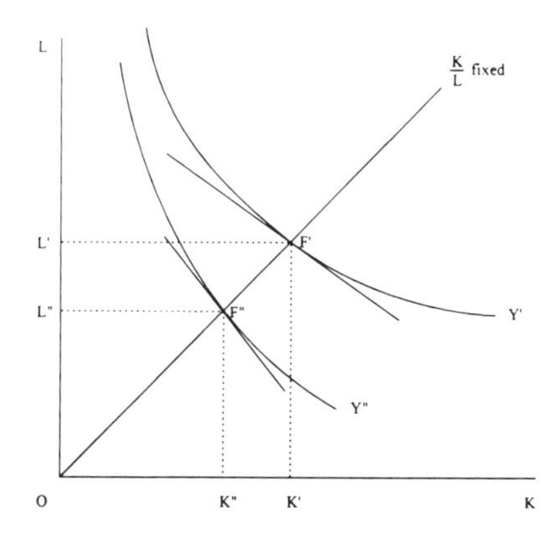

Fig. 9.2: A shift to a more capital intensive production.

technical change and productivity growth on the basis of production functions is provided in Appendix A to this chapter.)

Figure 9.3 illustrates the case of technical change where the firm is operating on isoquant DD' with input price ratio r^o/w^o where r is the price of capital and w the price of labour, implying the input allocation at point A. With technical change, the isoquant shifts to EE'. The short run effect is before the input prices have a chance to adjust, leading the firm to operate at point B. A labour saving innovation leads to lower demand for labour followed by a drop in wages leading to an increase in r/w. Similarly, a capital saving innovation is followed by a lower demand for capital leading to a fall in r/w. Consequently, the input price ratio becomes steeper, r^1/w^1, leading to the allocation of capital and labour to point C (for example).

The theory of production and productivity growth

1. Review of concepts in production decision making

Economic theory is based on a number of concepts, which also play a role in analysing productivity growth. These concepts are illustrated by means of Fig. 9.4, using one input and one output for simplicity. The top panel on the left hand side illustrates the standard production function, reflecting total physical product (TPP). The bottom left panel is directly related to the production function and gives the average physical product (APP) and the marginal physical product (MPP). The right hand side contains a top panel, only used for transformation and a bottom panel with the average variable costs (AVC)

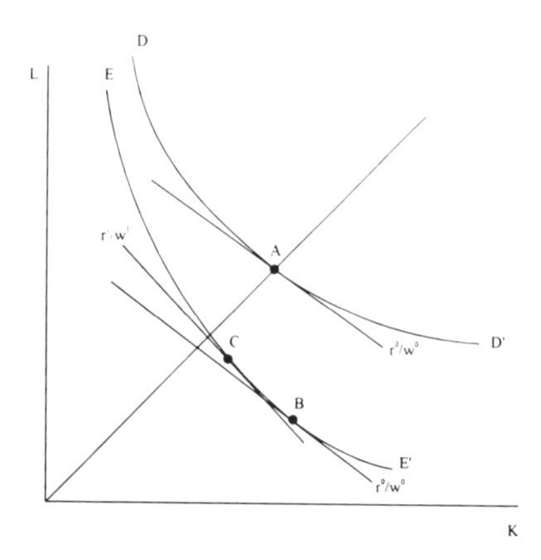

Fig.9.3: Technological change and relative changes of input prices.

and the marginal cost function (MC). The shift from stage I to stage II is exactly at the point where APP is at its maximum and equals MPP. Moreover AVC is at its minimum and equals MC. At this point, there are constant returns to scale, but with a lower input level, there are increasing returns to scale and at a higher input level, we observe decreasing returns to scale. (The relation between technological change, scale effects and productivity change is illustrated in Fig. 9.5.)

2. Productivity growth and technical change

Output may be growing over time because more inputs are being used as well as because of technical change. Measures of productivity growth seek to decompose the impact of scale effects of input changes (i.e. movements along the production function) and technical change (i.e. operating on a new production function).

Fig.9.5 presents the cases of increasing returns to scale (panel a), decreasing returns to scale (panel b) and constant returns to scale (panel c). Since productivity growth reflects changes in average physical product, we can geometrically describe productivity growth by evaluating changes in the average products. All cases presented in the three panels of Fig.9.5 exhibit positive productivity growth, or alternatively an increase in average product. Productivity growth is assessed by comparing the slope of the line OA to the slope of the line OC. The scale effect's contribution to productivity growth is the change in the slope of OA to the slope of OB. This reflects how average product changes

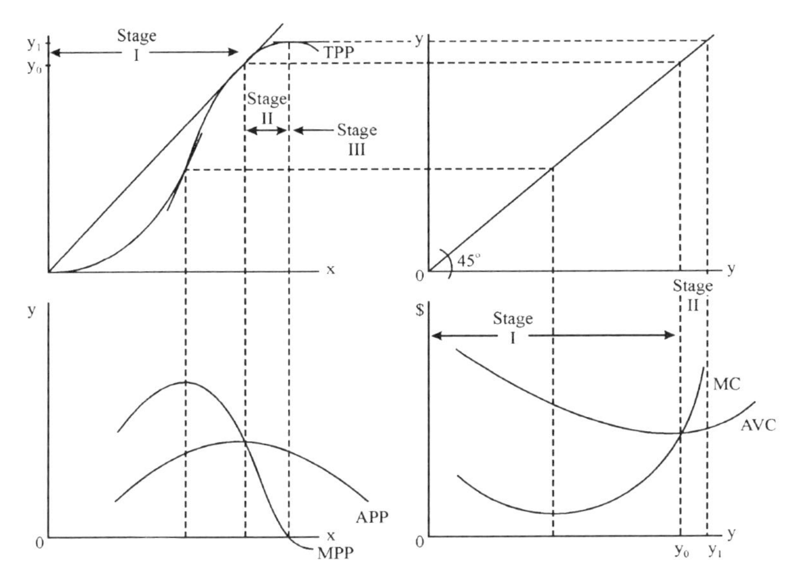

Fig.9.4: Production and cost relations in three stages of production.

as the firm increases input use using the same technology. The technical change effect is the change in the slope from OB to OC and is positive in all three cases.

Increasing returns to scale in Fig.9.5(a) are characterised by the production function exhibiting increasing average product with increasing input use. Thus, the scale effect's contribution is positive as illustrated by the increase in the slope from OA to OB. The positive technical change effect reinforces the positive scale contribution to productivity gain. Decreasing returns to scale in Fig.9.5(b) are characterised by the production function exhibiting decreasing average product with increasing input use. Thus, the scale effect's contribution leads to a negative productivity change and is illustrated by the decrease in the slope of OA to OB. The positive technical change effect overwhelms the negative impact of the scale effect leading to overall productivity growth. Constant returns to scale in Fig.9.5(c) are characterised by the production function exhibiting constant average product for all input levels. With no change in slope from OA to OB, there is no scale effect contribution to productivity growth. In this case, all productivity growth arises from technical change. In many situations in agriculture one might expect decreasing returns to scale in the short term, because of slow adjustments for some inputs (land, capital, family labour) and constant or slightly increasing returns to scale in the long term.

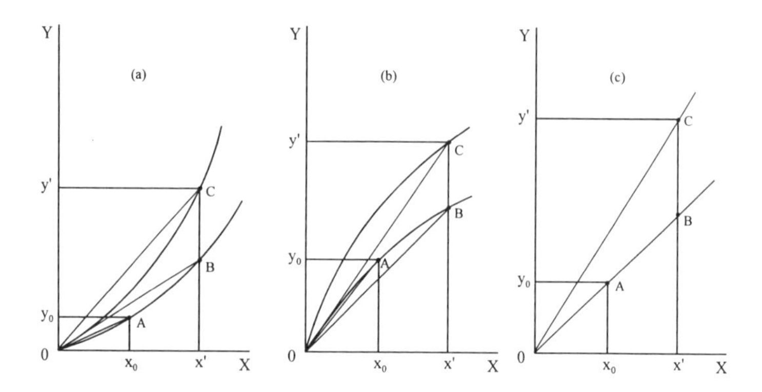

Fig.9.5: Productivity growth decomposition.

Technical change represents all the forces and activities leading to a change in the production technology. It may be disembodied from the inputs used in production and serves to make some (or all) of existing inputs more productive. Alternatively, technical change can be embodied in new equipment and thus transmitted into the sector as technical change. Another way technical change is frequently characterised is as either endogenous or exogenous to the sector's economic system. Exogenous forces include the efforts of research scientists pioneering new methods and materials to enhance production possibilities. Endogenous forces can be the direct efforts of research and development decisions by the producers (not likely in the case of production agriculture) but also include the role of forces stimulating the economic system. One of the primary stimuli is price relations. Consequently, changes in price levels can stimulate decision makers to adopt new technologies.

Technological change and productivity measurement: different ways to empirical results

The central definition of productivity is still the relation between inputs and the resulting outputs. In practice, both inputs and outputs change in quantity, quality and composition and this makes it more difficult to define productivity development. Therefore, productivity measurement is a clever way of handling different inputs and outputs. In Appendix B, a formal methodology of Total Factor Productivity (TFP) measurement is provided. Here it is sufficient to observe that index numbers of output and input are used. This methodology is often used in the literature.

A second line of productivity measurement is closely related to production economics. It is straightforward to consider productivity measurement in relation to the production function. Often one incorporates a time trend variable to represent technological change. Depending on specification, neutral or biased technological change is allowed. With several outputs there holds a transformation function. Transformation functions are often derived indirectly by means of cost or profit functions, including a technology variable (Antle and Capalbo, 1988).

Although the literature gives much attention to theoretical aspects of productivity measurement, the practical problems, such as the definition and grouping of input and output variables, the working methods to handle the data, etc. seems to be even more important. As will be clear from the beginning, technological change is no ‘manna from heaven’ and adding more input variables results in a smaller change of total factor productivity.

There seems to be a strong tradition in the types of input variables used in productivity analysis and in defining technology. However, variables like research and development (by governmental organisations), extension service, education, rural development projects, etc. are often not incorporated in the analysis.[3] These inputs explain a part of the change in technology and/or productivity. The same holds for the quality of output and input variables. Increased quality of input variables will take a part of the ‘unexplained’ difference between output and input development (Capalbo and Vo, 1988, p.105). The reverse effect might follow from an increase in the quality of output or a decrease in the quality of input. Although this area has been covered extensively by Griliches (1991), easy solutions are not available.

Some difficult issues in agriculture follow from the special position of land and family labour. Often land rents are regulated, while using wage cost rates for family labour can make total costs exceed total revenue. Another issue that often leads to difficulties is the price increase of capital goods (Ball, 1985). There is depreciation and interest on capital goods, which have been acquired several years before. All these problems have to be solved in a practical calculation of productivity development. Therefore one observes quite often that very crude methods are used in determining productivity changes.

Some caveats in measuring technological change

Without pretending to be complete on this issue, it might be helpful to consider a number of methods that are often used to represent the change of technology and the related productivity measures in agriculture. We give some of the more familiar methods and provide short comments.

Increase of production per hectare

Here in fact the basic assumption is that there is only one output (e.g. grain) and one input (land). If this were the case, the increased cereal production per unit of land is a correct measure of the increase in productivity. But several inputs are required to grow cereals (seed, fertiliser, pesticides, energy, machinery, labour, management, etc.). Although some of them might be constant others have to be changed to achieve a higher level of production.

Increase of the production per animal

This can refer to the production per animal unit, which is quite similar to production per hectare. However, here, this measure says nothing about e.g. the quantity of feed used per animal. Thus, within animal production feed conversion ratios are also used. These ratios focus completely on the output/input relation of animal production per unit of feed input. Both measures say nothing about other inputs and they often change at a different rate.

Production per unit of labour (labour productivity)

Both total production or gross/net value added are relatively easy to obtain. The same holds with respect to the number of persons or (even better) the equivalent labour units working in a sector of the economy. This explains a part of the popularity of labour productivity. Even Van der Meer and Yamada (1990) in their thorough study use this measure, and outside agriculture this is one of the most popular measures to compare productivity. Capital and land input play no role in the analysis, which might cause a serious bias.

The number of persons fed per farmer

This divides the number of people within an economy by the number engaged in farming. A correction for a net import/export position may be required. It is a very rough method, only using one aspect of productivity. Here a division between agriculture and the rest of agribusiness is completely neglected. Moreover, no attention will be given to e.g. general productivity development in the economy and capital intensity.

Technological change within the CAP; the theory

The Treaty of Rome and the Treaty of Maastricht

Article 39 of the Treaty of Rome reflects the basic ideas around 1958 on the effect of technological change in agriculture. Article 39 contains the following objectives: "to increase agricultural productivity by promoting technical progress and by ensuring the rational development of agricultural production and the optimal utilisation of the factors of production, in particular labour", "thus to ensure a fair standard of living for the agricultural community, in particular by increasing the individual earnings of persons engaged in agriculture". It has often been observed (e.g. Horring, 1971) that these articles contain an implicit contradiction, if the European Union speeds up technical progress without reducing the quantity of labour in agriculture. Otherwise technical progress stimulates production, gives lower prices and therefore depresses incomes. In the Treaty of Maastricht, these objectives have not been changed, although the more open character of the EU has been stressed, e.g. in article 3a. The cohesion within the EU, however, is presently one of the driving forces behind subsidised investments in agriculture and other parts of agribusiness. Given the limited growth in the demand for agricultural products, large subsidised investments will be a driving force in production growth.

General aspects

How can we analyse the consequences of different policy systems (market and price, quota, income support, structural policy) for technological change within the CAP? The relation between agricultural prices and technological change has been investigated in different ways and under different circumstances see Rutten (1989, Chapter 5) for an overview and Van der Meer and Yamada (1990, p. 63-67) for an application. It plays an important role in analysing the consequences of the protection of agricultural producers in rich countries and food consumers in poor countries (World Bank, 1986). As illustrated earlier, higher agricultural price levels mainly influence the optimal level of input use. In periods of increasing protection this might speed up the process of embodied technological change.

The theory of induced innovation, however, is not always clear about the direction of technological change because of price protection. Often, price protection is assumed to induce technical change and therefore productivity growth. Both Huffman and Evenson (1993) and Fulginetti and Perrin (1993) found weak and stronger confirmations of this hypothesis.

Van der Meer (1989) and Van der Meer and Yamada (1990), however, hypothesised that protectionism has a dual effect on technological change and productivity growth. With low agricultural prices the most efficient farmers remain, but their investment capacity in new technologies is very low. With high agricultural prices, resources do not shift to the most efficient farmers, but sufficient liquidities are available to introduce new technologies. This brings them to the inverse U-shaped curve between prices and productivity growth (see Fig.9.6). According to their hypothesis there is an optimal price level (P^M), where productivity change is at maximum (A^M), lower prices (reflected by P^L) or higher prices (see P^H) both lead to lower levels of productivity change. It is of course difficult to say whether the European Union price level is at P^M as suggested by Van der Meer and Yamada (1990).

The analysis of Kalaitzandonakes (1994) on productivity growth in the New Zealand beef/sheep industry came up with a quite different conclusion. The higher prices in New Zealand agriculture during the period 1975-1985, compared with the later period, had a negative influence on productivity growth. Because New Zealand moved later to the very low world market prices of beef, sheep and wool, the inverse-U-shaped curve seems to differ from these empirical results. This holds even more for the hypothesis that price protection speeds up technological change and productivity growth. The result of Kalaitzandonakes (1994) is also in line with the Porter theory (Porter, 1990), which indicates a high efficiency level in the situation of strong competition. Whether such a strong competition reflects also low price levels is not clear. The results of Mundlak *et al.* (1989) and the basic ideas behind the structural adjustment programmes in developing countries are not in line with the results of Kalaitzandonakes. A long period of relatively low prices is assumed to hamper efficiency in agriculture.

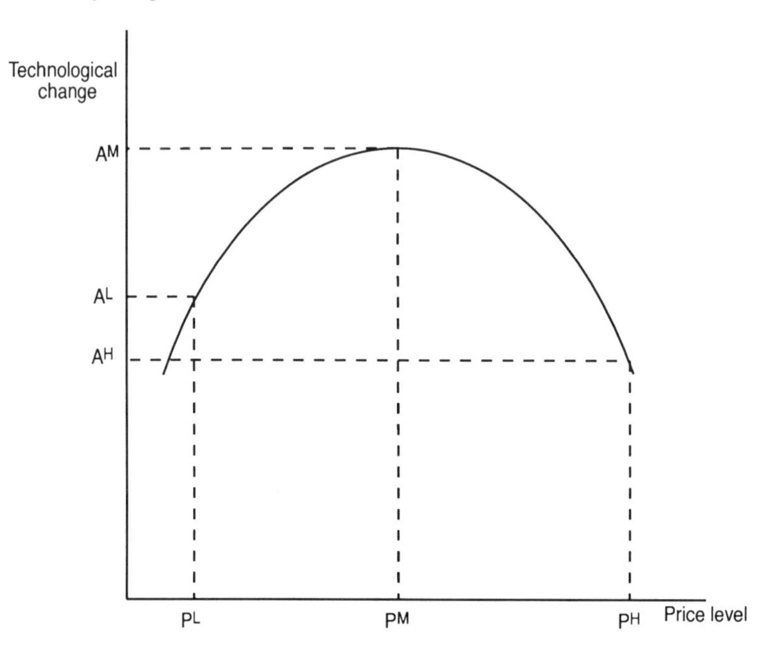

Fig.9.6: Inverse U-shaped curve of technological change.

Consequences under a market and price policy

The general analysis therefore does not lead to a completely clear picture, but we assume that the market and price policy of the CAP had at least the following consequences:

a) Price certainty on the domestic market will have increased agricultural production in the EU (and depressed agricultural production outside the EU).

b) Because high prices foster increased input use in agriculture, this speeds up embodied and also endogenous technological change.

c) Although there are no indications of important economies of scale in agriculture, this might play a role in upstream and downstream industries. Increased input use leads therefore to a positive effect on productivity because of the scale effect (endogenous growth effect).

d) Structural policies for agriculture largely remained national policies within the CAP. This implies that for each country it is profitable to speed up technological change, because the price effects are often limited. Although such investments might not be very productive for the EU-economy, this does not hold for the agricultural producers of individual countries.
e) In recent years a link has been made between EU-expenditures on structural policy and market and price policy. This link increased subsidies on investments and input use in agriculture, which might not be very productive.

All five consequences are output and productivity increasing for the producers who make the decisions, although for the EU total economy this might be different (because of the subsidies). According to the ideas of Rausser and de Gorter (1991) and De Gorter *et al.* (1992) there is a strong link between the expenditures on market and price policy and structural policy. The existing market and price policy is a type of 'insurance' for agricultural producers that technological improvements which lead to output increases do not depress income of agricultural producers too much. A shift from exogenous – government driven – technological change to endogenous technological change, which is generated within the sector, would be an additional reason to diminish price protection of the agricultural sector. If external effects of agricultural production are included, that might give even larger differences between the productivity increase for the individual producer and the total society.

A quota system

Under a quota system, output growth is restricted. Therefore, technological change should be compensated by reduced input use (see Appendix A). With increasing returns to scale, input reduction might be smaller. This implies that both embodied technological change (which depends on input use) and endogenous technological change (which depends on the scale effect) will be reduced. However, technologies which depend on price certainty and which are not output increasing might be very attractive within a quota system.

Two examples might clarify this position. The effect of Bovine Growth Hormone (BST) in milk production is less attractive in a situation of milk quota, compared with a market and price system. This holds for given market prices of milk (Giesen *et al.*, 1989). A new technology of storing and spreading manure requiring high investments, however, might be introduced much easier under a milk quota system, when price and income certainty play an important role.

Income policy

Income policy is a combination of no (or a relatively low) price support, together with income payments. Given the observations made in the case of market and price policy, the relatively low price of agricultural products hampers input use and embodied technological change, compared with market and price policy at higher price levels. If income payments depend on the continuation of the particular farmer, or on the continuous use of land under a base acreage system, this influences the long run competitive conditions and a labour or land using technological change results. (See Appendix A.)

To put it in another way: land remains the property of relatively old and inefficient farmers, leading to an additional restraint on productivity growth. It is a negative effect on productivity growth in agriculture, compared with free market prices without income allowances,[4] but might still be more efficient than market and price policy or a quota system.

Structural policy

Structural policy has been only a small part of EU agricultural policy, but it is of increasing importance. Originally its main intention was to stimulate productivity and to reduce costs (see Fig.9.5c for a situation with constant returns to scale). The effects of structural policy are illustrated by Fig.9.7. Because of structural policy, the supply curve (= marginal cost curve) of agricultural production shifts from S_0 to S'. If prices are fixed, this would increase production from Q_0 to Q_s. Assuming a constant demand curve D, structural policy without price adjustment would create surplus production. To clear the market again prices should decrease to P'. At the end, prices are much lower, production has increased and producers could have lower incomes. Often it is the consumers who gain most from this kind of adjustment. Recently, the structural policy of the EU has been extended. Important budgets are available for structural policy, both in general and on a more regional basis, as discussed in Chapters 10 and 13.

Technological change within the CAP; empirical results

It would be good if we could illustrate the theoretical results by means of sound empirical analysis. Productivity developments, however, are not readily available and we have to see what can be derived from the work of other researchers, in addition to the results provided earlier in this chapter.

It is interesting to observe that two different studies for the United States and the EC on the increase of total factor productivity came up with identical results: 1.7% per year (Ball, 1985 and Henrichsmeyer and Ostermeyer, 1988). Periods and methodology were slightly different but results are very homogeneous. This holds also with respect to the increase of total factor productivity in agriculture within different countries of the EC (see Table 9.1). Here one observes substantial differences in the growth rates of inputs and outputs, but only small differences in the development of total factor productivity: this would suggest that technological change is mainly exogenous.

But it is not very difficult to provide quite opposite results. Bernard and Jones (1993, p. 18) analysed data of thirteen OECD countries and their results

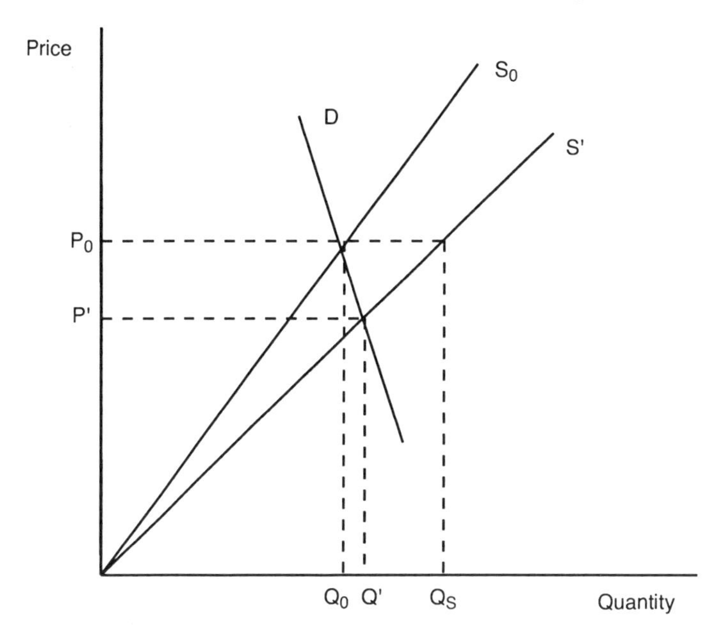

Fig.9.7: Effects of a structural policy.

on total factor productivity were not only far different in level, but often ended in quite different conclusions with respect to the level of productivity growth in agriculture for different countries: see Table 9.1. Trueblood (1996) analysed productivity changes by means of a Malmquist productivity index, using FAO data and came up with different results, although Japan and Australia show low productivity growth. The results of Bernard and Jones (1993) and Trueblood (1996) seem to be more in line with the hypothesis of Van der Meer and Yamada (1990, p. 63-67), that relatively low (Australia, USA) and relatively high agricultural prices (Japan) hamper productivity growth. Only the results for the Nordic countries (Finland, Norway, Sweden) with their high price level are difficult to interpret, unless one assumes that farming under such conditions is quite different. Bernard and Jones (1993) give no separate attention to land, an important input in agriculture. Because of their focus on all sectors of the economy, they only included labour and capital to determine total factor productivity. Given the relatively small change of land in agriculture (see Folmer *et al.*, 1995, p. 31), this leads to an overestimation of growth rates in total factor productivity. Trueblood has been limited by data availability, specifically for capital.

It is also possible to compare the results between different sources of partly overlapping periods. From a research perspective it is very disappointing that these sources often provide opposite results. The correlation between the TFPs of Henrichsmeyer and the OECD is 0.12; the correlation between the last two columns of Table 9.1 is equal to 0.64. This suggests a low level of reliability of the research, especially if TFPs are not very different.

Table 9.1: Annual percentage growth rates of agricultural production, input and total factor productivity from different sources.

Country	Annual change of output, input and total factor productivity, according to Henrichsmeyer (1965-85) and OECD (1973-1989)						Research of Bernard and Jones (1970-1987) and Trueblood (1962-1990)	
	Gross production		Gross input		TFP		TFP	
	H&O	OECD	H&O	OECD	H&O	OECD	B&J	True-blood
Bel/Lux	2.2	1.1	0.0	0.0	1.7	1.1	3.7	4.0
Den.	1.5	2.2	-0.1	-0.1	1.6	2.3	4.1	3.7
Fra.	1.8	1.8	0.1	-0.4	1.6	2.3	4.0	3.0
Ger.	1.7	1.2	-0.1	-0.4	1.4	1.6	4.3	4.6
Ire.	2.9	2.1	0.9	0.8	1.7	1.3	-	1.5
Italy	1.6	1.6	-0.6	-0.6	2.2	2.2	2.0	2.4
Neth.	4.2	2.9	2.6	1.3	1.5	1.6	4.4	1.7
UK	2.9	1.4	-0.3	-0.2	2.2	1.6	3.6	2.5
EC-9	1.9	-	0.1	-	1.7	-	-	-
US	2.2[1]	-	0.5[1]	-	1.7[1]	-	1.5	2.0
Aust.	-	-	-	-	-	-	1.8	0.9
Can.	-	-	-	-	-	-	0.9	1.4
Jap.	-	-	-	-	-	-	-0.2	0.7
Nor.	-	-	-	-	-	-	2.1	3.1
Swed.	-	-	-	-	-	-	2.0	3.1
Fin.	-	-	-	-	-	-	2.2	3.7

[1] Results of Ball (1985) over the period 1960-79.

Source: Henrichsmeyer and Ostermeyer-Schlöder (1988), OECD (1995), Ball (1985), Bernard and Jones (1993) and Trueblood (1996).

Table 9.2 provides results of the same researchers, using the same data set, but two different methods to determine productivity growth/technological development. Results of both methods are opposite for France and Germany, although the total factor productivity measure is quite similar to the OECD results shown in Table 9.1. Becker and Guyomard (1991) conclude on the basis of their analysis using a cost function approach that technological change is mainly exogenous.

Table 9.2: Detailed analysis of total factor productivity increase and technological change for France and Germany (1960-1985).

Country	Total Factor Productivity	Technical change
France	1.95	1.70
Germany	1.55	2.27

Source: Becker and Guyomard (1991), p.21, 22.

Table 9.3 shows volumes of inputs per unit of output within different countries of the EU. This is the inverse of productivity levels. Here one can observe rather large differences, although prices within the EU are quite similar. However, this does not hold with respect to typical input prices of farm family labour and land. Moreover, large differences in product quality might exist (e.g. soft wheat and durum wheat), while exporting countries should be more efficient than importing countries within an open internal market such as the EU. Table 9.3 suggests that there are still opportunities for technological change and a better allocation of production factors. From this one might deduce which countries/products are most promising for increasing agricultural production and/or for encouraging technological change.
The theory of exogenous technological change implies a convergence of productivity levels. Technology is freely available and countries (or farms) with a relatively high level of technology will be followed by the others. The theory of endogenous technological change (Romer, 1987) gives more attention to scale factors and ruling internal conditions, which would induce the countries with a high productivity level to show also a higher level of productivity growth. This hypothesis is not confirmed by Bernard and Jones (1993, p.19), who find opposite results in agriculture. A high productivity level gives significantly lower productivity growth.

Conclusions

The results of our – limited – overview are not very promising. No clear conclusions can be derived. Moreover, it seems that different researchers come up with quite different results, even if they focus on the same questions and analyse the same countries. If it were necessary to come to a conclusion on the basis of this information, then it seems probable that the CAP has on balance stimulated productivity growth in agriculture, although this conclusion is very weak.

Table 9.3: Input volume per unit of product within the EU; 1984-86. Average within the EU is 1.

Land	Wheat	Sugar beet	Pigs	Milk
Belgium	-	0.88	0.90	0.78
Denmark	-	-	0.98	0.88
France	0.86	0.77	0.98	1.01
Germany	1.02	1.07	1.10	1.16
Greece	2.10	1.26	1.13	1.73
Ireland	0.75	1.15	0.76	0.81
Italy	2.00	1.36	1.07	1.47
Luxembourg	-	-	1.17	0.94
Netherlands	0.91	0.93	0.89	0.80
United Kingdom	0.84	1.05	0.91	0.79

Source: Bureau and Butault (1992, p. 31).

Changes in the external effects of agricultural production

Over the past few decades agricultural production technology has changed considerably. Particular technologies have been displaced completely by new ones, based on the substitution of labour by capital equipment and purchased inputs and services. Often very limited attention has been given to the external effects (Oskam, 1991).

One could incorporate the external effects of agricultural production in a modified production function. Such a procedure elucidates the importance of disguising these effects of agricultural production, in the short as well as the long term. Various external effects might be included:

- air pollution, either by well known ingredients like nitrogen oxide, and sulphur oxide, causing the acid rain problem or by contributing to the greenhouse effect by means of carbon dioxide, methane and nitrous oxide, etc. or contributing to odour pollution (ammonia, but also other ingredients);
- surface water pollution, due to minerals (such as chemicals) or organic material in the surface water;
- groundwater pollution, due to leaching into the groundwater of minerals or other materials which have their origin directly or indirectly in agricultural production;
- soil pollution; this can mean that the soil attracts a certain amount of pollutants that could potentially come into the groundwater, the surface water or the air. Of course, there are two stages in this process, where the soil is still ‘attracting’ these materials and where there is already leakage due to saturation;
- the visual attractiveness of the landscape, due to composition and variety;

- the conservation of nature, leading to the availability of a wide diversity of different species. This can embrace a direct impact on flora but also indirect effects on fauna through impacts on wildlife habitat, including the effects of the level and quality of the groundwater and surface water. This illustrates also the interaction among different parts of the environment.

Several researchers have tried to tackle the problem of incorporating the environmental effects in a consistent framework of national accounts or sectoral accounts (Hueting, 1974; Mäler, 1990). The methodology in this area is still in development. Therefore, it is impossible to apply a 'ready-to-use' methodology. Here we provide some information of an empirical analysis to incorporate the environmental effects due to manure, fertiliser and pesticides usage resulting in pollution of air, groundwater and surface water and in global warming for the Netherlands (Oskam, 1991; 1995).

During the period 1970–1986, there was a strong increase in the environmental burden of agricultural production in the Netherlands. The results of Table 9.4 make clear that the negative external effects are limited compared with the annual growth of productivity. Of course one has to realise that these numbers depend very much on derived and assumed shadow prices of external effects. Basic theory makes clear that there is a long term development which makes the prices of negative external effects higher and therefore their influence in productivity developments larger.

Table 9.4: Output and productivity developments in % per year with and without including external effects of agricultural production in the Netherlands, 1970–1986.

Variable	No external effects	External effects included
Gross output	4.2	3.7
Total factor productivity	2.6	2.3

Source: Oskam (1995).

Besides external effects, food quality and food safety standards, animal rights issues, ecological production, workers protection in agricultural production and many other conditions influence the input output relation in agriculture. Although these issues are well known it would be too difficult to determine their effects on the level of technology. But let us take one example: ecological agriculture. Here it will be quite clear that on the basis of 'traditional measures' productivity is lower in ecological agriculture (Wijnands and Vereijken, 1992). However, considering ecological (organic) products as totally different products makes such a comparison invalid.

Conclusions

The definition and measurement of technological change is, although quite straightforward in economic theory, a very difficult issue in practice. Still one has to realise that the agricultural sector belongs to the parts of the economy where: (a) outputs are not changing too quickly, because nearly the same types of products are often still produced; (b) the scale effect in production technology is not very important due to typical characteristics of farming; and (c) for quite a number of products there is a clear price formation at international level, which provides basic information to make an intra country comparison of productivity levels or changes.

The most workable method to measure productivity increase in agriculture is total factor productivity, measured by means of index numbers. For international comparisons under quite different price regimes, the choice of the relevant prices is important for the final results.

The effect of the CAP on technological change is not completely clear. The hypothesis of Van der Meer and Yamada (1990, p. 63) of a relatively high level of productivity growth under the ruling conditions of the EU seems plausible, but has not been confirmed strongly and uniformly by empirical analysis. But the results indicate at least that ruling EU-prices did not influence productivity growth negatively.

If the CAP is characterised as a price support policy at EU level, together with structural policy, mainly financed by governments in combination with the EU, some conclusions follow clearly from incentive theory. Under the conditions of a large open market, it nearly always pays off for individual countries to stimulate production by means of structural policy. At individual country level, the prices are nearly given. An increase of production is therefore mainly paid by other countries and this is especially true for small countries within the Union.

Even though it is very difficult to measure productivity increase in agricultural production, it is clear from nearly all empirical studies that it is an important factor within the agricultural sector. This sector shows high levels of productivity growth, which — together with the related long run real price decrease of agricultural products — implies that agricultural incomes remain often relatively low. Here, quite opposite results follow from sectoral accounts and averages of farm families (Hill, 1996), but also from different researchers (Gardner, 1992). Although Hill and Gardner try to convince their readers that there is no income disparity in agriculture, this does not seem very plausible if one observes the large decrease of agricultural firms and agricultural labour force during many decades (Munk and Thomson, 1994, p. 45, 52).

The negative effect of the CAP on the environment has been mentioned quite often, including of course in Chapter 13 of this book. It is one of the driving forces behind environmental policy with respect to agriculture. A first indication for the Netherlands gives the idea that the positive effects of technological change are much larger than the negative environmental effects.

References

Antle, J.M. and Capalbo, S.M. (1988) An Introduction to recent developments in production theory and productivity measurement. In: S.M. Capalbo and J.M. Antle (eds) *Agricultural Productivity: Measurement and explanation*. Washington, Resources for the Future, 17-95.

Arrow, K.J. (1962) The economic implications of learning by doing. *Review of Economic Studies,* 29, 154-174.

Ball, V.E. (1985) Output, input and productivity measurement in U.S. agriculture, 1948-1979. *American Journal of Agricultural Economics,* 67, 475-486.

Barro, R.J. (1990) Government spending in a simple model of endogenous growth, *Journal of Political Economy* 98, 103-125.

Baumol, W.J. and Wolfe, E.N. (1983) Feedback from productivity growth to R & D, *Scandinavian Journal of Economics,* 85, 147-157.

Becker, H. and Guyomard, H. (1991) Bestimmung deterministischer Faktorproduktivitäten and Schätzung technischer Fortschritte für die Agrarsektoren Frankreichs und der BR Deutschland. *Agrarwirtschaft,* 40, 20-24.

Bernard, A.B. and Jones, Ch.I. (1993) Productivity across industries and countries: Time series theory and evidence. MIT-Department of Economics Working paper 93-17, Cambridge, Mass.

Bureau, J.C. and Butault, J.P. (1992) Productivity gaps, price advantages and competitiveness in E.C. agriculture. *European Review of Agricultural Economics,* 19, 25-48.

Capalbo, S.M. and Vo, T.T. (1988) A review of the evidence on agricultural productivity. In: S.M. Capalbo and J.M. Antle (eds) *Agricultural Productivity: Measurement and explanation*. Washington, Resources for the Future, 96-137.

Caves, D.W., Christensen, L.R. and Diewert, W.E. (1982) Multilateral comparisons of output, input and productivity using superlative index numbers. *Economic Journal* , 92, 73-86.

Chambers, R.G. (1988) *Applied production analysis: A dual approach.* Cambridge, Cambridge University Press.

Chavas, J.P. and Cox, T. (1992) A nonparametric analysis of the influence of research on agricultural productivity, *American Journal of Agricultural Economics,* 74, 583-591.

Cox, T.L. and Chavas, J.P. (1990) A nonparametric analysis of productivity: The case of US agriculture. *European Review of Agricultural Economics*, 17, 449-464.

De Gorter, H., Nielson, D.J. and Rausser, G.C. (1992) Productive and Predatory public policies: Research expenditures and producer subsidies in agriculture. *American Journal of Agricultural Economics*, 74, 27-37.

Diewert, W.E. (1976) Exact and superlative index numbers. *Journal of Econometrics*, 4, 115-145.

Diewert, W.E. (1981) The economic theory of index numbers: A survey. In: A. Deaton (ed) *Essays in the Theory and Measurement of Consumer Behaviour*. London: Cambridge University Press.

Färe, R., Grosskopf, S., Lovell, C. and Pasurka, C. (1989) Multilateral productivity comparisons when some outputs are undesirable: a nonparametric approach. *Review of Economics and Statistics,* 71, 90-98.

Folmer, C., Keyzer, M.A., Merbis, M.D., Stolwijk, H.J.J. and Veenendaal, P.J.J. (1995) *The Common Agricultural Policy beyond the Mac Sharry reform*. Amsterdam: Elsevier.

Fulginiti, L.E. and Perrin, R.K. (1993) Prices and productivity in agriculture. *Review of Economics and Statistics,* 75, 471-482.

Gardner, B.L. (1992) Changing economic perspectives on the farm problem. *Journal of Economic Literature*, 30, 62-101.

Giesen, G., Oskam, A. and Berentsen, P. (1989) Expected economic effects of BST in the Netherlands. *Agricultural Economics*, 3, 231-248.

Griliches, Z. (1991) *Technology, Education and Productivity*. New York, Basil Blackwell.

Hallem, D. and Machado, F. (1996) Efficiency analysis with panel data: A study of Portuguese dairy farms. *European Review of Agricultural Economics,* 23, 79-93.

Henrichsmeyer, W. and Ostermeyer-Schlöder, A. (1988) Productivity growth and factor adjustments in EC agriculture. *European Review of Agricultural Economics,* 15, 137-154.

Hill, B. (1996) Monitoring incomes of agricultural households within the EU's information system - new needs and new methods. *European Review of Agricultural Economics*, 23, 27-48.

Horring, J. (1971) European agricultural policy: A Dutch viewpoint. In: H. Priebe, D. Bergman and J. Horring, *Fields of conflict in European farm policy*. Agricultural Trade Paper 3, Trade Policy Research Centre, London.

Hueting, R. (1974) *Nieuwe schaarste en economische groei*. [New scarcity and economic growth]. Amsterdam: Agon.

Huffman, W. and Evenson, R. (1993) *Science for agriculture*. Ames: Iowa State University Press.

Kalaitzandonakes, N.G. (1994) Price protection and productivity growth. *American Journal of Agricultural Economics*,76, 722-732.

Koning, N. (1994) *The failure of agrarian capitalism*. London: Routledge.

Lucas, R.E. (1988) On the mechanics of economic development. *Journal of Monetary Economics*, 22, 3-42.

Lund, M., Jacobsen, B. and Hansen, L. (1993) Reducing non-allocative costs of dairy farms: Application of non-parametric methods. *European Review of Agricultural Economics*, 20, 327-341.

Mäler, K.G. (1990) National accounts and environmental resources. *Environmental and Resource Economics*, 1, 1-15.

Mansholt, S.L. (1986) Minder is moeilijk in de Europese Landbouw: Prijsverlaging, contingentering en areaalbeperking gewogen als middel tot produktievermindering [Less is difficult in European Agriculture: Price decrease, quota and set aside weighted as methods to reduce production]. *Spil* 55-56, 5-20.

Michalek, J. (1985) *Technical progress in West German agriculture: A quantitative approach*. Kiel: Vauk.

Mulder, M. and Poppe, K.J. (1993) Landbouw, Milieu en Economie. [Agriculture, environment and the economy]. Periodieke Rapportage 68-89, LEI-DLO, Den Haag.

Mundlak, Y., Cavallo, D. and Domenech, R. (1989) Agriculture and economic growth in Argentina, 1913-84. IFPRI Report 76, Washington.

Munk, K. and Thomson, K. (1994) EC agricultural policy for the 21st century. *European Economy*, Reports and Studies, 1994 (4).

Newbery, D.M.G. and Stiglitz, J.E. (1981) *The theory of commodity price stabilisation*. Oxford: Oxford University Press.

Oskam, A.J. (1991) Productivity measurement, incorporating environmental effects of agricultural production. In: K. Burger, *et al.* (eds) *Agricultural Economics and Policy: International challenges for the nineties*. Amsterdam: Elsevier, 186-204.

Oskam, A.J. (1995) Productivity measurement at sectoral level at different perspectives. In: M.J.G. Meeusen-van Onna and J.H.M. Wijnands (red.) De methoden van produktiviteitsmeting. [Methods of productivity measurement.] LEI-DLO-Mededeling 515, Den Haag, 1995, 84-104.

Pindyck, R. and Rotemberg, J. (1983) Dynamic factor demands and the effects of energy price shocks. *American Economic Review*, 73, 1066-1079.

Porter, M. (1990) *The competitive advantage of nations*. New York: The Free Press.

Rausser, G.C. and de Gorter, H. (1991) The political economy of commodity and public good policies in European Agriculture. *European Review of Agricultural Economics*, 18, 485-504.

Rebalo, S. (1991) Long-run policy analysis and long-run growth. *Journal of Political Economy*, 99, 500-521.

Romer, P.M. (1987) Growth based on increasing returns due to specialisation. *American Economic Review* 77, Papers and Proceedings, 56-62.

Rutten, H. (1989) Technical change in agriculture. LEI-Onderzoekverslag 45, Den Haag.

Sato, R. and Suzewa, G.S. (1983) *Research and Productivity: endogenous Technical Change*. Auburn Publishing Company.

Schäfer, A. (1985) *Zur Anwendung von Frontier- und Dualitätsansätzen in der Landwirtschaft*. Kiel: Vauk.

Trueblood, M. (1996) An intercountry comparison of agricultural efficiency and productivity. Dept. of Applied Economics, University of Minnesota, August 1996 (PhD thesis).

Van der Meer, C.L.J. (1989) Price level and agricultural growth performance. *Tijdschrift voor Sociaalwetenschappelijk onderzoek van de Landbouw*, 4, 295-312.

Van der Meer, C.L.J., Rutten, H. and Dijkveld Stol, N.A. (1991) Technologie in de landbouw. [Technology in agriculture.] WRR-Voorstudies en achtergronden T2, WRR, Den Haag.

Van der Meer, C.L.J. and Yamada, S. (1990) *Japanese Agriculture: A comparative Economic Analysis*. London: Routledge.

Wijnands, F.G. and Vereijken, P. (1992) Region wise development of prototypes of integrated arable farming and outdoor horticulture. *Netherlands Journal of Agricultural Science*, 40, 225-238.

World Bank (1986) *World Development Report*. New York: Oxford University Press.

---

[1] Here we focus on outputs; similar observations hold with respect to inputs.

[2] In comparison, the Harrod measure of technical change is a long-run classification looking at the impact of technical change during a period in which the capital-output ratio is constant. The Solow measure looks at the impact of technical change when the labour-output ratio is constant.

[3] The work of Van der Meer, *et al.* (1991) and the background study of Rutten (1992) form an exception. They include expenditure on education, extension and R&D. Oskam (1991) includes expenditure on land consolidation.

[4] We refer to Folmer *et al.* (1995, p. 298-302) for the effects of a free market system without income allowances for technological development. The authors argue that either large or small (part-time) farms will survive under such conditions.

Appendix 9A Basic principles of technological change presented in a formal way

Neutral and non-neutral technological change

Define the production function as $Y = F(K, L; t)$ where t denotes technical change and let:

$MPP_i(K, L; t_0) =$ the marginal product of input i $(i = K, L)$ before the onset of technical change, and

$MPP_i(K, L; t_1) =$ the marginal product of input i $(i = K, L)$ after the onset of technical change.

The Hicksian technical change is characterised as labour-saving (or capital-using), neutral or labour-using (capital saving), respectively, if:

$$\frac{MPP_K(K,L;t_1)}{MPP_L(K,L;t_1)} > \;,\; = \;,\; > \frac{MPP_K(K,L;t_0)}{MPP_L(K,L;t_0)}$$

Basic production theory: Cost minimisation

Let r be the price of capital and w the price of labour. Under long-run competitive conditions:

$$\frac{MPP_K}{MPP_L} = \frac{r}{w} \cdot \qquad (1)$$

The cost minimisation problem involves choosing input to minimise the cost of producing a specific level of output. Focusing on the case of a single variable input, we can express minimum cost, C, as a function of output, y, as $C(y) = w\,x(y)$. This leads to expressing average costs as:

$$\text{AC} = \frac{C(y)}{y} = w\frac{x(y)}{y} = \frac{w}{APP}$$

where APP is average physical product. The marginal cost of production is:

$$\text{MC} = \frac{\Delta C(y)}{\Delta y} = w\frac{\Delta x}{\Delta y} = \frac{w}{MPP}$$

where MPP is the marginal physical product $(=\Delta y/\Delta x(y))$. Observe that the marginal cost function is also the supply function (at least for quantities where marginal costs are not below average costs; see Fig 9.4).

Now, we define the elasticity of production ε as the percentage change in output given a percentage change in input:

$$\varepsilon = \frac{\frac{\Delta y}{y}}{\frac{\Delta x}{x}} = \frac{\Delta y}{\Delta x} \cdot \frac{x}{y} = \frac{MPP}{APP} \tag{2}$$

The production elasticity is also named the elasticity of scale. In standard theory there are regions of increasing, constant and decreasing returns to scale, depending on the level of input (Chambers, 1988, p. 20, 21).

Productivity growth and technical change

Define the production function as:

$$y = \mathrm{f}(x, \theta)$$

where θ represents the stock of technical knowledge. The total change in output is:

$$\Delta y = \frac{\Delta y}{\Delta x}\Delta x + \frac{\Delta y}{\Delta x}\Delta\theta$$

Dividing through by y leads to the percentage change in output:

$$\frac{\Delta y}{y} = \frac{\Delta y}{\Delta x} \cdot \frac{\Delta x}{y} + \frac{\Delta y}{\Delta \theta} \cdot \frac{\Delta \theta}{y}$$

$$= MPP \cdot \frac{\Delta x}{y} + \hat{A}$$

where $\hat{A}$ is the component related to a change in output resulting from a change in the stock of technical knowledge. Recall the production elasticity defined in (2). Thus:

$$\frac{\Delta y}{y} = MPP \cdot \frac{\Delta x}{y} \cdot \frac{x}{x} + \hat{A}$$

$$= MPP \cdot \frac{x}{y} \cdot \frac{\Delta x}{x} + \hat{A} \tag{3}$$

$$= \frac{MPP}{APP} \cdot \frac{\Delta x}{x} + \hat{A}$$

$$= \varepsilon \frac{\Delta x}{x} + \hat{A}$$

Productivity is defined as $APP = y/x$. The growth in productivity[1] is

$$\hat{G} = \frac{\Delta APP}{APP} = \frac{\Delta(\frac{y}{x})}{\frac{y}{x}} = \frac{\Delta y}{y} - \frac{\Delta x}{x} \quad (4)$$

which is the growth in output less the growth in input. Using (3) we find that productivity growth can be expressed as:

$$\hat{G} = (\varepsilon - 1) \cdot \frac{\Delta x}{x} + \hat{A}$$

Thus, the productivity growth is decomposed into a scale effect, $(\varepsilon\text{-}1) \cdot \frac{\Delta x}{x}$, and a technical change effect, $\acute{A}$.

Quota policy and technical change

Under a quota system, output growth is restricted. This implies that in a situation with scale effects and technical change effects (see equation (3)) there holds:

$$\frac{\Delta y}{y} = \varepsilon \frac{\Delta x}{x} + \acute{A} = 0 \rightarrow \frac{\Delta x}{x} = - \frac{\hat{A}}{\varepsilon}$$

Any increase of technology should be compensated by reduced input use, which is also influenced by the elasticity of production. With increasing returns to scale, input reduction might be smaller. This implies that both embodied technological change (which depends on input use) and endogenous technological change (which depends on the scale effect) will be reduced.

Income policy and input adjustment

Replacing a price policy by an income policy reduces product prices and the marginal value per unit of input at the original input level. Hence, input levels are adjusted to fulfil the long-run ratio of equation (1):

$$\frac{MPP_K}{MPP_L} = \frac{r}{w}$$

where L is either land or labour or both and w is the price of land or labour or both inputs. We assume that price of capital *(r)* is determined outside agriculture. Because opportunity return of land and/or labour is declining, w becomes small and (a) more labour or land is used with a declining rate of marginal physical product per unit of these inputs; (b) there is a tendency to use a more land and/or labour intensive technology.

Appendix 9B Measurement of total factor productivity and related approaches

A general definition of total factor productivity, applicable in all situations is:[2]

$$\text{Productivity} = f(Y_1,\ldots,Y_m)/g(X_1,\ldots,X_n) \tag{6}$$

where: Y_i = output generated by the process (i=1,...,m)
X_j = input used in the process (j=1,...,n)

It depends, however, in which way the functions f and g (mostly called aggregator functions (Diewert, 1981)) are defined as to what is the resulting productivity level. Recent developments in productivity measurement start with the implicit assumptions about production technology that are behind the aggregator functions.[3]

In productivity measurement, we mostly refer to a reference level: the absolute level of productivity depends on the dimension of input or output measurement. Given the definition of productivity in equation (6), the relative change in Total Factor Productivity (TFP) between two different time periods could be derived from:

$$\text{TFP}_t/\text{TFP}_{t-1} = [f_t(Y_{1,t},\ldots,Y_{m,t})/g_t(X_{1,t},\ldots,X_{n,t})/ (f_{t-1}(Y_{1,t-1},\ldots,Y_{m,t-1})/g_{t-1}(X_{1,t-1},\ldots,X_{n,t-1}))] \tag{7}$$

Note that this is a very general definition because even the aggregator functions between the two different units or periods might differ.

In literature, the optimal choice of index numbers, representing the aggregator functions, received substantial attention. The choice can be based on the type of production function of the underlying process (Diewert, 1976). Assuming a homogeneous linear translog production function and assuming competitive markets, the Tornqvist-Theil index number for the functions f and g of outputs and inputs are exact. These functions are defined, respectively, as:

$$\ln(Y_t/Y_{t-1}) = \tfrac{1}{2} \sum_{i=1}^{m} (S_{i,t} + S_{i,t-1}) \ln(Y_{i,t}/Y_{i,t-1}) \tag{8a}$$

$$\ln(X_t/X_{t-1}) = \tfrac{1}{2} \sum_{j=1}^{n} (R_{j,t} + R_{j,t-1}) \ln(X_{j,t}/X_{j,t-1}) \tag{8b}$$

where: $S_{i,t}$ = output share of output i in total output in year t
$R_{i,t}$ = input share of input j in total input in year t

Because the translog production function is a general second order approximation for any twice differentiable production function, the Tornqvist-Theil index can be considered as a 'superlative' index number (Diewert, 1981), which means that under the particular assumptions it gives a good approximation. Moreover, Caves *et al.* (1982) have shown that this index is also 'superlative' when the production process is non-homothetic (see Chambers, 1988, p. 37-41 for the definition of homothetic production functions). This means that the only important assumption that remains is competitive markets, which imply that input shares and output shares reflect marginal productivities.[4]

The development of *TFP* directly follows from the difference between (8a) and (8b):

$$\ln(\mathrm{TFP}_t/\mathrm{TFP}_{t-1}) = \tfrac{1}{2}\Big[\sum_{i=1}^{m} (S_{i,t} + S_{i,t-1}) \ln(Y_{i,t}/Y_{i,t-1}) - \sum_{j=1}^{n} (R_{j,t} + R_{j,t-1}) \ln(X_{j,t}/X_{j,t-1})\Big] \qquad (9)$$

It will be clear that from one year to another no new outputs or inputs can be incorporated (undetermined logarithms). This brings researchers often to the practice of using rather broad input categories.

Other methods of productivity measurement

As has been mentioned already in the main text, productivity measurement can be based directly on the production function (Antle and Capalbo, 1988). Under certain conditions, technological development can be related to productivity measures (Chambers, 1988). Moreover a 'complete' description of the production technology, either in the form of a production (or transformation) function or in the form of a cost or a profit function makes it possible to define several interesting productivity measures, because one can keep particular variables constant. Productivity measures, however, depend on estimated functions, which is both very demanding on research and risky with respect to estimation results.

A third approach in productivity measurement focuses on frontier production functions (see e.g. Färe *et al.*, 1989). Although frontier functions have been applied at sectoral level (Cox and Chavas, 1990; Michalek, 1985; Schäfer, 1985), the most promising route seems to be at farm level (see e.g. Hallem and Machado, 1996; Lund *et al.*, 1993). This approach determines directly the transformation function between inputs and outputs and can also include undesirable and/or restricted inputs and outputs. The methodology is much easier to apply than the 'production function approach' but is highly dependent on the quality of the data.

---

[1] By taking natural logarithms of $APP = Y/X$ we obtain:

$$\ln APP = \ln Y - \ln X$$

where ln denotes the natural logarithm. In general, $\Delta \ln z = \Delta z / z$ Thus,

$$\Delta \ln APP = \Delta \ln Y - \Delta \ln X = \frac{\Delta Y}{Y} - \frac{\Delta X}{X}.$$

[2] Antle and Capalbo (1988) give a very good overview of the relation between production theory and productivity measurement.
[3] The very common Laspeyres index number is either exact for a Leontief type of function with no substitutability between inputs or outputs, or a linear function with perfect substitution possibilities (Antle and Capalbo, 1988, p.53 and 54).
[4] This is still an important assumption in agriculture because of many quasi-fixed production factors (land, family labour, buildings). Moreover, many long-term analyses come up with the result that farming gives on average negative profits.

PART III

The CAP and the European Union

Chapter 10

The CAP and the Farmer

Michael Keane and Denis Lucey

This is an extremely broad topic, as both the CAP and the farm population to whom the policy applies have been continually changing in the period of over thirty years since 1962 when the original common policy for a range of products was established. The CAP itself has evolved to take account of both internal and external factors. These include market balance, budgetary pressures, reorientation to give greater prominence to such issues as environmental preservation and rural development, and recently the need to accommodate the provisions of the GATT Uruguay Round agreement. For much of the time this evolution has been gradual. However, on occasion, there have been more radical steps such as the Mac Sharry CAP reforms. The farm population has also been changing steadily, both in geographic terms as EU membership expanded, and also numerically within countries as long-term structural decline continued. Against this wide-ranging background, it is proposed to confine discussion to a limited range of topics, as follows:

employment in agriculture;
changing farm structure;
output and productivity;
income from farming;
pluriactivity;
co-operation between farmers;
the future for farmers.

Employment in EU agriculture

Increasing EU membership has radically altered the numbers employed in agriculture, with profound effects on the structure and diversity of farming activity. The original northern European expansion in 1973 (Denmark, Ireland, UK) added about 15% to total numbers, the southern European expansion in 1981 and 1986 (Greece, Spain, Portugal) added close to 60% to total numbers, and the recent middle-northern European expansion (Austria, Sweden, Finland) added around a further 8%. Likely future expansion to central-eastern Europe will have similar profound effects.

Decline in numbers employed in agriculture within countries has continued, with a reduction of over 55% from 1970 to 1993 for the EU (12) (Table 10.1). There has been considerable variation within countries, however, with highest rates of decline in Spain, Germany, Italy, France and Denmark. The numbers employed in agriculture thus nowadays constitute a much smaller proportion of the EU (12) civilian labour force, falling from 13.4% in 1970 to 5.6% in 1993 (Table 10.1).

Table 10.1: Persons employed in agriculture.[1]

	'000		Percentage change	Agriculture as % of total civilian employment	
	1970	1993	1970 to 1993	1970	1993
Belgium	176	100[2]	-43	5.0	2.5[2]
Denmark	303	140	-54	12.9	5.4
Germany	2262	849	-62	8.6	3.0
Greece	1279	794	-38	40.8	21.3
Spain	3310	1198	-64	27.1	10.1
France	2647	1101	-58	12.8	5.1
Ireland	283	144	-49	27.0	12.7[2]
Italy	3878	1504	-61	20.1	7.5
Luxembourg	12	6	-50	8.7	3.0
Netherlands	289	289[2]		6.2	4.6[2]
Portugal	984	805[2]	-18	29.2	11.7
UK	806	547	-32	3.3	2.2
EUR 12	16,230	7,198	-56	13.4	5.6

[1]inc. hunting, forestry, fishing.
[2]pre-1993.

Source: CEC, 1994.

The economic fundamentals that explain the decline in numbers employed in agriculture in developed countries are discussed in many texts (including Chapter 8 of this one) and need not be elaborated on at length. Briefly, the relatively static demand for farm output in expanding developed economies, unlike the increasing demand in other sectors, combined with steady advancements in technology (which expand farm supply) result in agriculture accounting for a steadily declining proportion of the overall economy. The emerging income disparity between farming and non-farming is alleviated by the steady decline in numbers employed in agriculture. The interesting question is the extent to which this inevitable adjustment is modulated and made less painful by policies such as the CAP.

For much of the period since 1962 the CAP has been dominated by a market support policy involving relatively high food prices supported by variable import levies, with budgetary provision to subsidise exports and some internal consumption. While price support policy under the CAP undoubtedly eased farm income pressures in general, such provisions have probably aggravated income and wealth disparities within agriculture, one of the issues underlining the CAP reform debate discussed in Chapter 4. Five general factors are involved (OECD, 1994). Support policies have concentrated on commodities typically produced on large farms. Given significant economies of scale, support levels based on median costs have provided bigger margins to larger producers. As the elasticity of supply tends to be higher in the larger commercial part of the farm sector, larger suppliers have been better placed to expand their output in response to revenue incentives. As the equity base per hectare rises with size, the size of farm is an important determinant of its capacity to secure finance to expand. With the benefits of price support partially or wholly capitalised into land values, large farms obtain a disproportionate share of the benefits arising.

In EU terms the disproportionate large farm advantage from the CAP was succinctly expressed in the widely discussed estimate that 80% of CAP spending was going to about 20% of farmers (EC, 1994), primarily those with least need of extra funds in welfare terms. What has been called in this book the 'new' CAP, post Mac Sharry reform and GATT Uruguay Round, involves an era of support based much more on administrative measures rather than the market. With administration structures down to individual farm level in place, it is potentially possible to achieve much better targeting of aid. While too early yet to make an overall judgement, it will be interesting to compare the 'new' and the 'old' CAP in terms of their comparative ability to alleviate hardship in the inevitable process of continuing structural adjustment in farming.

A number of model-based studies have attempted to simulate the effect of CAP reform on employment in EU agriculture. To take one example, the EC Agricultural Model (ECAM) provided estimates of agricultural employment under three scenarios, the reference run i.e. pre Mac Sharry reform, the Mac Sharry reform and a decoupled[1] Mac Sharry reform. It was estimated that the Mac Sharry reform would not change the rate of decline in agricultural employment relative to the reference run, with each estimated to result in an annual decline of 2.6% in the decade to 2002, compared with 2.8% in the previous decade (Folmer *et al*., 1994). However, it was estimated that a decoupled Mac Sharry scenario would result in a greater rate of annual decline of 2.9%.

Changing farm structure

Total EU utilisable agricultural area has declined very slowly over time, with some loss of land to urbanisation and related use in wealthier regions and with increased afforestation and land abandonment occurring mainly in poorer regions.

While the size of agricultural holdings (over 1 ha) has increased gradually to over 17 ha (EU 12), the dominant feature in the EU is the enormous variation in size of holding between countries. At one extreme is the UK with over three times the EU (12) average, which in turn is up to ten times the average holding size in Greece, Italy and Portugal, which range from 5 to 10 hectares. In the southern member states generally, around four-fifths of all holdings are less than 10 hectares, while in some northern states 20-40% of all holdings are greater than 50 hectares (Table 10.2).

Table 10.2: Size structure of agricultural holdings.

	% of holdings in each size group		
	<10 ha[1]	10 - 50 ha	>50 ha
Germany	46	47	7
France	32	49	19
Italy	84	14	2
Netherlands	44	51	5
Belgium	45	48	7
Luxembourg	29	41	30
UK	24	42	34
Denmark	17	64	19
Ireland	25	64	11
Greece	89	11	-
Spain	73	21	6
Portugal	88	10	2
Total Twelve	66	27	8

[1]excluding holdings <1 ha, some countries <2 ha.

Source: CEC, 1994.

The change in average holding size since 1975 has tended to aggravate the unequal structure in the EU with the biggest increases in size occurring in Denmark and the Netherlands and the smallest increase occurring in Italy (Larsen and Hansen, 1994). Overall, the data clearly indicate the limited land resource available from which to obtain a satisfactory livelihood from agriculture in much of the EU and the consequential need for increased farm pluriactivity and rural development if a policy objective of achieving regional balance and limiting migration to urban areas is to be achieved.

Output and productivity

Overall EU agricultural output has continued to grow, with about a 20% increase in the total volume index over the last 20 years. With the steadily declining labour force there has therefore been a continuing growth in the capital/labour ratio. In reviewing EC and other developed countries' agricultural sectors, OECD estimates show that this process was more rapid in the 1960s and 1970s than in more recent times (OECD, 1994). Farming practices have been revolutionised by technological progress (as discussed in the previous chapter) with the result that agriculture is now generally a capital intensive sector. There is, however, great disparity, with developed economies generally displaying a growing divergence between two groups of holdings. One comprises the large, commercially oriented farms, while the other small-farm sector generates a relatively small volume of output and income from farming, with the households relying on non-farm sources, commercial or welfare, for supplementary income.

The uptake of new technology and workforce performance are very much influenced by the age and educational status of farm operators. Detailed comparisons show that the agricultural workforce is far less formally educated than the industrial or service sectors (OECD, 1994). Notwithstanding that informal learning and on-the-job experience may be particularly important in farming, savings from reduced EU budgetary expenditure on price support over time could generate good returns if invested instead in human capital. Demographic patterns show that the average age of the agricultural workforce is considerably older than the overall workforce in the EU and other developed countries, with over half of all farm operators over 55 years (Eurostat, 1991). The continuing development and active promotion of trained young farmer establishment, allied with the farm retirement scheme under CAP reform, may play a vital role in ensuring that a cohort of farmers emerges which is fully equipped to compete internationally in the highly knowledge-intensive activity which modern farming has become.

Income from farming

Sectoral income is commonly measured by net value-added (NVA) per agricultural work unit (AWU). There has been a long term average growth rate of close to 1.5% in real NVA per AWU in European agriculture since 1973, with a somewhat slower and more fluctuating growth rate of 1.2% from 1980–1982 to 1992–1994. Overall, it can be concluded that, for about a 20 year period until the 1990s, the decline in the agricultural workforce has been sufficient to keep real agricultural incomes on a slight and rather unsteady upward trend. However, long term average GNP per person employed has been increasing at a higher rate of about 2% per annum, so that incomes from farming per person employed in agriculture have been falling in relative terms (EC, 1994).

With regard to the future, agricultural value added, which can be taken as a rough indicator for farm income, was estimated using the ECAM model mentioned earlier for three policy scenarios, the reference run (pre-CAP reform), the Mac Sharry reform and a decoupled Mac Sharry reform. It was estimated that the Mac Sharry reform would result in a lower rate of decline in real agricultural value added relative to the reference run in the decade to 2002, with the reference run estimate of -0.9% per year reduced to -0.5% per year under Mac Sharry reform (Folmer *et al.*, 1994). Indeed, a decoupled Mac Sharry reform was estimated to result in a marginally increasing rate of agricultural real value added to 2002. Finally, in sectoral terms, the estimated positive effect on real value added in agriculture of the Mac Sharry reform was associated with the livestock sector rather than the crop sector (Folmer *et al.*, 1994).

Variation in income by country is very substantial, with average NVA per AWU in 1992/93 for Portugal, the country with lowest values, being only about one-tenth that of the highest countries, Belgium and the Netherlands (Table 10.3). Variation by country is little changed from that found by Kearney (1991) for 1987/88, with some northern member states having twice or more the EU average range of incomes, and with Ireland, Italy, Greece and Spain below average and Portugal very far down the scale. While some changes in rank occur when income is defined in terms of family farm income per unit unpaid labour, overall variation by country remains very considerable (Table 10.3).

The Farm Accountancy Data Network (FADN) provides a more detailed breakdown of income from farming in terms of NVA per AWU. In 1992/93 the NVA per AWU of the least productive 46% of EC (12) farms was just over 3000 ecu while it was 40,000 ecu (13 times higher) in the class of most productive farms which account for 6% of total farms (Table 10.4). An even larger spread of labour productivity per unit is found in family farm unpaid labour. This variation is little changed from that discussed by Marsh for the year 1987/88 (Marsh, 1991). The distribution of holdings between countries under the FADN classification scheme shows the extent of the regional variation in the EU, notwithstanding the methodological problems. Portugal, for example, has 87% of all holdings in the lowest income class, 0–5000 ecu, in contrast with Belgium which has only about 10% in this class.

A more detailed regional breakdown may be derived from the 166 separate regions identified in the nomenclature of territorial units used in the EU for regional development purposes. The subset of regions with greatest problems, lowest incomes, highest unemployment, etc. are very dependent on agriculture. Kearney has summarised this very detailed regional breakdown in terms of a strongly dualistic pattern involving:

a) the large scale competitive regions including the field crop farming areas of northern France, parts of England and Denmark, the intensive agriculture and horticulture of the Netherlands, Flanders, Emilia-Romagna, parts of northern Germany, Provence and the Mediterranean coast of Spain;
b) the problem areas of Greece, Mezzogiorno (except Abruzzi), Portugal, Spain, Ireland and Northern Ireland (Kearney, 1991).

Table 10.3: Income variation by country, 1992/93.

	Net value added per annual work unit		Family farm income per unit unpaid labour	
	'000 ecu	EU12=100	'000 ecu	EU12=100
Belgium	24.8	213	20.3	220
Denmark	20.9	280	10.6	115
Germany	14.9	128	10.3	112
Greece	5.0	43	4.8	52
Spain	10.7	92	10.9	118
France	19.2	166	14.3	155
Ireland	13.2	114	11.3	123
Italy	9.3	80	8.7	95
Luxembourg	20.7	178	18.5	201
Netherlands	24.0	207	12.8	139
Portugal	2.2	19	1.7	18
United Kingdom	22.4	193	20.3	220
EU 12	11.6	100	9.2	100

Source: CEC, 1994.

Table 10.4: EC holding size and family income 1992/93.

Class of income in '000 ecu	No. FADN holdings '000	% holdings	NVA per AWU '000 ecu	Family f'rm income 1000 ecu
0 – 5	1908	46	3.1	0.8
5 – 10	907	22	8.7	7.3
10 – 20	814	20	16.2	14.2
20 – 30	261	6	25.7	24.3
>30	267	6	39.7	52.4
All holdings	4157	100	11.7	9.2

Source: CEC, 1994.

In relation to policy, the 'old' (pre Mac Sharry reform) CAP, while maintaining average income primarily through price support policy, did comparatively little to alleviate the relative disadvantage of the weaker regions as listed above. A major challenge for the 'new' reoriented CAP, with its administrative capability of precisely targeting regions of greatest disadvantage, is to help reduce regional disparity. It will be very interesting over the coming years, when the 'new' CAP will have been in operation for sufficient time to have had a real effect, to examine whether reduced regional disparity is in practice being achieved.

Sarris has concluded that in agricultural terms the southern EU will lose relative to the northern EU under the Mac Sharry reform. In particular, he indicated that southern products, such as sheep and tobacco, would lose relative to the large farm northern products such as milk, livestock and cereals. However, it was concluded that the reforms would result in proportionately higher consumer gains in the south, so that the estimated overall net gains for both regions are almost exactly proportional to both regions' total GNP (Sarris, 1994).

Pluriactivity

While the above discussion involved income from farming, many of those who practice farming are also involved in pluriactivity (engaged in off-farm activities which increase total income) or part-time farming (dedicating less than a conventionally defined number of hours/days per year to the operator's farm). While noting the definitional problems in making comparisons between countries or in combining data and allowing for the unavailability of recent data, available estimates show that up to 30% of farmers in the EU (12) have other gainful employment, with a particularly high estimate of 43% for Germany (CEC, 1993). For many regions, and particularly in low income regions with a high dependence on agriculture, off-farm employment often involves employment on another farmer's holding or in some activity associated with agriculture.

Pluriactivity can make a major contribution to alleviating the low farm income problem. Gasson (1988) cites numerous studies which show that part-time farmers, on average, earn high incomes compared with full-time farmers and compared with other rural residents. The availability of non-farm income to a farmer and/or spouse may reduce the opportunity cost of capital to the farm household, thus enabling farm investment activities to complement labour-substitution by other household members and adjustments towards less labour-intensive activities so that increases rather than decreases in farm output and income can actually be associated with pluriactivity (Lucey *et al.*, 1987). Other benefits listed by OECD, especially if it involves employment outside agriculture, include the lowering of exposure to farm sector events (OECD, 1994). This may be quite important with respect to adjustments stemming from policy reform. Part-time farming can generate income to enable households new to farming to establish themselves as farmers and to increase their size of holding, with pluriactivity being a feature of households new to farming (Bryden *et al.*, 1990), or ease the transition out of farming.

Two features of farming are the dominance of owner-operators and the overlap of place of work and place of residence. The owner-operated family farm is the EU norm. While the data are somewhat outdated, less than 8% of the total employment in agriculture in the EU are regularly employed non-family members, with the UK at about one-third being the only major exception (CEC, 1994).

The organisational arrangement of units of farms and households in agriculture has been seen as a consequence of limited economies of size relative to the size of the family's labour capacity. In an interesting discussion drawing on the transaction costs approach, it was noted that economies of size were increasing; however, due to labour saving technological innovations, labour capacity was also increasing. The persistence of smaller than technically optimal farm sizes was explained by lower transaction costs of family farms *vis-à-vis* hired labour farms (Schmitt, 1991). Given that the family owner-operator normally resides in a house located on the farm, the perceived costs of leaving the sector are magnified. Studies in the US show that farmers, when faced with severe economic hardship forcing them to abandon farming, demonstrate a great reluctance to abandon their rural communities (Saupe and Bentley, 1990). Several low income farming areas do not have many non-farm employment opportunities within reasonable commuting distance. This indicates the desirability of supporting off-farm employment where existing farmers reside. This is of particular importance in those depressed low-income regions discussed earlier, where large further outflows from agriculture are inevitable.

Moreover, selectivity factors such as age, education and availability of substitute household labour may influence which farm households may be in a position to seek or to be offered whatever level of employment is available in any local area (Lucey and Kaldor, 1969). Special training/educational programmes oriented at enhancing the capacity of low income farmers to engage in pluriactivity may need to become key components of enhancing the vitality of rural communities.

At policy level it is fully recognised that the diversification of economic activity in rural areas into other sectors, whether or not connected with agriculture, is an essential condition for job maintenance and/or new job creation and therefore for slowing down the flight from the land. Thus the new regional measures have "widened the scope of the CAP Guidance section into such fields as improvement of rural living conditions, the renovation of villages, a policy of quality products and product promotion, support for applied research and financial engineering measures" (EC, 1996). The implementation of these measures puts considerable emphasis on the need for farmers and others who live in rural areas to work co-operatively to achieve rural development goals. The LEADER II Programme is a good example.

Co-operation between farmers

Farmers are very small, isolated rural entrepreneurs who have seen their numbers shrink rapidly over time, as discussed earlier. Like all parties engaged in similar activity, they have combined successfully into associations to exert political influence, an influence which is still widely acknowledged to be considerable in most EU countries, and in the EU in total through the Committee of Agricultural Organisations in Europe (COPA). The continuation of this influence, despite their decline in numbers, may be due to their continuing common need to mitigate the consequences of free market economics. Thus in terms of public choice theory, political 'rent-seeking' through organisation in an interest group gives an economic pay-off which counteracts the difficulties in the market. The focused and concentrated issues and representational structures of farmers contrast markedly with the much weaker and more amorphous consumer lobby. While this political dimension to farmers in terms of representation to government is not further discussed here, readers are referred to an interesting recent article on the subject drawing on public choice theory by Nedergaard (1994) and to the discussion of this issue in Chapter 8 of this book.

As farms are extremely small businesses involved in the market in buying inputs and selling often perishable and poorly timed and located food raw materials, the emergence of powerful suppliers and buyers through growth and merger is a development of profound consequence for farmers. Traditionally, farmers have formed producer co-operatives to seek some control over their market activities, and the need would now seem to be even greater from their viewpoint due to the changing structure of the markets in which they operate.

Farmers' co-operatives have attempted to redress the market power imbalance between farmers and proprietary food processors whose activities have become concentrated so that they act like oligopolies or even locally like monopolies. Farmers' co-operatives have also integrated their activities forward in the food business chain by processing and distributing food products manufactured by themselves from the farm produce of their members. They thus try to attract to the co-operative as much as possible of the added value in the food business chain. By doing so, the strategic position which the co-operative adopts in the food business chain is strengthened, which may be crucial to retain access to markets, on behalf of its members. In addition, a strong position in the chain facilitates a clearer profile in the market and may even result in capturing higher profit margins in some cases (Verheijen and Heijbrock, 1994). Farmers have, for similar reasons, through their co-operatives, organised themselves to exert market power relative to suppliers of farm inputs and have also integrated backwards into the manufacture and distribution of some of these inputs. Contrasting sets of favourable and unfavourable competitive scenarios for farmers in the food marketing channel from consumers to farm inputs have been discussed by Keane (1986).

The proportion of agricultural products sold through co-operatives in the member states of EU (12) in 1992 is shown in Table 10.5. The importance of co-operatives varies considerably among member states, being generally strongest in Denmark, France, Ireland and the Netherlands. Similarly, milk, fruit and vegetables generally are the farm products with highest co-operative involvement, while cereals are also high in some countries. The position regarding meat is variable with a high co-operative presence in pigmeat in Denmark, France and Ireland, a variable presence in beef/veal ranging from 60% in Denmark downwards and a low presence in poultry meat ranging from 30% in France downwards. Of course, it must be remembered that co-operatives do not have to dominate any national market to have a significant influence on price formation. Any group with 15-20% market share can have such an influence.

The key role of co-operatives in co-ordination of transactions along the food business chain is stressed by Ollila (1989), who shows that a well functioning co-operative is able to solve many marketing problems that neither the market nor an administration can. This refers particularly to a co-operative's role in market intermediation and relaying of signals to its members to guide their farm production decisions regarding volume, seasonality, quality or composition of farm produce.

Co-operatives can, of course, also play a strong role in the general development of rural areas, not merely the agricultural and food aspects. This is especially the case for smaller rural communities. A Canadian study concluded that co-operatives in smaller communities play a more important role in providing competitive prices and in providing services that otherwise would not be provided than do co-operatives in larger communities (Fulton and Hammond Ketilson, 1992). If such is also the case in Europe, then co-operatives of various kinds may have a key role in future rural development in low-income farming communities.

Table 10.5: Percentage of agricultural products sold through co-operatives in the EU-12 (1991 or 1992).

	Pig-meat	Beef/ Veal	Poul -try	Eggs	Milk	Sugar-beet	Cer-eals	All fruit	All veg.
Belgium	15	1	-	-	65	-	25-30	60-65	70-75
Denmark	95	59	0	59	92	0	48	90	90
Germany	23	25	:	:	56	:	:	20-40	55-65
Greece	3	2	20	3	20[1]	0	49	51	12
Spain	5	6	8	18	16	20	17	32	15
France	80[2]	30[2]	30	25	50	16[3]	70	45	35[4]
Ireland	55	9	20	0	98	0	26	14	8
Italy	10	13	0	8	33	0	20[5]	40	13
Lux.	35	25	-	-	81	-	79	10	-
Neth.	25	16	21	17	82	63	65	78	70
UK	20	5.1	0.2	18	4	0.4	19	31	19

(1) Cows', ewes' and goats' milk.
(2) Finished animals: young cattle not included 70%: Store animals not included 40%.
(3) Processed into sugar.
(4) Excl. potatoes (seed potatoes, 65%; early potatoes and ware potatoes, 25%).
(5) 28% maize.

Source: The Agricultural Situation in the Community (1994).

The future for farmers

While overall decline in numbers will inevitably continue as part of the economic adjustment process, policy change involving a gradual erosion of price support may alter the strategic approach to their business for many farmers. In the past, with substantial price supports applying to a limited number of staple products, a low cost strategy predominated. Applying the approaches to industrial development promoted by Porter (1990), the future may involve much more emphasis on diversification away from standardised products towards price responsive market opportunities. These could include, for example, higher value-added regional food products or very high quality environmentally sensitive products. Despite the price supported staple product ethos, some regions, particularly in France and Italy, have maintained traditional regional production based on a favourable product image sufficient to maintain a substantial premium over basic standardised products. Thus the components of the creation of competitive advantage for non-farming businesses, such as specialisation and vertical integration along the value added chain as suggested by Porter, may also increasingly apply to farming. Farm families may also become steadily more involved in non-farming activity as rural development is fostered, and participate in the many aspects of rural tourism, craft industries and related activity. Because individual farmers are small isolated rural entrepreneurs, it will be necessary to encourage this development through the growth of co-operative or mutual support structures and through the provision of appropriate educational and training modules. As government price support for stable products lessens, increased investment in research, education and extension to underpin rural development may be the key to maintaining a viable farming sector within vibrant rural communities in the future.

References

Bryden, J.M., Bell, C., Fuller, A.M., MacKinnon, N., Salant, P. and Spearman, M. (1990) Emerging Responses of Farm Households to Structural Change in European Agriculture. *Paper to VI European Congress of Agricultural Economists*. The Hague, Netherlands.

Commission of the European Communities (CEC), various years. *The Agricultural Situation in the Community*. Brussels.

European Commission (1994) EC Agricultural Policy for the 21st Century. *European Economy Reports and Studies*. No. 4. Brussels.

European Commission (1996) *The Agricultural Situation in the European Union*. Brussels.

Eurostat (1991) Farm Structure, 1987. *Survey: Main Results*. Luxembourg.

Folmer, C., Keyzer, M.A., Merbis, M.D., Slovhvijk, H.J.J. and Veenendaal, P.J.J. (1994) Common Agricultural Policy reform and its differential impact on member states. In: *European Economy, Reports and Studies*. No. 5.

Fulton, M. and Hammond Ketilson, L. (1992) The Role of Co-operatives and Communities: Examples from Saskatchewan. *Journal of Agricultural Co-operation*, 7, 15-42.

Gasson, R. (1988) *The Economics of Part-Time Farming*. Longman Scientific and Technical, Harlow, United Kingdom.

Keane, M. (1986) Food Marketing, Competition and Co-operatives. *Working Paper No. 10*. Centre for Co-operative Studies, University College Cork.

Kearney, B. (1991) Rural Society - Disparities in Incomes and Alternative Policies. In: *The Changing Role of the Common Agricultural Policy*. Belhaven Press, London, 126-137.

Larsen, A. and Hansen, J. (1994) Agricultural support and structural development. In: *European Economy, Reports and Studies*. No. 5.

Lucey, D.I.F. and Kaldor, D.R. (1969) *Rural Industrialisation*. London. Geoffrey Chapman Ltd.

Lucey, D.I.F., Walker, S., Cuddy, M.P., Doherty, M. and Lyons, D. (1987) *New Jobs in Mayo - A Study of Recent Major Employment Developments*. Report prepared for Mayo County Development Team, Castlebar, Ireland.

Marsh, J. (1991) Initial Assumptions. In: *The Changing Role of the Common Agricultural Policy*. Belhaven Press, London, 2-24.

Nedergaard, P. (1994) The Political Economy of CAP Reform. In: Kjeldahl, R. and Tracy, M. (eds) *Renationalisation of the Common Agricultural Policy?* APS, Belgium, 85-103.

Ollila, P. (1989) Co-ordination of Supply and Demand in the Dairy Marketing System. *Journal of Agricultural Science in Finland*, 61(3), 135-317.

OECD (1994) *Farm Employment and Economic Adjustment in OECD Countries*. Paris.

Porter, M. (1990) *The Competitive Advantage of Nations*. MacMillan, London.

Sarris, A.H. (1994) Consequences of the proposed Common Agricultural Policy reform for the southern part of the European Community. In: *European Economy, Reports and Studies*. No. 5.

Saupe, W. and Bentley, S. (1990) Adaptation of Family Farms to Economic Shocks in the Midwest of the USA: A Longitudinal Farm Household Survey. *Paper to VI European Congress of Agricultural Economists*, The Hague.

Schmitt, G. (1991) Why is the Agriculture of Advanced Western Economies Still Organised by Family Farms? *European Review of Agricultural Economics*, 18(3/4) 443-458.

Verheijen, J.A.G. and Heijbrock, A.M.A. (1994) Co-operatives in Changing Market Conditions. In: Hagelar, G. (ed) *Management of Agrichains*, 168-174. Wageningen Agricultural University.

---

[1]i.e. where compensatory payments can be made in a manner fully decoupled from producer decisions and are, as a consequence, not contingent on any set-aside requirement.

Chapter 11

The CAP and the Consumer

Christopher Ritson

Introduction

In view of the considerable media attention devoted to the impact of the Common Agricultural Policy on consumers, it is perhaps surprising that such a small proportion of the academic work on the CAP should have been directed specifically at the consumer interest. It is therefore instructive to begin by quoting Adam Smith:

> Consumption is the sole end and purpose of all production; and the interest of the producer ought to be attended to, only so far as it may be necessary for promoting that of the consumer. The maxim is so perfectly self-evident that it would be absurd to attempt to prove it. But in the mercantile system, the interest of the consumer is almost constantly sacrificed to that of the producer; and it seems to consider production, and not consumption, as the ultimate end and object of all industry and commerce.
>
> (Adam Smith, 1776).

Substitute 'CAP' for 'mercantile system' and the author could be an agricultural economist writing 200 years later!

In this chapter, I first briefly consider the nature of the 'consumer lobby' relating to the CAP. Then I attempt to identify various aspects of consumer welfare which, in principle, might be affected by the CAP. Third, I consider the objectives of the CAP in the context of the consumer. Finally, the CAP market policy is assessed relative to the various consumer interests.

It is generally accepted that the consumer voice in Europe, insofar as agriculture is concerned, is weak. A former Permanent Secretary in the British Ministry of Agriculture, who was previously a senior official in the European Commission Agricultural Directorate, comments:

> The consumer lobby which has in this country been fairly vocal about the demerits of the CAP has never been a potent force on the Continent. ... When I joined the Commission in 1973, DG-VI made no pretence to be other than a Directorate General concerned with Agriculture. There was — and still is — a division in DG-III concerned with the food industry, but the notion that it could have any significant impact on the policies of DG-VI was, and I suspect still is, laughable. Over the years, the Commission has gradually built up a directorate of consumer affairs, but it has chosen to concentrate its efforts on other consumer issues and its peashooters have little effect on the armoured divisions of the CAP.
>
> (Franklin, 1994).

Only belatedly can one identify the consumer interest as having a major impact on agriculture at a European Union (EU) level, and this has been in the context of food safety — in particular decisions relating to beef hormone implants, the use of BST in milk production, listeria in cheese, salmonella in poultry, and most dramatically, the 1996 BSE beef crisis. However, where decisions have been taken by the Council in these areas, although they may have had a major impact on agriculture, and indirectly on the CAP, they have usually not been CAP policy decisions as such. Indeed, it is difficult to cite major examples of CAP decisions which have been influenced predominantly by the consumer interest.

One reason for this is perhaps that, even when 'vocal' about the CAP, the consumer voice in Europe has been poorly focused. One can criticise the approach adopted by the consumer lobby with respect to the CAP in three specific ways. First, often it seems to be the wrong thing which is criticised. For example, much attention is devoted to 'food mountains', and in particular to the destruction of food, and also to budgetary expenditure. But, arguably, visible food surpluses are only symptoms — often rather minor symptoms — of the real consumer cost of the CAP. Just what this is I explore below, but it is worth giving one example of this particular problem at this point.

The front page headline story in a popular Scottish Sunday newspaper in April 1994 went as follows:

> **WHAT A WASTE**
> The scandalous destruction of vast quantities of fruit and vegetables in the European Common Market has shocked MPs. In a single year, European growers destroy enough fruit and veg to keep a city the size of Glasgow going for years.

It is not difficult to provoke outrage in an audience on this issue. Pictures of cauliflowers or lemons piled up, waiting to be destroyed, raise strong emotions — particularly when it is pointed out that 'they can only be destroyed if a "Common Market inspector" certifies that the produce is of a sufficiently high quality to be destroyed!'

However, in my view, the destruction of 'surpluses' of fruit and vegetables can be justified, even from a consumer perspective, rather more than many other aspects of the CAP. Production of fruit and vegetables is highly seasonal and varies substantially from year to year. Demand can be quite inelastic. Produce cannot be stored (or at least is very costly to store). In these circumstances it does not make good sense to allow market prices to fall to ruinously low levels for producers in times of glut. It is quite sensible, from the point of view of *all* sections of society, if small quantities are removed from the market in such periods, and the most efficient course of action may well then be to destroy the produce; and the produce destined for destruction must be of an acceptable quality, or the EU would find itself compensating farmers for produce which would never have found a commercial market (and consequently would not have depressed prices).

Second, and following from the above, it seems that in its more popular manifestations the 'consumer view' of the CAP becomes obsessed with the idea that the policy represents some kind of great evil, which has an adverse effect on the lives of 'ordinary people' in every conceivable way. This view has particularly manifested itself in the context of interest in diet and health, where the popular media view of the CAP seems to be that, not only does it mean expensive food, but that it also has a severe adverse nutritional effect. For example, the Sunday newspaper story mentioned above went on to quote an MP as pointing out:

> the health loss involved, since fruit and vegetables are essential to healthy eating.

I shall return to this issue below.

Third, I think that the consumer interest in the CAP is often defined far too broadly to be helpful. For example, the following is an extract from the terms of reference of a report published by the British National Consumer Council entitled 'Consumers and the CAP':

> The CAP affects people primarily as consumers of food but also as consumers of the countryside and as taxpayers.
>
> (National Consumer Council, 1988).

Thus, often the so-called consumer lobby seems to be interested in the impact of the CAP on consumers, not just as consumers of food, but also as taxpayers and as 'consumers of the countryside'. In practice this means regarding the consumer interest in the CAP as comprising virtually every aspect of national welfare influenced by the CAP, other than the producer interest. The argument then becomes very diffuse. It is much more helpful, in a discussion of the CAP and the consumer, to restrict the issue to that of consumers *as consumers of food*. That is the approach taken in this book, where other chapters are concerned with aspects such as the CAP and the environment, and taxpayer and budgetary aspects of the Policy.

The consumer interest in European agricultural policy

It is possible to identify four main aspects of (food) consumer welfare which are potentially influenced by the CAP.

Food prices and expenditure

Most obviously, the policy influences the prices paid by consumers for individual food items and thus the proportion of total consumer expenditure devoted to food.

Food availability

Consumers now expect a wide range of foodstuffs to be available in the shops throughout the year.

Food security

As consumers, we are interested in minimising the possibility that basic foodstuffs may suddenly become unavailable, or so expensive that a substantial proportion of the population cannot afford an adequate diet.

Food quality

This is perhaps the most difficult of the four areas to define and it embraces a variety of aspects of consumer welfare. The provision of a range of qualities of individual food items should be covered by the issue of food availability. But there is also a need for consumers to be informed adequately concerning what they purchase, and it is usually accepted as a responsibility of government to legislate to control food safety, in terms of health and hygiene, and food additives. Finally, there is the question of the overall nutritional balance of the diet, and what have become known as the 'diseases of affluence'.

The objectives of the CAP

Table 11.1 lists the objectives of the CAP as specified in Article 39 of the Treaty of Rome and often referred to in this book. The right-hand column makes an informal estimate of the extent to which these objectives might be regarded as either producer-oriented (P-O) or consumer-oriented (C-O).

In the case of the first objective, agricultural economic analysis shows that, typically, the characteristics of agricultural product markets are such that the major part of the benefit of productivity improvement will be passed on to consumers. This issue is discussed in more detail in Chapters 8 and 9, but I have allocated the first objective as being 75% consumer — and only 25% producer-oriented.

Table 11.1: The objectives of the CAP.

		C -O	P-O
(a)	To increase agricultural productivity by developing technical progress and by ensuring the rational development of agricultural production and the optimum utilisation of the factors of production, particularly labour:	75%	25%
(b)	to ensure thereby a fair standard of living for the agricultural population, particularly by the increasing of the individual earnings of persons engaged in agriculture;	0%	100%
(c)	to stabilise markets;	40%	60%
(d)	to guarantee regular supplies;	100%	0%
(e)	to ensure reasonable prices in supplies to consumers.	100%	0%

Source: Article 39, Treaty of Rome (HMSO, 1962).

The second objective is, however, unambiguously a producer-oriented objective. The error made by the architects of the CAP, of course, was to believe that objective (b) could be achieved via objective (a). In practice, as we saw in Chapter 9, the contrary has been the case. The very success of European agriculture in improving its productivity has meant downward pressure on market prices. This, in turn, has made it more difficult to achieve a fair standard of living for the agricultural population — certainly when viewed in the context of rising incomes elsewhere in the economy.

It is generally accepted that the severe instability which characterises many agricultural markets, if left uncontrolled, is undesirable. Therefore looking at the objective of stability, the benefits are shared between producers and consumers. There are, however, theoretical arguments which suggests that, in certain circumstances, consumers might benefit from fluctuating food prices. This thesis, in my view, is rather weak (Ritson, 1985) but, because of it, I have allocated this objective 60–40 in favour of producers.

The objective of guaranteeing regular supplies is closely associated with what I have specified as 'food security' and 'food availability'. It is thus a consumer objective, as is, of course, the objective 'to ensure reasonable prices in supplies to consumers'.

Adding these figures up and dividing by five leads to the conclusion, albeit rather crudely, that the objectives of the CAP are biased, by nearly two to one, in favour of consumers. It is therefore difficult to criticise the CAP from a consumer perspective, merely by looking at its objectives. To assess the Policy genuinely we must consider what it does, not what it states it would like to achieve.

The CAP market policy

In practice, the CAP consists predominantly of a complicated set of regulations which control the marketing of agricultural produce, mainly the prices at which produce is marketed at the wholesale level. A description of these mechanisms was given in Chapter 1 (illustrated in Fig. 1.1), but to summarise, typically the controls have involved:

a) A tax on imports to prevent them selling below a predetermined price — a tax which varies so as to bridge the gap between international prices and this predetermined (or minimum import) price.
b) Intervention in the market to prevent internal prices in Europe from falling beneath a floor level, usually set a little below the minimum import price.
c) Subsidies on exports to bridge the gap between the floor (intervention) level and the prices at which produce can be disposed of on world markets.

The 1994 GATT agreement has constrained somewhat the size and scope of import levies and export subsidies, but the upshot of all three of these mechanisms taken together is that prices of agricultural products within the EU — and thus food prices to consumers — are kept above, usually substantially above, prevailing international prices. Let us now consider how these regulations, and some other aspects of the CAP, influence the four aspects of consumer welfare identified above.

Food prices and consumer expenditure

The CAP does not directly control the prices paid in the shops by consumers for food products, but influences the cost of agricultural raw materials — and most relevant academic work has concentrated on the gap between international and EU agricultural commodity prices. Table 11.2, however, provides some comparisons of UK and international food prices published by British consumer organisations.

The products listed in Table 11.2 have of course been selected by consumer organisations because they are foods for which, typically, the gap between UK wholesale prices and international prices has been rather high. They are also all products where what the consumer buys is recognisably similar to the agricultural commodity for which CAP mechanisms, such as intervention, import levies and export subsidies, apply. This means that price gaps can be related to the size of levies and subsidies on the related agricultural commodities, but selective examples like these do not provide a complete picture of the impact of the CAP on food prices and consumer expenditure.

According to the Consumers Association (1988), 'the CAP costs every man, woman and child in the EC £110 per annum as a consumer and £59 per annum as a taxpayer'. In his study for the Trade Policy Research Centre, Hubbard, using the 'Newcastle CAP Model', estimated that:

Table 11.2: Examples of food prices.

Food Product	UK Wholesale Price		International Price	
	1988	1993	1988	1993
Butter 250g pack	47p	62p	19p	22p
Beef (Topside) kg 1988 (Rump) lb 1993	178p	216p	112p	170p
Cheddar Cheese kg	na	245p	na	135p
Rice (American Long Grain) kg	65p	na	38p	na
Sugar kg	37p	53p	12p	18p

na = not available.

Source: 1988: (Which, April).
1993: (Consumers in Europe Group, 1994).

> ...consumer expenditure on food in the United Kingdom in 1983 totalled £32,000 million. The consumer cost of the CAP, estimated at £3,890 million, effectively represents therefore an explicit tax on food at around 14%.
>
> (Hubbard, 1989)

Even allowing for inflation during the 1980s, at about £70 per person this is rather lower than the Consumers' Association estimate. More recently, the very substantial work on agricultural policy effects undertaken by OECD (1993) has produced estimates which are a little higher — a consumer cost of about 200 ecu per head in the European Community of Twelve in 1993. However, all these estimates broadly substantiate the 'rule of thumb' that others have sometimes applied — that the CAP is equivalent to a value added tax on food of about 15%.

How do we know? Figure 11.1 represents a highly simplified version of the kind of model used in making these estimates (and covered in detail in Chapter 7). Conceptually the problem is quite easy; we simply revalue consumer expenditure on food products at alternative 'world prices' to obtain the consumer cost of the CAP. (We can also estimate the subsidy required to dispose of surplus production, which is the main component of the taxpayer cost.)

In practice, of course, the calculation is complicated and presents difficult theoretical problems. Information on the level of consumption and production is not too difficult to come by and therefore does not provide too many problems. The 'EU price' also can be obtained. There is the problem (already referred to) that the price consumers pay is at retail level, whereas these calculations will usually be undertaken further up the marketing chain, to be comparable with international prices, and they often refer to agricultural commodities (e.g. wheat)

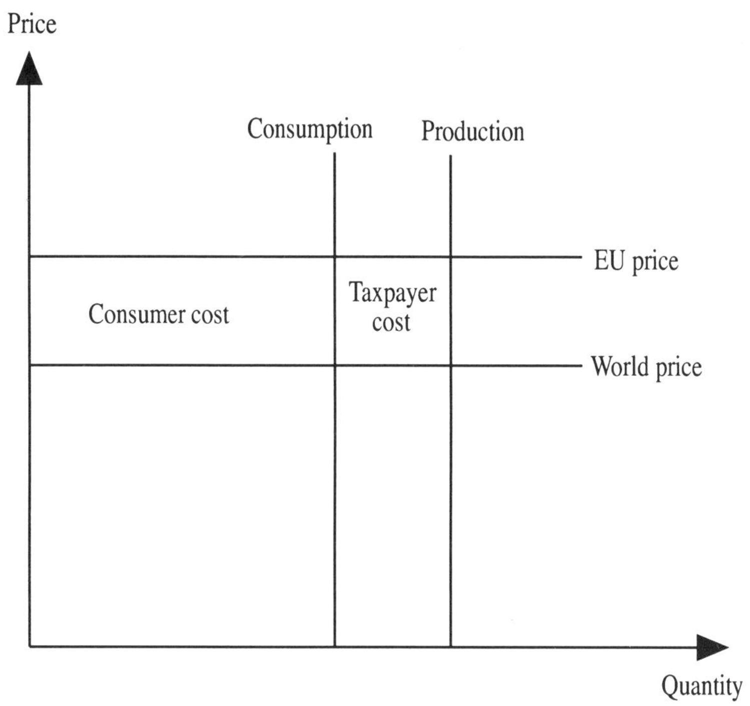

Fig. 11.1: Simplified model for estimating consumer cost of the CAP.

rather than recognisably food products (e.g. bread). It is not certain that the marketing margin (the gap between the retail and wholesale prices) will be independent of the wholesale price, so the whole cost of higher wholesale prices will not necessarily be borne by consumers — but this is the assumption which is usually made. It is also necessary to be cautious in using EU administered support prices, particularly intervention prices, as indicating the actual prices on EU markets, as increasingly they are different. Nevertheless, estimating the level of EU agricultural commodity prices at the wholesale level is a manageable task.

The real problem is the so-called 'world price'. In order to estimate the consumer cost of the CAP we must postulate an alternative scenario against which the present policy can be judged; and, irrespective of this alternative scenario, the so-called 'world price' is likely to be different from the prices currently prevailing on international markets. It is most unlikely to be correct simply to take the prevailing gap between European market prices and international prices, multiply this by the level of consumption, and call this the consumer cost of the CAP.

I have identified five possible alternatives, all of which might legitimately be used to measure the consumer cost of the CAP. These are listed in Table 11.3.

Table 11.3: Alternative scenarios to the present CAP.

1	Abolish agricultural policy in the EU.
2	Country considered withdraws from EU (or at least CAP).
3	Free trade, but sustain producer prices.
4	International movement towards free trade.
5	'Realistic' reform of the CAP.

The most usual alternative scenario is to view the CAP against a situation in which no EU or member state policy influences agricultural markets. Under this alternative, less would be produced in Europe, more would be consumed, and this would have the affect of raising world prices. (This is the assumption which was used in the Newcastle CAP model referred to above in estimating the overall effect of the CAP on producers, consumers and the European Community (EC) budget, and it is also the assumption which underlies Table 8.1.)

The question then arises as to whether we are talking about the consumer cost of *the* CAP or simply the consumer cost of *an* agricultural policy. Looking at the position from that of any individual member state, one can envisage the CAP no longer applying; but it is clearly quite unrealistic to regard the alternative as being no agricultural policy at all. Thus another scenario is the second in the table; that the cost of the CAP is viewed against the alternative policy that a country might be expected to adopt if it withdrew from the EU (or at least from the CAP).

A third possibility is to free trade between the EU and the rest of the world (that is to abolish levies on imports and subsidies on exports), but to sustain prices to farmers by direct product subsidies. This is the scenario which the simple calculation of the current excess of internal prices over world prices most closely resembles. This is because, under scenario 3, producer prices (and thus the level of production) are sustained, and international prices would only be affected by the increase in EU consumption, which would be likely to follow from allowing consumer prices in the EU to fall to world levels.

My fourth alternative scenario is described as 'international movement towards free trade'. Here I have in mind the kind of policy changes which are negotiated under the General Agreement on Tariffs and Trade (GATT) (and now the World Trade Organisation (WTO)), and in particular reciprocal 'decoupling' of agricultural policy and price support (see Chapters 5, 16, 17). If the alternative is one in which prices are allowed to move towards world levels in the course of international agreement in which other major trading countries do likewise, we might expect international prices to rise substantially, thus leading to a lower estimate of the consumer cost of the CAP than that obtained under any of the other alternatives.

The fifth alternative is described as 'realistic reform of the CAP' by which is meant an alternative CAP which might in some sense be regarded as politically feasible. This is the most difficult of the five scenarios to specify, as will be evident from the discussion of CAP reform in Chapter 4, but represents perhaps the most realistic alternative. Indeed the 1992 Mac Sharry CAP reform and the related changes in the CAP as part of the 1994 GATT Agreement can be viewed as steps in the direction of scenarios 4 and 5. The benefits to consumers of food resulting from these changes to the CAP are however modest and this indicates the problem in using scenarios 4 or 5 as ways of measuring the consumer cost of the CAP. If one takes the view that any politically feasible reform of the CAP will retain an element of supporting agricultural product prices above internationally traded levels, is it defensible to use that price level, rather than an estimate of 'world' prices, as a baseline for measuring the consumer cost of the present CAP?

Thus the consumer cost of the CAP (in terms of food prices and consumer expenditure) depends very much on the alternative scenario postulated. However, there is one point on which we can be reasonably certain. All five alternatives imply a net benefit to consumers. In other words, irrespective of which alternative scenario is used to judge the consumer cost of the CAP, all lead to the conclusion that there is a cost; but they all also lead to the conclusion that simply taking the prevailing gap between European market prices and world prices exaggerates that cost.

Food choice and availability

Under 'Choice of Foods' the National Consumer Council report referred to earlier begins:

> So it is clear that the CAP, by distorting the relative prices of various foodstuffs, distorts the choice consumers make when they buy food. However, the CAP also operates in such a way that it reduces the *absolute* amount of choice available to consumers.

The report then proceeds to quote a number of examples of the way consumer choice is restricted, in all cases by distorting prices. It defends this on the grounds that:

> It is meaningless to argue that people have a choice of foods if many of these foods are priced beyond the reach of their pockets.

It is obviously correct that a policy which pushes up food prices will lead some low income consumers to eliminate certain items from their diet. But to make 'food choice' merely another aspect of the effect on food prices and consumer expenditure does not seem very helpful. For 'food availability' to have any sensible meaning, we must be able to argue that the CAP influences the *range* of products actually *available* for sale.

First, it is worth noting that the CAP might have a positive effect. There is the problem, identified when considering the impact on food prices, of the benchmark for comparison. A country which left the EU might adopt a restrictive import policy which prevents supplies from some other European countries reaching its domestic market. The principle of 'Free Inter-Community Trade' (Chapter 1) has certainly facilitated the range of foods available to EU consumers which originate from other member states.

For example, before Portugal joined the EC in 1986, Portuguese agriculture was highly protected. In particular, imports were subject to very restrictive quotas and were only allowed when government officials decided there was a market need (Josling and Tangermann, 1987; Avillez, 1992). Since then Portuguese consumers have progressively had access to a wider range of produce which has generally been cheaper, and often of a higher quality.

Thus the main way in which the CAP influences the range of choice available to consumers is by substituting EU produce for third country produce. However, this usually applies to agricultural commodities in which the nature and quality of the consumer product is not significantly affected by the source of raw material supply. There may be some examples where consumers are adversely affected by having the range of, for example, cheese or rice from non-European supplies restricted. But it is difficult to argue that most EU consumers do not now have available to them a range of produce unparalleled in history.

However, two typical characteristics of fruits and vegetables, compared to most other agricultural products, suggest that the CAP might affect the range of these products available to food consumers, as well as the prices they pay. The first is the very seasonal nature of most production which, together with the fact that the produce is often very perishable, means that these products are particularly susceptible to intermittent seasonal lack of availability. Second is the very heterogeneous nature of the produce, with consumers not necessarily regarding produce from one supply source (because of variations in quality, size and variety, for example) as an acceptable substitute for supplies from an alternative source. These factors mean that it is much more likely that the CAP policy mechanisms will influence the *range of choice* of fruit and vegetables available to consumers than it is with respect to other CAP products.

Consider first the internal policy mechanisms. Earlier, I defended the policy of destroying small quantities of certain fruits and vegetables as not being significantly against the consumer interest. Figure 11.2 shows by way of evidence the proportion of produce withdrawn from the market for three typical products — tomatoes, cauliflowers and apples — for which the withdrawal mechanism has been very active. The fact that the amounts withdrawn from the market average less than 5%, and that there are intermittent peaks between which very little produce is withdrawn, suggests that the withdrawal mechanism is significant only when there is a potential glut of produce on the market. Figure 11.3 illustrates the point for apples, as it can be seen that peaks in withdrawal are associated with peaks in production, with the early 1980s and early 1990s experiencing extraordinarily large swings in the size of the harvest.

Indeed, it is quite probable that, in the occasional years of substantial withdrawal, much of the withdrawn produce would never have been harvested, as experience in a free market suggests that some producers no longer find it profitable to cover the cost of harvesting when a surge in supplies forces down market prices. Thus, it is in the nature of the products and their markets, rather than the extremity of the policy, that these withdrawals occur. For most fruits and vegetables it is similarly the case that only relatively small quantities of produce are withdrawn (Ritson and Swinbank, 1988).

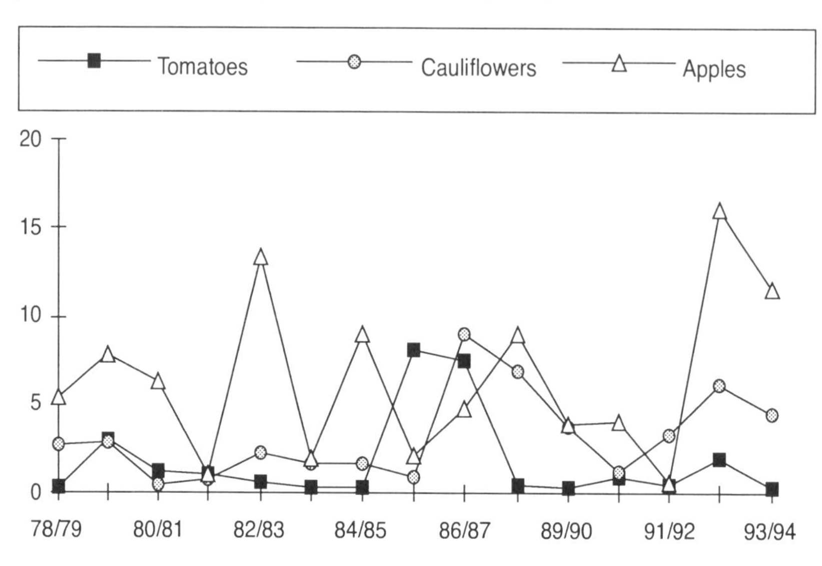

Fig. 11.2: Percentage of harvested production withdrawn — tomatoes, cauliflowers and apples.

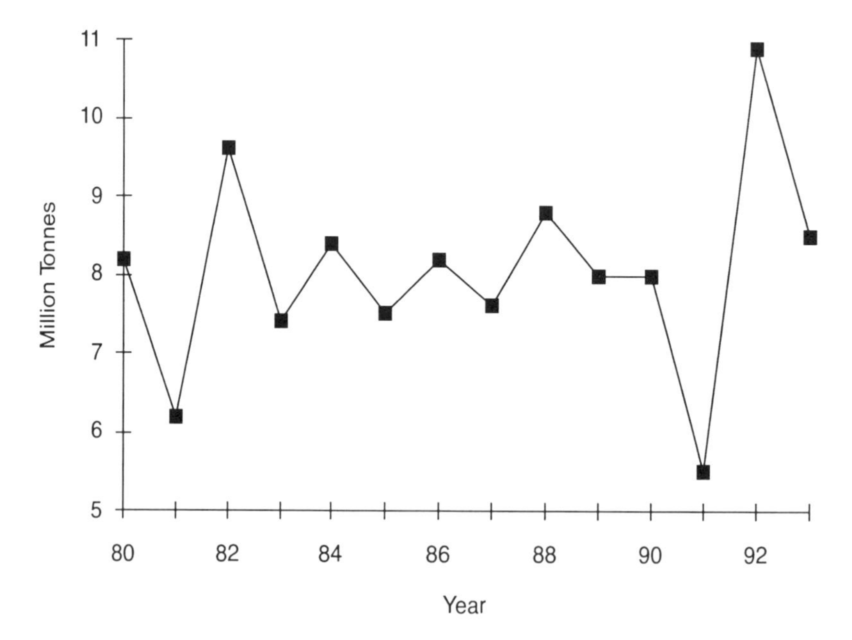

Fig. 11.3: Total production of apples in the EC (1980–1993).

There is one exception to this general observation — peaches (Fig. 11.4). In recent years, persistently, about 20% have been withdrawn from the market and the proportion in Greece has risen progressively to about 60%. It is difficult to argue that availability of *Greek* peaches is not being influenced by the policy — though whether this represents a major problem for consumers in northern Europe is another matter! The other classic case (now apparently ended) of what might be termed 'production for withdrawal' was Italian mandarins (Fig. 11.5). With 80% of Italian mandarins withdrawn from the market during the mid-1980s we clearly have a most peculiar market policy. Even here, however, arguably this was not really significant for consumers. In general citrus fruits are plentiful, but this particular variety had little consumer appeal and the withdrawal from the market was concerned with budgetary expenditure and producer interests, not really denying consumers the benefit of the consumption of a valued product.

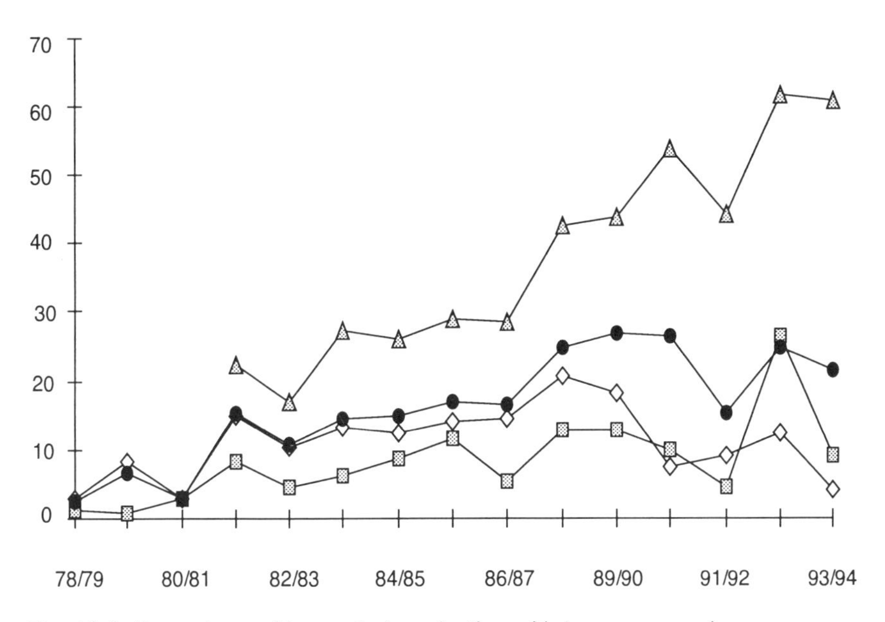

Fig. 11.4: Percentage of harvested production withdrawn — peaches.

Thus the verdict must be that the fruit and vegetable withdrawal system has not had a major impact upon the range, and indeed volume, of foods available to consumers within the EU. It has been necessary to deal with this particular aspect of the Policy carefully, as there is a certainly a popular perception, fuelled by media publicity of the destruction of fruits and vegetables, that this is *the* major way in which availability is affected.

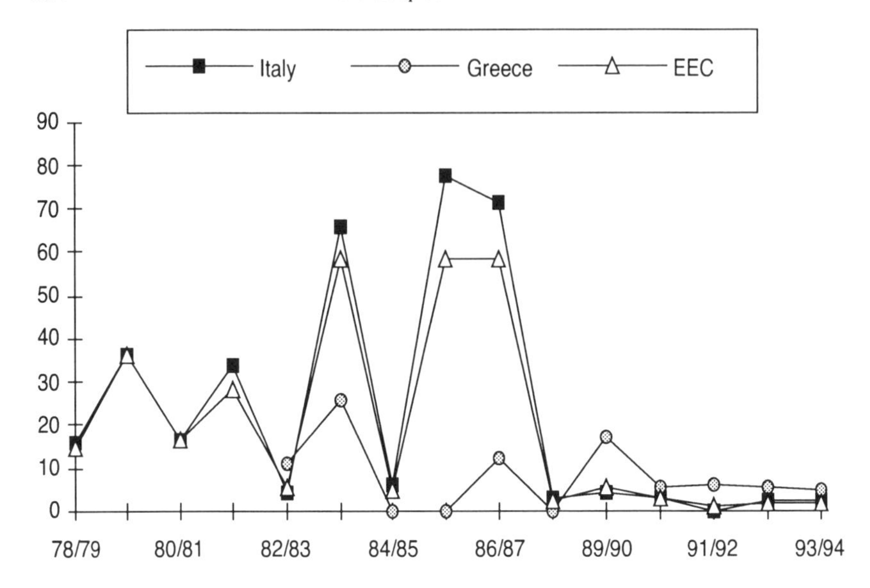

Fig. 11.5: Percentage of harvested production withdrawn — mandarins.

In contrast, there *is* evidence that the fruit and vegetables Policy influences the availability of produce which originates in non-member states. However, because the way the Policy achieves this is complex and subtle, and not conspicuous like the withdrawal mechanism, this appears to be hardly known at all, at least outside the narrow confines of the trade. The fact that produce from specific destinations can be completely driven from the European market was first drawn attention to by Williams and Ritson (1987) in their study of the impact of reference prices on the marketing of fruit and vegetables in the UK.

They pointed out that the minimum import price applied under the fruit and vegetable reference price system was different from that applying for most other agricultural commodities under the CAP, in two important respects. First, whereas with most agricultural commodities, imports from third countries receive common treatment (that is, are subject to the same import levy) the reference price system was country-specific. Second, the system operated *ex post facto*: special import taxes (known as countervailing charges) were applied after produce from a particular third country had been observed to be selling on EC wholesale markets at less than the reference price (plus any import duty applicable).

Once countervailing charges had been applied the minimum price which must be respected was increased by the amount of the countervailing charge. The nature of competition on EU markets, or the inability of exporting countries to control the pricing of their produce, might make it impossible for suppliers to react to countervailing charges by raising their selling prices to the level required by the reference price system. In these circumstances countervailing charges could quickly rise to prohibitive levels until produce from the offending country was driven from the market. This development was described by the European fruit and vegetable importers association (CIMO) as 'the spiral effect'.

From a consumer point of view the important aspect of this is that purchasers *are not* indifferent between the supplies of a particular fruit or vegetable coming from different countries — in the way, arguably, they are for most of the agricultural products covered by the CAP. Thus they may be denied access to varieties or qualities of produce during certain periods of the year because of the import controls exercised under the CAP — which are indifferent to variety or quality differences.

Williams and Ritson showed that in the early 1980s, the main products to suffer in this way were tomatoes and cucumbers from Spain (including the Canary Islands) but since then the range of products and countries affected has increased considerably (Ritson and Swinbank, 1993). At first it was still Spain which was most affected (about one third of all cases of countervailing charges being applied), but following the end in 1992 of the restrictions imposed on Spain during its transition to full adoption of the CAP, the effect has been to increase the problems for other countries.

For example, the countervailing charge on imported tomatoes from Morocco in April 1994 began at 32 ecu/100 kg and rose progressively to reach 83 ecu/100 kg on 25 May. This is rather more than tomatoes were selling for on wholesale markets at the time; in other words, in order to have the privilege of selling tomatoes within the EU, Moroccan producers were being asked to pay Brussels a sum of money which was more than they were receiving for the sale of their produce — an invitation they understandably declined. Again it needs to be emphasised that this only represents a significant example of the CAP affecting food availability if consumers regard tomatoes from Morocco as a distinct product.

It might have been expected that the GATT Agreement, under its tariffication clause (see Chapter 17), would have ended this system because, in principle, all import restrictions were supposed to be converted to conventional tariffs. In the event, the EU has replaced the Reference Price system with an even more complicated set of regulations, which nevertheless seem to be very similar in effect. A detailed discussion of the new import system will be found in Swinbank and Ritson (1995), but from the point of view of consumer food choice there are only two changes of significance. Reference Prices have been replaced by Entry Prices, and Countervailing Charges by Tariff Equivalents, but the latter are now subject to a cap, called the Maximum Tariff Equivalent (MTE). As a consequence, when Moroccan tomatoes faced marketing difficulties in 1995, an MTE applied of 36 ecu/100 kg, rather than spiralling upwards indefinitely. However, this is still equivalent to an import tariff of more than 50% and thus still high enough effectively to prohibit imports. Second, the MTE is now applied as soon as market prices for imports from a particular country fall below 92% of the Entry Price. This means that incidents of the CAP for fruit and vegetables eliminating sales of produce from specific countries may become more frequent in the future.

In addition to the intermittent application of countervailing charges, imports of fruits and vegetables are also subject to conventional import tariffs. These may also influence availability of supplies during certain periods of the year, as the duty concessions granted to supplying countries are often linked to seasonal calendars — that is, duties are highest during the EU production seasons. It has, for example, been argued by the government of Cyprus, in its trading negotiations with the EU, that the date at which the full import duty on table grapes becomes applicable in Cyprus is several weeks before there are plentiful supplies from EU countries. The consequence, it is argued, is that prices are depressed on the European market as third countries attempt to export as much as possible before the duty becomes applicable, and that there is then a period of relative scarcity until EU produce is available to replace the imports. Even then the varieties are different. Consumer choice *is* being restricted.

In conclusion then, there are instances in which consumer choice and availability are affected by the CAP at least with respect to fruit and vegetables; but on balance the policy does not really come out too badly in this respect.

Food security

The CAP influences the security of our food supplies principally by raising the level of self-sufficiency in individual agricultural commodities. The most important thing to recognise in this context is that food security and self-sufficiency in food supplies are not the same thing. Self-sufficiency is thought to contribute to food security, because the larger the proportion of produce which comes from domestic supplies, the greater control the country apparently has over its food supplies.

There are two problems with this rather simplistic approach. The first is that there is insecurity attached to domestic supplies, as well as to imported supplies. In fact, typically, a greater degree of insecurity will emerge if a country is forced to rely solely on domestic production, because any individual country's production fluctuates proportionately more than the food supplies coming on to world markets. However, a country is able individually to use the world market to offset domestic fluctuations and so the potential availability of imports provides security against fluctuations in domestic food supplies. This is of course what the Portuguese government was doing in the example mentioned above — granting import quotas only when a potential 'shortage' was perceived to apply on the domestic market.

Relying on imports to deal with domestic-orientated food insecurity is, however, only valid from the perspective of an individual country. If all countries choose to aim to be self-sufficient in food supplies, then there can be no international trade in agricultural products and the world trading system can no longer provide security against the failure of domestic harvest. Thus there is a paradox; if all countries aim to increase food security simultaneously by policies of self-sufficiency, they will collectively reduce security for all.

The second problem with equating self-sufficiency with food security is that there is insecurity associated with imported inputs and the use of non-renewable domestic resources, as well as with imported agricultural products. If a country boosts its self-sufficiency, but is dependent on imported inputs or is using up scarce domestic resources, it is reducing its food security in the longer term. Thus an agricultural policy directed genuinely towards food security would not be concerned with increasing the level of self-sufficiency using modern intensive methods of agricultural production, but be directed more towards the capacity of the farming system to sustain an adequate supply of food for the population in times of emergency using, perhaps, low-input techniques.[1]

Turning to the CAP, the verdict must be that it has increased self-sufficiency (as shown in Table 1.1) beyond the level that is necessary to attain food security — in the sense of ensuring the EU against over-reliance on imported supplies; and that it has tended to encourage a kind of agriculture which might not always be consistent with food security. Nevertheless, it would be difficult to argue that the policy had *decreased* our food security, though we could have achieved the same degree of security at much less cost in different ways. But as far as food consumers are concerned, the policy has perhaps had a mildly positive effect on this aspect of consumer welfare.

Food quality and nutrition

As indicated earlier, this category embraces a variety of, often rather elusive, aspects of food consumer interests. It is possible, however, to distinguish three different ways in which 'quality' might be interpreted. These are, in the sense of, food safety; product characteristics; and the quality (or nutritional balance) of a nation's diet.

The incidence of so-called 'food scares', in which purchases of a particular food plummet following publicity over some aspect of safety underlines the importance of this aspect of consumer concerns, and at both a national and EU level there is widespread government monitoring and control. For example, in the UK, £2.2m per annum is spent on surveillance of pesticide residues covering 6000 pesticide/commodity combinations (MAFF, 1995). The Food Minister is quoted as saying:

> Despite the fact that the monitoring programme is specifically targeted at foodstuff which are most likely to contain residue, 99% of the samples analysed had no residue or were below the recommended maximum residue level.

However, the link between the CAP and the 'safety' of the EU food supply is tenuous, based on no more than the claim that CAP price support

> has encouraged intensive methods of production, including the use of growth hormones, pesticides, antibiotics and other chemicals
>
> (National Consumer Council, 1988).

Second, in principle, some of the policy mechanisms might also influence the 'quality' of individual foods, in terms of product composition. Within the meat sector, two examples have often been quoted as evidence of the CAP affecting product quality. First, that fat class 4 qualified for the sheepmeat variable premium in the UK, (whereas 'the market requires a less fat product') implied that the lamb consumed in the UK was of a higher fat content that would otherwise apply. (1991 saw the end of the premium.) The other example concerns the intervention standards for beef, which ensure that it is high quality beef which is removed from the market. Thus, it is argued, the overall average quality of beef consumed within the EU is reduced.

Otherwise, it is difficult to find specific examples, except perhaps with fruit and vegetables, which again seems to provide the most interesting (and ambiguous) case. Withdrawal of fruit and vegetables from the market might be regarded as analogous to the beef example given above — as we have already commented that produce must be of a sufficiently high quality before it can be destroyed. In fact this is rather different, as it is a minimum quality standard which applies, rather than the very best produce being withdrawn, and some traders argue that withdrawal induces EU growers to produce to minimum quality standards, rather than meeting the requirements of the market (reported in Ritson and Williams, 1987).

When it comes to imported produce, in contrast, there are strong grounds for believing that the Reference (minimum import) Prices may have improved quality. Ritson and Williams (1987) quote one Covent Garden importer who was quite explicit on the subject, claiming that British membership of the EC had provided a clear signal to third countries to improve their marketing in the UK. Their reaction was:

> to improve their quality standards — thus securing the highest prices by having a quality premium, and so avoiding countervailing charges. The event of Reference Prices has had many bad effects but one good thing is this quality upgrading.

Another way in which the CAP is supposed to affect quality of fruit and vegetables is by quality standards — imposed rigorously on produce imported from non-EU countries, but much less so for produce sold locally in southern Europe. Ideally, such a system should be consumer driven. Kohls and Uhl (1990) comment:

> The food grading system sets up a channel of communication between food producers and consumers. Ideally, grades will result in a perfect match between the diverse wants of consumers (according to their means and preferences) and the heterogeneous quality of commodities produced and marketed. This sorting and matching process has the potential of increasing both consumer satisfaction and farm profits.
>
> However:
>
> [A] criticism of present [US] food grades is that they are convenient for traders, but are not consumer-orientated.

The problem is, of course, that many of the properties of such products which consumers value are difficult (or at least expensive) to measure objectively and reliance is placed on colour, size, shape, and so on. This only has an *adverse* effect on food quality if it leads to a suppression of other quality characteristics — in particular 'taste'. But this can happen if the officials responsible for food grades are insensitive to genuine consumer attitudes. There is a view that the European Commission is obsessed with the need for straight cucumbers and uniform, blemish-free (but tasteless) apples, and that the grading system has a negative (and not as intended, a positive) impact on food quality.

Barker (1985) refers to a study which showed that only 3% of strawberries chosen by consumers at a self-pick farm would have met EEC standards.

> There appeared to be very little relationship between grading standards and consumer preferences.

Nevertheless, these are all really isolated examples, and it is very difficult to sustain an argument that CAP quality rules for intervention and marketing standards have had a major adverse impact on the quality of foods available for consumption within the EU.

However, the CAP does influence *patterns* of food consumption and thus the nutritional balance of diets in member states. The overall level of food consumption is reduced somewhat by higher food prices, but the main impact is derived from the fact that the policy raises the prices of certain products (relative to whatever alternative is applied) much more than others. Therefore it must have a significant impact on the *balance* of food products consumed.

Table 11.4 shows a 'league table' of the major food products according to the extent to which the CAP has raised prices (and thus discouraged consumption). It would be possible to dispute some of the ordering, but not that the Policy has had a very substantial effect on the price of those near the top; and a broadly neutral effect for those at the bottom. Easily in first place is butter and other dairy products, where typically prices have been two or three times international levels. In second place is sugar, where world prices are usually very much below EU prices, though interspersed with short periods, every six or seven years, in which world prices rise, and have even exceeded EU levels. Next comes beef, where again estimates of EU prices have usually been well in excess of international levels. This is followed by lamb, though the lack of major international exporters other than New Zealand makes this a difficult product to rank.

Table 11.4: Agricultural products ranked according to the price raising effect of the CAP.

Milk and Dairy Products
Sugar
Beef
Lamb
Pork
Poultry and Eggs
Bread and Cereal Products
Most Fruit and Vegetables
Vegetable Oils
Potatoes

The next group of products — essentially cereal products and the cereal-fed livestock — are those in which the CAP has had a 'moderate' impact on food prices. In the case of vegetable oils and most fruits and vegetables (for different reasons), European consumer prices are not significantly above those which would apply in the absence of the Policy.

Anyone who has followed the recent debate over the relationship between diet and health, and particularly the possible link between heart disease and diet, will not fail to notice that the Policy seems to have had the effect of discouraging, in the main, consumption of those products which medical experts tend now to regard as 'less healthy', and encouraging consumption of products more favoured from a health viewpoint.[2]

For example Fig. 11.6 shows the development of the relative prices between butter and margarine and Fig. 11.7 the development of consumption levels in the UK since EC membership. Prices start off close together in the early 1970s; reach a position where the butter price was double that of margarine by 1980 (and the UK had completed its transition to adoption of the CAP); since then, the price relativity has remained roughly constant (and the sustained decline in butter consumption has been driven by other factors). But this does not mean that price is no longer important. The CAP has in a sense 'cocooned' agricultural product prices from world price movements; but the consumption distorting effect is still present (and could be reversed).

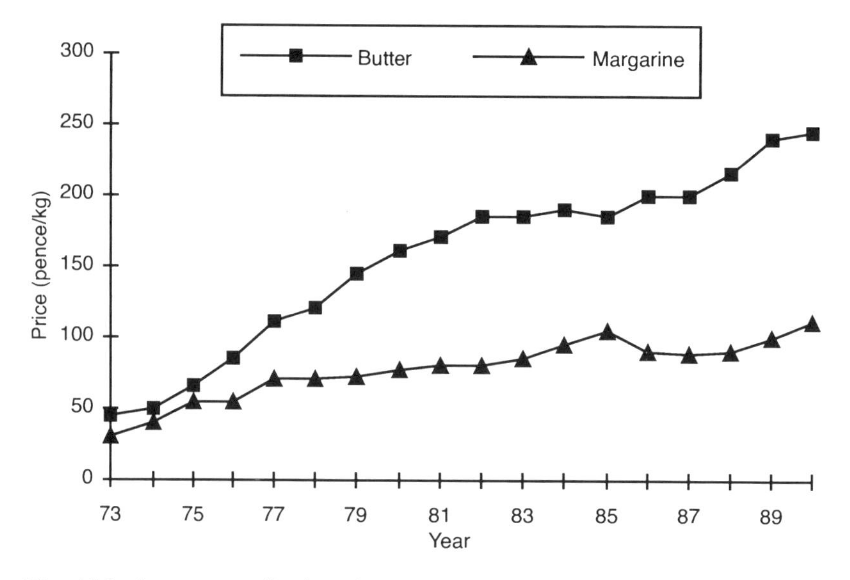

Fig. 11.6: Average retail price of butter and margarine in the UK, 1973–90.

Source: Henson (1992).

It is on this aspect that the popular conception of the CAP, as displayed in newspapers and television programmes, most blatantly misrepresents the consumer interest in the Policy. When the issue of diet and health is raised typically, as we have seen, what is presented will be pictures of fruit and vegetables being destroyed (thus 'the CAP destroys those foods which are thought healthy') and pictures of large stocks of dairy produce, sugar and beef (thus 'the CAP encourages the production of less healthy products'). There is a complete failure to distinguish between a policy which encourages the *production* of a product and a policy which encourages its *consumption*. In fact, by virtue of its reliance on price support, the CAP tends to discourage the consumption of those products for which it most encourages production. Coincidentally, it has succeeded in pushing the diet of EU consumers in the direction that the medical profession would now regard as 'more healthy'. For example, a study of the impact of the CAP on the consumption of food products in Greece shows a decline in consumption of sugar, meat and dairy products, and increases for citrus fruit, vegetable oils and vegetables (with no change for bread and cereals) — broadly consistent with the implications of Table 11.4 (Georgakopoulos, 1990).

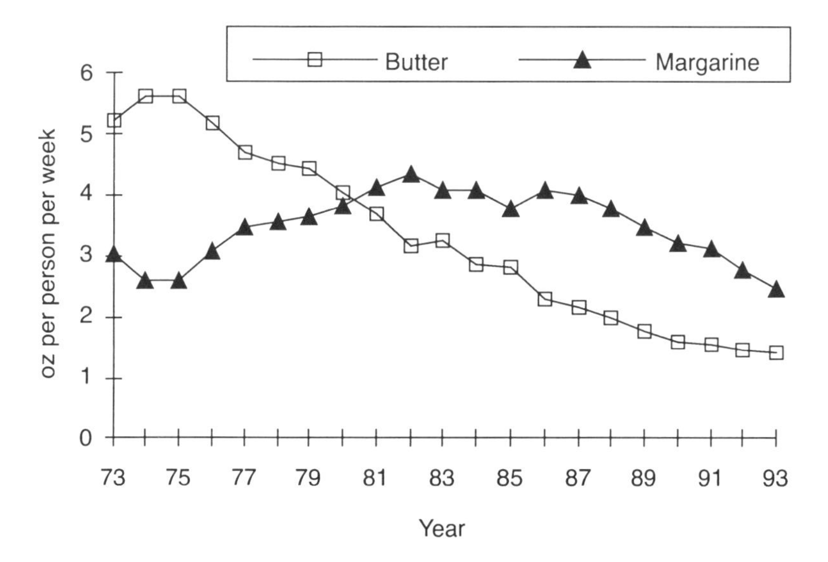

Fig. 11.7: Butter and margarine purchases 1973–93.

Source: Calculated from National Food Survey data.

The relatively small number of studies which have looked specifically at the impact of the CAP on the UK diet (Colman, 1987; Ritson, 1983, 1992; Henson, 1992; Swinbank, 1992 and Cawley *et al.*, 1994) have all broadly come to the same conclusion — that the main effect has been via food prices, and that this effect has been (at least mildly) positive with respect to healthy eating.

Recent changes to the Policy are, if anything, strengthening this effect. For example, during the 1980s the way the sheepmeat Policy worked in the UK could reasonably be described as subsiding consumption; but that has now changed. More significantly the Mac Sharry Reform of the CAP, and the changes necessitated by the GATT Agreement, will have the effect of reducing somewhat the price raising effect for cereal products and grain-fed livestock; but not for milk and dairy products.

Conclusion

The verdict must be that the CAP has had an adverse impact on consumer welfare when viewed solely in terms of food prices and consumer expenditure. Simple estimates of the impact of the CAP on food prices will, however, almost certainly overestimate this. When viewed against a broader perspective of the consumer interest, the verdict is rather better. There are isolated examples of the Policy adversely affecting the range of choice of food available, but on balance the Policy cannot be said to have a major widespread impact in this respect. In the case of food security the Policy has probably had a positive effect, although at much greater cost than could alternatively have been achieved, and at the cost of weakening the international trading system, which ultimately provides food security for all. Finally, although it has certainly never been any part of the intention of the Council of Ministers, the Policy, curiously, has probably had a positive impact on the overall nutritional quality of the European diet. Nevertheless, it remains the case that the CAP results in an implicit tax of food consumption, which has a regressive effect because lower-income households devote a higher proportion of their incomes to food than do the better off. Because of this most politicians argue that food taxes should not be contemplated; yet this is precisely the effect of the current CAP.

References

Avillez, F. (1992) *A Agricultura Portuguesa Face à Política Agrícola Comum*, Instituto Superior de Agronomia, Lisbon.

Barker, J.W. (1985) *Agricultural Marketing*, Open University Press, Milton Keynes.

Cawley, D., Lee, A. and Lund, P. (1994) The Common Agricultural Policy and the UK diet. Paper presented at the 36th EAAE Seminar, Reading.

Colman, D.R. (1987) Consequences of national and European pricing policy for nutrition and the food industry. In: Cottrell, R. (ed) *Food and Health*. Parthenon, Carnforth.

Consumers in Europe Group (1994) *The Common Agricultural Policy: How to spend £28 billion a year without making anyone happy*. CEG, London.

Franklin, M. (1994) Food policy formation in the UK/EC. In: Henson, S. and Gregory, S. (eds) *The Politics of Food.*. Proceedings of an Inter-disciplinary Seminar held at Reading University 1993.

Georgakopoulos, T.A. (1990) The impact of accession on food prices, inflation and food consumption in Greece. *European Review of Agricultural Economics*. 17(4), 485–493.

Henson, S. (1992) The CAP and healthy eating. *Nutrition and Food Science* January/February 1992, 2–4.

HMSO (1962) *Treaty Establishing the European Economic Community, Rome 1957, Article 39,* London.
Hubbard, L.J. (1989) *Public Assistance to UK Agriculture* Report for the Trade Policy Research Centre (unpublished).
Josling, T.E. and Ritson, C. (1986) Food and the nation. In: Ritson, C., Gofton, L. and McKenzie, J. (eds) *The Food Consumer*. Wiley, Chichester.
Josling, T.E. and Tangermann, S. (1987) Commodity policies. In: Pearson, S.R., Avillez, F. and Bentley, J.W. *et al*. (eds) *Portuguese Agriculture in Transition*. Cornell University Press, Ithaca, New York.
Kohls, R. and Uhl, J. (1990) *Marketing of Agricultural Products*. 7th edn. Macmillan, London.
MAFF (1995) Press release no 328/95 *Latest Pesticide Residues Results Published*.
National Consumer Council (1988) *Consumers and the Common Agricultural Policy*. HMSO, London.
OECD (1993) *Agricultural Policies, Markets and Trade*. Monitoring and Outlook, Paris.
Ritson, C. (1980) *Self-Sufficiency and Food Security*. Centre for Agricultural Strategy, University of Reading.
Ritson, C. (1983) A Coherent Food and Nutrition Policy: the Ultimate Goal. In: Burns, J.A., McInerney, J. and Swinbank, A. (eds) *The Food Industry: Economics and Policies*. Heinemann, London.
Ritson, C. (1985) Some observations on price instability, agricultural trade policy and the food consumer. In: De Haen, H., Johnson, G. and Tangermann, S. (eds) *Agriculture and International Relations*. Macmillan, London.
Ritson, C. (1992) Non-price effects of the CAP on food markets. In: Henson, S. (ed) *Proceedings of the CAP and Healthy Eating Workshop*. University of Reading.
Ritson, C. and Swinbank, A. (1988) EEC Fruit and vegetables policy in an international context. *Agra-Europe,* Special Report, No. 32.
Ritson, C. and Williams, H.E. (1987) Reference prices and the marketing mix for fruit and vegetables. *Food Marketing,* 3(1), 61–76.
Ritson, C. and Swinbank, A. (1993) *Prospects for Exports of Fruit and Vegetables to the European Community after 1992*. Food and Agriculture Organisation of the United Nations, Rome.
Smith, Adam (1776) *An Enquiry into the Nature and Causes of the Wealth of Nations*. Dove, London. 1st ed.
Sturgees, I.M. (1992) Self-sufficiency and food security in the UK and EC. *Journal of Agricultural Economics*. 43(3), 311–326.
Swinbank, A. (1992) The CAP and food prices. In: Henson, S. (ed) op cit.
Swinbank, A. and Ritson, C. (1995) The impact of the GATT agreement on EU fruit and vegetable policy. *Food Policy*. August.
Which (1988) Consumers Association.

Williams, H.E. and Ritson, C. (1987) *The Impact of the Reference Price System on the Marketing of Fruit and Vegetables in the UK*. Department of Agricultural and Food Marketing Report No. 31, University of Newcastle upon Tyne.

---

[1]For a fuller discussion of agricultural self-sufficiency and food security, see Ritson (1980) and Sturgees (1992).

[2]There is, of course, the associated question of the extent to which a government may be justified in manipulating food prices, to alter the diet of the nation, by adopting an explicit food policy. This issue is discussed in Ritson (1983) and Josling and Ritson (1986).

Chapter 12

The CAP and the Food Industry

Simon Harris and Alan Swinbank

Introduction

It is perhaps not stretching the point to claim that farmers produce raw materials for the food industry, for few farm products are sold directly to households: most undergo some form of preservation or transformation before ultimate sale. In just the same way that agriculture's input industries are bound in to the food chain and are necessarily affected by the changing commercial fortunes of the farm sector brought about by agricultural policy, so too is the food industry. In the EU, however, the food industry's concerns extend beyond those shared by food manufacturers the world over, relating to the availability of their agricultural raw materials, because of the characteristics of the CAP. Thus, the CAP affects not just the sourcing of the food industry's raw materials, but also the scale of operation of particular sectors of the industry. It impacts upon the price of raw materials, and thus has the potential to affect not only the competitive position of one branch of the EU's food industry in comparison with another, but also the relative competitiveness of the EU's food industry on world markets. Furthermore, to a large extent the CAP's support mechanisms are operated through the food industry, so causing the industry to be involved in the implementation of agricultural policy in a way which would not be necessary were only direct support payments to be made to farmers, as used to be the case under the deficiency payments scheme in the UK prior to EC entry in 1973.

Economic importance of the food industry

In this chapter, the term 'the food industry' should be taken to embrace all those economic activities concerned with the transport, processing and storage of food and drink products from the farm gate to the retail outlet. Thus we exclude catering from our purview, for here the effect of the CAP is rather diffuse.

To give some perspective on the economic importance of the food industry, comparisons have been made with the rest of manufacturing industry and with agriculture. The key conclusion to be drawn from Table 12.1 is that the food, drink and tobacco industries, taken as a whole for official statistical purposes, are very much more significant than consideration of any of their component parts would suggest. Thus the EU's food industry makes great play with the finding that it is the largest, in terms of production, in the manufacturing sector. But it is clear that the food industry is also extremely significant in terms of employment and total value added. Its net trade position, however, is less favourable than other more export oriented manufacturing sectors, although still positive.

Table 12.1: The significance of the food industry compared with the rest of EC 12 manufacturing industry. Top 10 sectors ranked by volume of production (1992).

	Production (million ecu)	Employment (million people)	Value Added (million ecu)	Trade Balance (million ecu)
Food, drink & tobacco	337.8	2.4	101.4	2.8
Chemicals and man made fibres	295.8	1.7	103.9	13.2
Electrical engineering	181.9	2.7	106.8	3.5
Motor vehicles and parts	181.9	1.8	81.2	20.4
Mechanical engineering	166.4	2.3	86.9	35.8
Metal products	124.6	2.0	74.2	7.1
Paper, printing and publishing	110.9	1.4	64.2	-6.3
Rubber and plastics	71.2	1.0	42.4	4.1
Non-metallic mineral products	69.9	1.0	41.8	4.9
Other means of transport	53.6	0.8	26.9	2.1

Source: European Commission (1995, pp. 18 and 20).

Perhaps most surprising, as generally its principal domestic markets are mature, is the finding that the growth in food industry production has been significantly positive, at an average annual rate of 3% over the period 1986 to 1992[1] (European Commission, 1995, p.5). Although less than the obvious high flyers (e.g. pharmaceuticals at 7.3%, rubber and plastics at 5.3%, and computers and office equipment at 5.1% annually), it seems that the food industry is continuing to gain significant growth from the development of new convenience products with a higher value added than its traditional offerings.

The comparison with agriculture in Table 12.2 is made more difficult because the data do not completely match. The annual EU inquiry into industrial activity, which provides results for the food industry, covers only enterprises with 20 or more employees (except in Portugal and Spain). This is an unfortunate cut-off as it excludes the 'craft' or 'artisan' businesses so characteristic of parts of the EU food industry. By contrast the data for agriculture includes small businesses and self-employed farmers.

Table 12.2: Agriculture and the food industry compared (1992).

	Employment		Value Added	
	Agriculture (a)	Food Industry (b)	Agriculture (c)	Food Industry (b)
	(1,000 people)		(billion ecu)	
Belgium	98	71	2.3	2.1
Denmark	142	64	2.0	3.5
Germany	914	520	9.1	22.4
Greece	793	56	6.9	1.0
Spain	1,253	339	12.4*	12.7
France	1,142	360	21.4	16.8
Ireland	154	44	2.6	3.0
Italy	1,749	212	21.6	10.3
Luxembourg	6	2	0.1	...
Netherlands	305	130	5.8	6.7
Portugal	522	65	1.9	...
UK	548	513	7.4	21.3
EC 12	7,626	2,378	93.5	101.4

Notes: *Changed by the authors from the figure of 21.4 reported in the original text
(a) Agriculture, forestry, hunting and fishing
(b) Food, drink and tobacco
(c) Net value-added at factor cost for agriculture

Sources: European Commission (1995, pp. 13-1 and 13-6). European Commission (1994, tables 3.1.4 and 2.0.1).

Despite these differences in data coverage a clear picture emerges. Although agriculture employs three times the number of people recorded for the food industry, this proportion is significantly less than in 1985, when agriculture provided four times the employment of the food industry. The continuing outflow of labour from agriculture, by comparison with the food industry's ability to maintain aggregate employment levels (despite the pace of change within the industry), has led to this result.

The proportionately much greater contribution of the food industry to national economic activity is also clear. Although employing only one third the number of people in agriculture, the food industry's value added is about the same as that of agriculture. Over time, as development of the food industry in regions of the EU where it is currently weak catches up with countries such as Germany and the UK, where it is already well developed, the food industry's value added is likely to exceed that of agriculture. This is already the case for the UK where food manufacture and food distribution each contribute more to national income than does agriculture (Table 12.3).

Table 12.3: Value added and employment in the UK's food sector, 1989.

	Contribution to GDP	Employment* June	
	(%)	('000)	(%)
Agriculture	1.4	279	1.2
Food & Drink Manufacturing	2.5	544	2.4
Food Wholesaling	0.5	241	1.1
Food Retailing**	2.5	781	3.4
Catering	1.9	748	3.2

* In addition to people in paid employment, there are a large number of self-employed workers in these industries. For example, the addition of farmers and their spouses to the figure for agriculture reported above produces a total comparable with that reported by the European Commission for the UK in 1989 on the basis given in Table 12.2. Employment figures are expressed as a percentage of the employed workforce.
** Includes confectioners and tobacconists.
Source: Burns and King (1993, p. 4), based on the Census of Production 1990, and personal communications with the Ministry of Agriculture, Fisheries and Food.

It is worth pointing to the differences between EU countries. France, Italy and Greece have larger agricultures, in terms of value added, than their food industries. The reverse applies for countries such as Denmark, Germany and the UK with very well developed food industries, whether largely first-stage processing of agricultural products (Denmark) or a mix of first-stage processing and the manufacture of more complex food products (Germany and the UK).

In the future there is likely to be further growth in the relative importance of the food processing, distribution and retailing activities, at the expense of the farmer's share of retail food expenditure. This increased activity beyond the farm gate reflects the changing lifestyles associated with modern society. In particular, the increasing proportion of women working outside the home, and the trend towards casual eating rather than formal family meals, have meant an emphasis on food preparation by the manufacturer, so that the time spent preparing food at home is minimised. Not surprisingly, the result is to enhance the importance of food preparation, packaging and distribution at the expense of on-farm production.

The proportion of consumer spending on food taken by food manufacturing and distribution reflects the contribution these sectors make before food reaches the consumer and the associated costs. The convenience factors built into food, ranging from the presentation of raw materials in ready-to-cook forms to the creation of complete meals ready for the microwave oven, require a range of economic activity generating employment and value added activities. Furthermore the role of distribution, and its associated costs, has become more important as consumers have sought out 'fresher' food presentations, such as chilled foods, where a whole food chain capable of keeping products at a constant temperature from factory to retail display cabinet is necessary. In this dynamic market, food manufacturers and retailers brand and advertise their wares as they strive to maximise their share of the consumer's purse and, in the case of manufacturers, their place on the retailer's shelves. These marketing activities contrast sharply with those of the farmer selling into a protected and relatively static market, and these changes underlie the complaints from the farm sector that the food industry expands its share of consumer spending on food at the expense of the farmer.

First and second-stage processed products

The body of EU legislation generally referred to as the CAP, as explained in Chapters 2 and 3, is derived from articles 38 to 46 of the Treaty of Rome establishing the European Economic Community. Despite periodic revisions to the Treaty, these articles have remained unaltered since first applied on 1 January 1958. In article 38 'agricultural products' are defined as "the products of the soil, of stockfarming and of fisheries and products of first-stage processing directly related to these products". The items subject to these provisions are then listed in Annex II to the Treaty. Flowing from the definition adopted for agricultural products, Annex II covers not only basic agricultural products (e.g. cereals, animals, grapes), but also first stage processed products. For example, Annex II includes not only cereals, but also "products of the milling industry; malt and starches; gluten; inulin". Thus the first-stage products of the cereal processing industry are included in the CAP as are the first-stage products of abattoirs (fresh, frozen and processed meat), creameries (dairy products), oilseed crushers (vegetable oils), renderers (animal fats), sugar beet and cane processors and refiners (sugar), fruit processors (canned fruit, jams), and wine makers and distillers (wine and ethyl alcohol). Second-stage processed products (typically manufactured food products: for example pasta, bread, lemonade, chocolate bars) are excluded from the CAP's purview as they are not included within Annex II. Such products in Euro-jargon are known as non-Annex II goods.[2]

The split between first-stage and second-stage processed products, reflected in the cut-off in the CAP's coverage, may be thought of as a distinction between 'agricultural processing' and 'food manufacture'. Obviously such a distinction is not rigid, but the characterisation is nevertheless useful. Agricultural processing takes raw materials direct from the farm, and either converts them to raw basic ingredients for further processing (e.g. flour, vegetable oils) or products ready for final consumption (e.g. canned fruit, wine). Some, such as butter and sugar, can either be directed to retail outlets or used for further processing. As a generalisation the agricultural origin of first-stage processed products remains clear. Food manufacturing, however, takes raw material ingredients and incorporates them into goods ready for the consumer in which indications of their farm origin may well be lost. Schematically the difference between first and second-stage processed food products may be represented as in Table 12.4.

Table 12.4: Characteristics of processed food products.

First-stage	Second-stage
Supply driven, and strongly influenced by the CAP	Market driven, and produced to fill a market demand
Bulk, undifferentiated products	Individual, branded product lines
Price takers (prices set by the market or the CAP)	Price setters (manufacturers set prices, subject to competitive forces, and retailer power)
Low value added	Higher value added
Declining share of consumer spending on food	Growing share of consumer spending on food

An equivalent distinction, but by no means the same as that between Annex II and non-Annex II goods, is made in the GATT Uruguay Round Agreement on Agriculture. This refers to "basic agricultural products" and "incorporated products". Basic agricultural products are defined "...as the product as close as practicable to the point of first sale" (GATT, 1994, p.40). Incorporated products are not defined as such, but in relation to the commitments on export subsidies, article 9 refers to "subsidies on agricultural products contingent on their incorporation in exported products" (GATT, 1994, p.48).

The difference between first and second-stage processed products should not be over emphasised in commercial terms. Many food companies' activities cover both levels of processing, especially where there has been a strategy of vertical integration. For example companies may be both flour millers and bakers, or abattoir operators and meat pie manufacturers. Nevertheless, an absolute distinction tends to exist in government administrative structures. At the level of the European Commission, Annex II products are the responsibility of the directorate-general for agriculture (DG VI) whereas non-Annex II goods are the responsibility of the much less well resourced Food Production and Biotechnology division of the directorate-general for industry (DG III). With this split responsibility, involving not just the directorate-generals, but also their respective commissioners, the EU's food industry has repeatedly complained that its interests are not well represented. DG VI is seen by the food industry as a directorate-general for agriculture which pays scant regard to the interests of their industry (let alone the interests of consumers, taxpayers, the environment or third country producers).

Paradoxically, the Commission's day-to-day detailed management of agricultural markets on its own competence or through the management committee procedure (fixing intervention buying and disposal activities, setting the level of export refunds, and so on), which is largely the responsibility of DG VI, has a more profound effect upon the commercial fortunes of food traders and manufacturers than it does on farmers. The Commission's management role is, of course, important in maintaining agricultural market prices, which are of prime concern to farmers; but the day-to-day variations in export subsidy rates and intervention activities which sustain market prices are of little direct concern to farmers, whereas they have immediate implications for traders and manufacturers.

The food industry feels strongly that a body (DG VI) which has such power over their industry should at least display more knowledge of, and sympathy for, their interests. But the Commission's preoccupation is one of managing markets for bulk CAP commodities as it is charged to do under the common organisation of the market for each basic agricultural commodity, whereas the food industry's concern is the development of markets for manufactured food products. Frequently these two concerns conflict, as is apparent in the dilemma the EU faces over implementation of the GATT Agreement with respect to food products, as discussed later in this chapter.

It may be that the recent (1995) creation of a new management committee for non-Annex II goods, under the auspices of DG III, may help with this dilemma. This seems unlikely, however, as DG III will not be able to alter the division of EU budget resources between export refunds on non-Annex II goods for which it is responsible, and the Annex II products for which DG VI is responsible, as under the GATT Agreement there is a separate constraint on the level of spending on export subsidies on 'incorporated products' (i.e. non-Annex II goods). The DG III responsibility is to manage the non-Annex II allocation in the EU Budget, as well as ensuring that the EU's GATT constraint on export subsidy spending for non-Annex II goods is achieved. Currently, however, DG VI is continuing to set the unit rates of export refunds[3] for the basic agricultural commodities incorporated in processed food products, albeit with DG III representatives sitting-in on the various agricultural management committees when export refund rates for non-Annex II goods are being discussed. Such a split responsibility does not seem likely to be maintained in even the medium-term, let alone the long-term, as a separation of the budget-holding authority from that involved in fixing export refunds is an artificial one.

First-stage processing

First-stage processors are the most directly involved with the CAP. In some sectors, the CAP sets the levels of price they must pay for their raw materials and for all it determines the scale of operation for their industries. It is through first-stage processors' activities that the majority of CAP support is given to farmers. But a close involvement with the CAP is not an unmixed blessing.

On the positive side, from the industry's point of view, as the CAP encouraged a growth in EU agricultural production, so a commensurate growth in the scale of processing activity was necessary. Allied with the expansion in processing, there was a growth in the ancillary activities necessary to handle the volumes of increased production. This has ranged from increased storage capacity to take intervention stocks, through an expanded transport infrastructure, to enlarged or specially built port facilities to handle the export of bulk agricultural raw materials. For example Hallam, Midmore and Peston (1994, p.49) found that for every two jobs in agriculture related to sugar beet cultivation, one was to be found in the transport industry hauling beet and sugar.

First-stage processors have also been helped in that the CAP supports the price of the first-stage processed product as a result of support buying at the intervention price when applicable, the payment of export refunds and other support mechanisms. This has also meant that first-stage processors have often had a guaranteed market in which to sell their products: many simply produced for sale to intervention. With a good measure of policy continuity since the CAP's introduction in the mid-1960s, the protection given to processors' activities and the growth in volumes to be processed, the result has inevitably been that the EU's first-stage processing industry is larger than it otherwise would be.

The downsides of such a close involvement with the CAP are the lack of protection against policy change, and a burdensome managerial involvement with the operation of the CAP. The food industry would argue that the dichotomy between the treatment of agriculture and the food industry is nowhere more clearly shown than when there are important changes in EU agricultural support. In the case of agriculture, 'compensation' will frequently be paid when support is reduced. Thus the CAP reform measures agreed in 1992 (see Chapter 5) led to a 30% cut in support prices for cereals and the introduction of a wide-ranging system of direct payments to farmers as compensation. When milk quotas were introduced in 1984 with little warning, there was no explicit recognition of the burden imposed on processors in adjusting to the reduced level of milk throughout. This was particularly relevant for the butter and skim-milk powder processors, and triggered significant redundancies and asset write-downs.

The link between CAP support and farm-gate prices

The mechanism by which the benefit of CAP support prices is transmitted back to farmers is the payment of inflated prices by first-stage processors for agricultural raw materials. There are wide variations between products, however, in the closeness of the link between CAP support levels and farm-gate prices. At one extreme is the sugar sector where, by law, sugar beet processors are compelled to pay *minimum prices* for sugar beet set by the EU. Not surprisingly there is little variation in the sugar beet prices received by farmers. Processors may pay market-based premia on top of the minimum prices, but the levels of any such premia are small. For sugar beet processors, the EU requirement that they pay the minimum price for beet is seen as a quid pro quo for the EU's support of the price for the first-stage processed product — white sugar — which in effect builds in a guaranteed processing margin.

In the dairy sector, although the aim of the CAP policy mechanisms for milk is to generate a return to farmers which approaches an annually determined *target price*, there are, in fact, no CAP mechanisms which directly support the price of raw milk. However, import duties and export refunds are important mechanisms for maintaining the market prices of processed products; and in the past intervention buying of butter and skim-milk powder has been a significant source of market support. The policy implicitly assumes that the benefits of such price support will be passed back to producers. In fact, the European dairy industry is characterised by its co-operative structure. On the Continent, in particular, dairy farmers are usually owners, through their co-operative, of the processing capacity; and thus as long as the co-operative remains efficient they can expect to retain the majority of the CAP's largesse. However, if milk producers in remote areas are faced with a localised monopsony buyer, as could conceivably be the case in the UK following the demise of its milk marketing boards, then under present CAP arrangements farmers would have little bargaining strength.

In the cereals sector, in addition to area payments, support is still given through the standard CAP mechanisms of intervention buying, the granting of export refunds and the charging of import duties. Farmer co-operatives are much less well developed than for milk, however, and so the degree to which CAP support gets passed back to farmers depends on the market situation in particular seasons. Even the CAP's support buying activities for cereals are themselves increasingly of a 'safety-net' nature. Thus intervention buying is not available before November in each marketing year, leaving the harvest flush of supplies free to depress market prices as necessary to ensure commercial uptake. The farmer without access to grain storage capacity will have to take the going market price at the time of harvest. Even when intervention buying is in operation it only applies at an intervention price set for a basic feed-grain quality. Higher quality cereals (e.g. milling wheats and malting barleys) have to obtain market-based price premia as best they can, depending on the circumstances of each particular season.

For some other crops such as oilseeds and pulses, where EU border protection is slight, prior to the GATT Uruguay Round farm support was effected by paying a subsidy to the processor in the expectation that the financial benefit of the subsidy would be reflected in higher farm-gate prices, and for some products a *minimum producer price* was set. The payment of the subsidy to the processor was conditional upon the payment of the minimum producer price to the farmer. But this could lead to difficulties, and a farmer would not necessarily be in a position to insist upon his rights. For peas and beans, for example, it has been suggested that on occasions animal feed compounders were unable, or unwilling, to pay the stipulated minimum price because of weak market prices and an inadequate incorporation subsidy:

> ...market movement has been dependent upon trade taking place at prices effectively below the minimum price, albeit that this had to be disguised.

In the absence of intervention arrangements for these products:

> ...sticking to the letter of the regulation and having no sales below the minimum price could mean that growers would be severely penalised by the rule that is intended to protect them.
> (Abbott, 1990, p.73).

The form of surpluses

The Commission, through its market management activities can influence very markedly the form an EU 'surplus' takes. The surpluses in the sugar sector, for example, are substantial, and the CAP for sugar provides for automatic intervention purchases of sugar in much the same way as the CAP for milk products used to provide for automatic intervention purchases of butter and skim-milk powder. Despite this, the sugar surplus has never materialised in intervention stocks, whereas the butter surplus frequently has. In large part this is because the Commission, through its weekly export tenders, is able to sell subsidised sugar on world markets in an orderly fashion, whereas large export sales of bulk dairy products have always been more problematic. A second and rather important factor is that, prior to subsidised export, the storage costs of the EU's privately held sugar stocks are met from the EU's budget, with these budget expenditures in turn matched by revenues generated by storage levies charged on the production of all EU sugar, and ultimately paid for by the consumer. Thus the incentive to sell to intervention, to shift to the taxpayer the costs of storage, is removed.

For the dairy sector, the main mechanisms for supporting market prices in an over-supplied market have been intervention buying of butter and skim-milk powder, and subsidised exports. With an over-supplied world market, however, EC milk surpluses were frequently diverted to intervention. Indeed certain factories produced principally for intervention. The dilemma for factory managers was clear: if your farmers are producing ever larger quantities of milk that the market cannot readily absorb, then more milk must be turned into butter and skim-milk powder for sale to intervention if the EC's price guarantees are to be realised; and this in turn may mean investment in new butter and skim-milk powder manufacturing capacity. A private company might be unwilling to undertake such a politically risky investment; a co-operative would be less able to resist. Indeed, a justification that was advanced for the purchase of Unigate's factories by the Milk Marketing Board for England and Wales in 1979 was the MMB's fear that the private sector would be unwilling to maintain a sufficiently large processing capacity to cope with farm production.

Subsequently, the imposition of milk quotas in 1984 and further reductions in the level of quota once set, rendered much of that butter manufacturing capacity redundant. First-stage processors (the creameries) benefited from the expansion of the CAP in the 1970s and early 1980s, but suffered uncompensated cut-backs from 1984 on. A knock-on effect of the dairy 'reforms', and similar measures in other sectors, was that the cold storage industry, which itself provided a service to the EC's intervention agencies, also faced over-capacity. Intervention (and private storage) stocks of butter, which had stood at 1.3 million tonnes at the 1986 year end, had dwindled to less than 0.2 million by the end of 1988 for EC 9 (Milk Marketing Board, 1989, p.63).

Throughout the 1980s, EC milk and milk product prices remained well above world market prices and, despite the quota-induced cut-backs, EC milk production continued to exceed consumption by a sizeable margin. Paradoxically, by the late 1980s certain sectors of the EC's dairy industry were reporting a milk 'shortage'. In the absence of quota transfers, all factories had been affected in a similar fashion by the quota-induced supply shortfalls. Butter-making factory managers, faced with weakened price support measures for butter, saw this as evidence of excess capacity in butter manufacture; whereas managers of factories producing products with buoyant EC demand, or selling to profitable export markets on the back of ample EC export subsidies, bemoaned the 'shortage' of milk, induced by quota rigidities, which limited the throughput of their plant. An export-based strategy is not, however, without its risks, for export refunds which today guarantee profits can be reduced tomorrow. McClumpha (1989, p.262), of Nestlé, complained that the "regular export business of added value products to meet genuine consumer demand, has been subject to sometimes capricious change in export [refunds] interfering with orderly long term market development and promotion".

The CAP and second-stage processing

A biscuit-maker would be a typical second-stage processor. All his principal ingredients — flour, sugar and milk powder — are CAP products and have to be purchased at EU-supported prices. Butter may, however, be available at reduced prices from intervention stores or new production. If 'subsidised' butter is obtained, a series of checks and controls have to be fulfilled in order to satisfy the authorities that the butter was used for the stated purposes, and was not fraudulently diverted to the full-price market, so requiring an increased administrative effort.

Thus, the major part of the raw materials required by food manufacturers are only available at EU, and not world, market prices. This simple fact would render second-stage food manufacturing in the EU unprofitable if it were not protected from manufacturers elsewhere with access to lower-priced raw materials. Consequently the whole protective mechanism of the CAP and, in particular, its import duties have to be applied to imported processed foodstuffs.

Similarly, it is argued, if processed food products are to be exported from the EU, then they must be able to benefit from export refunds of a comparable value to those granted on the CAP basic agricultural products which are the food manufacturer's raw materials. The trade regime for non-Annex II goods provides for an import duty reflecting the cereals (and rice), sugar and milk products incorporated into the product, plus an *ad valorem* component protecting the processing activity itself. On export, the export refund is payable on the cereals, sugar, milk and egg products incorporated in the product. The managerial task of dealing with these cumbersome rules — which also necessitates a thorough understanding of the appropriate conversion rate from ecu to national currency on any particular day and circumstance — is considerable. Small companies are consequently disadvantaged by the operations of the CAP.

The EC's payment of export refunds on non-Annex II goods used to be problematic in international terms, for the previous GATT rules prohibited the granting of export subsidies on *processed* products, and in May 1983 a GATT panel of inquiry had ruled that pasta was a processed food product and concluded that EC export subsidies were being granted in breach of GATT rules. Despite this, and reflecting the GATT's then weak powers, export refunds continued to be paid, as the EC was able to block the adoption of the panel report by the GATT contracting parties.

The outcome of the GATT Uruguay Round negotiations was to settle this issue. Under the terms of the Agreement on Agriculture, export subsidies are now formally permitted on the agricultural products incorporated in manufactured foodstuffs provided:

a) the unit rates of subsidy do not exceed those on the basic agricultural products exported as such; and

b) the spending on such export subsidies in aggregate is cut by 36% over the six years of the GATT Uruguay Round implementation period.

An export orientation?

On a number of occasions in this chapter, reference has been made to the real or imaginary constraints the EU's food industry feels it faces on international markets. There is a widespread view, certainly among second-stage processors, that international niceties and GATT regulations have limited the sales of processed, 'value-added', products on world markets. Instead of exporting bulk commodities, the food industry would like to see greater export volumes of processed products which, they argue, would aid employment and the balance of payments (and of course enhance their profits).

Their view is that world markets for primary agricultural products show relatively slow growth and are subject to alternating periods of 'glut' and 'famine', depending upon the vagaries of weather-induced fluctuations in production, and governments' disposal programmes. They are cut-price markets, subject to trade wars as the major exporters jostle for market share. In contrast, the food manufacturers maintain that markets for manufactured food products show strong growth and higher prices. Here manufacturers can create and sustain markets by their use of brands and advertising, thereby maintaining margins and, indeed, setting prices in a way that is not possible for primary commodities.

It is clear that world markets for processed food products have been growing much faster than for basic agricultural commodities. The FAO (1994, p.4) estimated that between 1979–81 and 1992, world agricultural commodity exports, as a whole, grew in value by 2% annually in real terms, whereas world exports of manufactured food products grew by 7% annually in real terms. In recognition of this change in world markets, the claim is increasingly being heard that the USA must move away from its reliance on bulk commodity exports. Some years ago, Handy and MacDonald (1989, p.1253) reported that

> According to unpublished statistics provided by the Foreign Agricultural Service of the USDA, the United States accounts for 21% of world exports in bulk agricultural products, but only 5% of world exports in consumer oriented processed food products. Moreover, since the late 1970s, the United States has been running relatively large trade deficits, of $5 to $6 billion annually, in consumer oriented processed food products, while the European Community has shifted from trade deficits to trade surpluses, of around $2 billion annually, in those products.

For the future, it seems likely that the trend of the past two decades will be maintained for faster growth in processed food products, as against bulk agricultural commodities, in world agri-food markets. Whether the EU will be able to maintain its proportionately greater share than that of the USA will depend, in part, on the continued effectiveness of the EU's export refund arrangements and future developments in the CAP. It will also depend on decisions taken by the multinational food companies, with plants in both the USA and the EU and a global view.

The GATT constraints

A key outcome of the GATT Uruguay Round negotiations was the introduction of GATT disciplines into agriculture trade. Constraints were agreed on countries' agricultural support policies to reduce their trade-distorting effects. Over the six year implementation period (1995/96 to 2000/01), developed countries have to:

- cut levels of domestic agricultural support by 20%;
- convert all mechanisms for import protection to import duties (the process of 'tariffication');
- cut import duties by an average of 36%;
- provide minimum levels of import access equal to 3%, rising to 5%, of domestic markets;
- cut spending on export subsidies by 36%;
- cut the volume of subsidised exports by 21%.

Because of the way the negotiations were conducted, the food industry's concerns were never separately identified from those of agriculture (Harris, 1994). The upshot is that the GATT agriculture constraints apply also to the food industry — the key ones being those on levels of import protection and, even more importantly, on subsidised exports. The export constraints raise major issues for the food industry and its links with the CAP.

For agricultural (broadly speaking Annex II) products, the twin constraints of a cut in the spend on export subsidies, and a cut in the volume of subsidised exports, apply. This is on a product specific basis. For example, for milk products, there are four product categories: butter and butteroil; cheeses; skim-milk powder; and 'other' dairy products. No flexibility is allowed between these product categories. For dairy farmers, it is of comparatively little importance whether the milk 'surplus' is exported in the form of cheese or butter; but for product-specific dairy processing companies the distinction is of crucial importance. As it happens, there is considerable slack in the constraints on butter exports, as the base period (1986–1990) covers the major run-down of intervention stocks built-up in the early 1980s. Indeed, compared with more recent export performance, the EU can increase its exports of subsidised butter without contravening the GATT constraints. For cheeses, however, quite the reverse applies. Here, subsidised exports had increased in the late 1980s/early 1990s. Thus the GATT constraints imply a significant cut-back in exports of subsidised cheese.

Even within product categories, and within the volume constraints, policy makers have a choice between favouring the export of basic agricultural products, or of the export of the higher value added processed products included in each basic commodity regime. For example, wheat and wheat flour fall within the same GATT export constraint heading; and 'pigmeat' embraces pork, bacon, sausages and canned pigmeat. With limited budget resources, the archetypal DG VI response would be to encourage the export of the basic agricultural products and pay scant regard to the higher value-added processed products which would have access to whatever GATT permitted funds were left. With the development of major first stage processing operations (e.g. pigmeat canning in Denmark), this could have major implications for domestic employment. The choices need not be quite as stark if, for example, world price levels are on a rising trend (because, say, of a strengthening of the US dollar) so allowing the available export subsidy resource to go further. Nevertheless it is clear that policy makers have difficult decisions to take.

For 'incorporated' products (cereals, sugar, milk and egg products incorporated in non-Annex II goods) only the GATT-induced cut in spending on export subsidies applies given the difficulties involved in attempting to identify the volume of agricultural products being exported in manufactured food products. Furthermore, 'incorporated' products form a single entity in the GATT constraints. There is no attempt to separate the spend on cereals from that on sugar or dairy products for example. If, as seems likely, the GATT constraints do limit the payment of export subsidies on non-Annex II goods, then clearly choices have to be made as to how the limited funds are to be rationed. The lower CAP support prices for cereals, following the Mac Sharry reforms of 1992 do, however, provide some relief, in that subsidy expenditure which in the past would have been paid on cereals incorporated in non-Annex II goods can now be spent on other incorporated products.

One crucial question is what will happen to the market prices of the CAP raw materials used by food manufacturers? Should domestic market prices remain relatively unchanged, then the GATT constraint would appear to require a major cut-back in exports of processed food products. Not surprisingly, EU food manufacturers are concerned about the possibility of such an outcome, particularly given their ambitions to continue expanding their exports, building on the export growth achieved in recent years. A cynical observer might be justified in suggesting that export refunds are probably unnecessary if, as claimed, non-Annex II goods enjoy buoyant demand on world markets as a consequence of their brand appeal and high added-value. Nonetheless, for many manufacturers the export refund is an important revenue source.

The options for allowing the food manufacturing industry to continue to expand its extra-EU exports, despite the GATT constraints, would appear to include:

a) lowering CAP market support prices;

b) introducing limited special schemes to allow food manufacturers access to cheap EU agricultural raw materials;

c) restricting the availability of export subsidies to those manufactured food products which, in the Commission's eyes, most need them;

d) allowing a greater use of third country raw materials bought at world prices and processed under customs control for re-export in the form of manufactured food products to world markets;[4]

e) relocating factories outside the EU and/or expanding export production from existing non-EU factories.

The outcome will presumably be some mix of these options. The implications of each are different, but all will impact on EU agriculture as well as EU food manufacturers. Apart from the implications for EU employment and economic activity were food manufacturers to relocate some of their production for world markets outside the EU, there would be significant implications for EU agriculture. Not only would important (and growing) markets for EU agricultural production be lost, but the EU's ability to meet its GATT export disciplines on the basic agricultural commodities would be made more difficult. Agricultural production not exported in the form of manufactured food products merely adds to the surplus of basic agricultural commodities, so compounding the EU's difficulties.

At the time of writing, EU policy makers and food industry leaders were still debating how the GATT Accords should be implemented for the food industry. Whatever the outcome of these discussions, it is clear that the different interests of the EU's food manufacturers and its farmers has been emphasised by the Uruguay Round Agreement on Agriculture.

CAP or CFP?

The focus of Community policy making in the agricultural and food sectors has been the CAP since the EC's inception in 1958.[5]

There have been various stages in this process, which could be characterised as follows:

- devising and putting in place the CAP (to 1968).
- operating an open-ended support mechanism (to 1984).
- attempts to 'reform' the CAP, reducing support levels and limiting its open-ended nature (to 1992).
- use of direct payments, lower market support prices and supply control measures (1992 to date) — the 'new' CAP described in Chapter 5.

Nevertheless, despite this evolution, the CAP remains, as it has always been, an agriculturally oriented set of policies.

This constant focus on agriculture and its problems, however, has ignored a fundamental structural change in the European economy: the decline in the importance of agriculture, on the one hand, and the rise in the relative importance of the food manufacturing, distribution and retailing sectors, on the other. It is this economic change, allied with a political change as consumers have become of greater political significance, which has underlain the calls for a common food policy (CFP).[6] The excesses of the CAP have allied consumers, outraged at having to pay unnecessarily high prices for unwanted surpluses, with 'free traders' concerned to reduce the trade-distorting effects of EU policies. Both camps can unite in a demand for a CFP which is consumer-oriented; and the food industry lobby (or at least second-stage processors) has found it politic to be associated with such appeals.

Whether or not consumers and food manufacturers are natural allies is a moot point: the food industry lobby has itself often faced internal divisions as some sectors (the first-stage processors) have seen their interests tied in with farming and a continuation of the CAP, whilst other sectors (the second-stage processors) have espoused free trade, at least in raw materials. These divergent interests help explain the relatively weak organisational structure of the CIAA (*Confédération des Industries Agro-alimentaires*), the food industry's equivalent of the farmers' Brussels-based lobby, COPA (*Comité des Organisations Professionelles Agricoles*).[7] And, in turn, this inability or unwillingness to present a united front weakens the food industry's case for a CFP, and for more sympathetic recognition of the industry's needs.

The single market

In the mid-1980s the EC decided that it wished to renew its efforts to eliminate all barriers to intra-Community trade by 31 December 1992. Discussion on the effects of the Community's single market programme — under which an internal market "without internal frontiers in which the free movement of goods, persons, services and capital is ensured" — for the EC's farm and food sectors was limited.[8] For agriculture this was not too surprising as the existence of a common Community policy for agriculture, for more than 20 years, meant that the single market programme did not have much left to cover. Even so, the few agricultural issues raised were contentious in themselves, and of considerable importance to the food industry: for example, the question of whether or not MCAs should be abolished (as we saw in Chapter 6, they were); whether national production quotas should be abolished (they were not); and the prospects for a successful harmonisation of veterinary and phytosanitary provisions.

For the food industry, it is not so much the detail of what was contained in the single market programme that was of importance,[9] but rather the change in industrialists' patterns of thought. Because businessmen were bombarded by governments with '1992' propaganda, telling them that the single market was coming and that they must respond, they started to take the message on board and to adjust their behaviour. It is the actions that businessmen take that make a reality of politicians' rhetoric. At the food manufacturing level, the principal changes will be a rationalisation of production facilities and a further concentration of ownership, although parts of the food industry are already highly concentrated. According to Shaw *et al.* (1989, p.15) 35 firms accounted for more than 40% of total food industry sales in the UK, by as early as 1979.

Production facilities will be rationalised as companies having plant in more than one member state, made necessary when each market was thought of as a separate entity, reorganise so that a single larger plant serves a region of the EU, comprising two or more member states, or even the entire EU market. This may lead to abrupt changes in demand for agricultural raw materials as plants in some countries are shut down while those in other countries are expanded. The structural concentration of ownership is also a major concern, as previously nationally-based companies seek to organise themselves on a EU-wide scale. A rash of take-overs marked the food industry in the closing years of the 1980s, as manufacturers jockeyed for market share by buying up their equivalents in other member states. The result is likely to be a repetition, at an EU level, of the structural concentration seen already in several national markets.[10]

Another feature of food manufacturers' behaviour resulting from the single market was increased attempts to adopt single-purchasing and, if possible, sourcing strategies covering plants in more than one country. The implication is that, where possible, manufacturers will buy agricultural raw materials where they are cheapest in the EU. The harmonisation of transport provisions under the single market programme will make such strategies more feasible. At the retailer level the effects of the single market are likely to be slower in showing through. This is essentially because of the deep-seated differences in national cultures and tastes, and hence the difficulty of transplanting a successful retailing concept from one country to another. Nevertheless, some concentration of ownership does seem likely. Furthermore, retail chains are beginning to form alliances, or buying groups, across national frontiers which have more muscle than their individual company members, and which may in time develop joint 'own-label' brands.

What is clear is that the single market programme should not be treated as a one-off event, but as a process. The changes triggered by the single market, and all that goes with it, will indeed speed up the pace of change in the EU's food industry but they will take time to work through. This is a process which will last throughout the 1990s and, probably, a good deal longer.

Conclusions

The EU's food industry is a disparate group of businesses in terms of their size, their ownership and their links with agriculture. Collectively they form one of the most important economic sectors of the EU with sizeable contributions to employment, national income, foreign trade and the price of food. Farmers (and their input suppliers), food manufacturers and food retailers make up a complex and interdependent food chain which is very characteristic of modern society. Farmers do not produce food, but raw materials: they are but one cog in a complex food system. Any attempt to mould part of the chain, such as the CAP, is bound to influence the whole; and, in the CAP, policy-makers devised a series of price support mechanisms which can only be deployed through the food industry. The CAP is complex and bureaucratic: and any trader or food company that fails to understand the subtleties of policy, have timely access to accurate information, or predict the effect of changes in agricultural conversion rates or export refunds can face severe financial penalties. The risks associated with trade, and the demands on management's time which a thorough understanding of the CAP requires, are bound to disadvantage the small company in competition with the large. Given that the CAP so strongly influences the commercial fortunes of food companies, the price of food to consumers, the environment, and so on, it is surprising that it is still seen in many quarters as a purely agricultural concern on which farmers and *their* ministers can adjudicate in splendid isolation.

References

Abbott, M. (1990) *Combinable Crops and the EEC*. A National Farmers' Union Handbook, BSP Professional Books, Oxford.

Burns, A. and King, R.D. (1993) *Training in the UK Food and Drink Sector*. Report for the European Commission's Task Force Human Resources (FORCE), Reading.

European Commission (1994) *The Agricultural Situation in the Community: 1993 Report*. Office for Official Publications of the European Communities, Luxembourg.

European Commission (1995) *Panorama of EU Industry 94*. Office for Official Publications of the European Communities, Luxembourg.

FAO (1994) *Commodity Review and Outlook 1993–94*. FAO, Rome.

GATT (1994) *The Results of the Uruguay Round of Multilateral Trade Negotiations; The Legal Texts*. GATT Secretariat, Geneva.

Hallam, D., Midmore, P. and Peston, Lord (1994) *The Economic Impact of the British Beet Sugar Industry*. Department of Agricultural Economics and Management, The University of Reading.

Handy, C. and MacDonald, J.M. (1989) Multinational Structures and Strategies of US Food Firms. *American Journal of Agricultural Economics*, 71(5).

Harris, S.A. (1989) Agricultural Policy and its Implications for Food Marketing Functions. In: Spedding, C.R.W. (ed) *The Human Food Chain*, Elsevier Applied Science, London.

Harris, S.A. (1994) The Food Industry Perspective. In: Ingersent, K.A., Rayner, A.J. and Hine, R.C. (eds) *Agriculture in the Uruguay Round*. Macmillan, London.

Holmes, P. (1989) Economies of Scale, Expectations and Europe 1992. *World Economy,* 12(4).

McClumpha, A.D. (1989) International Trade Implications. In: Spedding, C.R.W. (ed) *The Human Food Chain*. Elsevier Applied Science, London.

Milk Marketing Board for England and Wales (1989) *EEC Dairy Facts and Figures 1989*. MMB, Thames Ditton.

Shaw, S.A., Burt, S.L. and Dawson, J.A. (1989) Structural change in the European Food Chain. In Traill, W.B. (ed) *Prospects for the European Food System*. Elsevier Applied Science, London.

Swinbank, A. (1990) Implications of 1992 for EEC Farm and Food Policies. *Food Policy,* 15(2).

---

[1]At constant 1985 prices.

[2]The term 'non-Annex II goods' also includes products of other industrial sectors which are based on Annex II products, for example pharmaceuticals based on sugar.

[3]The EU uses the term 'export refund' for the subsidies paid on products exported to third countries. They are referred to as export subsidies in the GATT Uruguay Round Agreement on Agriculture.

[4]Inward processing relief is a long-standing arrangement under which, under specified conditions, certain raw materials can be imported into the EU without payment of import duties, provided they are re-exported from the EU within a specified period. Clearly, no export refund is payable, and complex administrative structures have to be put in place to guard against fraud.

[5]Food law harmonisation and completion of the internal market have, however, been a continuing preoccupation of the Community since the early 1960s and the single market programme has emphasised this importance.

[6]Presumably the acronym for a Common Food Policy would be CFP, except that this has been appropriated by the Common Fisheries Policy.

[7]On this see Harris (1989, pp. 302-303).

[8]See, however, Swinbank (1990) and other articles in this special edition of *Food Policy*.

[9]The detailed arrangements concerning food law and the concept of mutual recognition are not discussed here in a book on the CAP. For further details see Swinbank (1990).

[10]As Holmes (1989, p.533) cautions, however, "The aim of public policy must be to facilitate the realisation of scale economics without reducing competition too much".

Chapter 13

The CAP and the European Environment

Philip Lowe and Martin Whitby

Introduction

Agriculture and forestry do not only produce food and fibre; they also help shape the rural environment. A managed environment that ensures the productivity of the land is of course essential to the maintenance of primary production. Increasingly, also, modern society values the environmental resources which arise as joint outputs with primary land use, including water supply, semi-natural habitats, wildlife, the historic pattern of land settlement and its associated cultural artefacts, rural landscapes and open spaces. For these resources to be conserved and made available, the land must be managed and this mainly entails the continuity of certain farming and forestry practices. However, rapid changes in primary land use and technology have jeopardised the supply of these resources. The CAP has been criticised for helping to drive those changes and for not taking sufficient account of the environmental consequences, and recently European policy makers have begun to respond to such criticisms.

This chapter reviews these issues. First, it surveys the primary land use pattern within Europe and its associated environmental attributes. Then it reviews the contemporary pressures on the environment arising from changes in agriculture and considers the responsibility of the CAP for rural environmental change. Recent reforms to the CAP addressing environmental problems are described. The chapter ends with an initial evaluation of these reforms and a brief discussion of more fundamental solutions.

Land use in Europe and its environmental attributes

The broad pattern of land use in Europe is summarised in Table 13.1 in terms of the percentage share of primary land between arable, pasture and forestry in EU member states. The rows in the table summarise the pattern within the member states but, by expressing land use as a percentage of the EU total, each cell also reflects the importance of each use relative to the EU total primary land area of some 310 million hectares. Agriculturally, these diverse patterns of use are determined predominantly by the distribution of soils and climate.

Table 13.1: Primary land use in the European Union (percentage of total).

	Agricultural Land		Wooded Area	Other Land	Total
	Arable and Permanent Crops	Permanent Meadows and Pastures			
Austria	0.5	0.6	1.0	0.5	2.6
Belgium	0.3	0.2	0.2	0.3	1.0
Denmark	0.8	0.1	0.2	0.3	1.4
Finland	0.8	0.0	7.4	1.5	9.7
France	6.1	3.6	4.7	3.1	17.6
Germany	3.8	1.7	3.3	2.3	11.2
Greece	1.2	1.7	0.8	0.4	4.1
Ireland	0.3	1.5	0.1	0.3	2.2
Italy	3.8	1.6	2.2	1.9	9.4
Luxembourg	0.0	0.0	0.0	0.0	0.1
Netherlands	0.3	0.3	0.1	0.4	1.1
Portugal	1.0	0.3	0.9	0.7	2.9
Spain	6.4	3.3	5.1	1.2	15.9
Sweden	0.9	0.2	8.9	3.1	13.1
UK	2.1	3.6	0.8	1.3	7.7
Total EU	28.4	18.6	35.8	17.1	100.0

Source: Stanners and Bourdeau (eds), 1995.

Arable production and permanent cropping require dry harvesting conditions and soils of innate or enhanceable fertility: they account for 28% of total land use. Major European shares of arable area are found in France, Germany, Italy, Spain and the UK, together accounting for more than three quarters of the EU total arable area. States climatically influenced by the Atlantic are mainly wet and temperate and therefore show more emphasis on grass production: one fifth of the total land use is under grass. France, Germany, Spain, Ireland and the UK account for most of it. The natural climax vegetation across most of Europe would be woodland, and indeed the largest land use is forestry, with 112 million hectares in total. This includes both natural and semi-natural woodland much of it on marginal land and, especially in Scandinavia, extensive areas of commercial forestry where the climate and the terrain limit the scope for animal husbandry. Austria, Finland, Sweden, France, Germany, Italy and Spain, together account for nearly 90% of EU forests.

Focusing on the rows of Table 13.1 reveals the distribution of land uses within member states, indicating their degree of specialisation. Thus Denmark, Italy and Spain are particularly dominated by arable and permanently cropped land; and Ireland, the Netherlands, the UK and Greece by grass. Forestry is the dominant land use in Finland, Sweden and Austria.

However, because there are only four main categories of land use in Table 13.1 and the presentation is aggregated at the national level, it tells us little about the environmental attributes associated with land. The diversity of conditions is arguably much more significant than that between categories, and is determined as much by the patterns of land occupation and the history of cultural practices as by the distribution of soils and climate. For example, arable production in the UK is mainly cereals, with root crops providing the most frequent alternative, but the category of 'arable and permanent crops' also includes extensive areas of orchards, vineyards and olive groves in the Mediterranean countries. The table also tells nothing about the intensity of production which, above certain levels, is generally inversely related to ecological diversity. Within the grassland category, for example, there is a wide range of variation from the ecologically rich semi-natural rough grazings, found in varying proportions in most member states, to the intensively managed grasslands, particularly in northern and western areas.

By focusing at the national level and on only a few categories of use, Table 13.1 elides the rich local variation in soil types and climate. Moreover, land use is determined not only by these two factors, but by the pattern of ownership and management also. This is not simply a matter of present cultivation rights and obligations. The current pattern of ownership embodies a mixture of historical traditions and events. Likewise, the present capacity of soils reflects their cultivation history, itself dependent on the way in which they have been owned and managed for centuries. To examine this heritage requires focusing in more detail on land use systems.

The different species and sylvicultural regimes operating in forests, for example, produce quite different environments. Table 13.2 breaks down the wooded area of Table 13.1, showing the distribution between coniferous and deciduous plantings and the extent to which coppicing is practised. Whilst conifers are very important in northern and mountainous regions, broadleaves are found on richer, lower lying soils. The majority of the EU's deciduous trees are found in France, Germany, Italy and Spain. Nearly all of the coppicing reported is in broadleaved species and this is concentrated in these same countries. In France and Italy it takes up virtually half of the forested area and in Spain it approaches one third. The management categories in Table 13.2 still conceal a great deal regarding different species, practices and environmental consequences. For example, the category of 'high forest' includes the full range from preserved, natural and semi-natural woodlands to naturally regenerated, commercially exploited forests, to single-species, single-age plantations. Virgin forest is rare: Peterken (1996) reports significant numbers of virgin sites only in Sweden (38) and Finland (28) and few in other EU member states (23 in total). In general, there is much more semi-natural forest of conservation value. Equally, the category of coppice with standards includes different degrees of intensity of management and commercial exploitation. In many regions traditional coppicing practices have lapsed leaving derelict woodland which nevertheless usually retains substantial ecological interest.

Table 13.2: Distribution of EU woodland use by sylvicultural system, species and member state: 1980s (percentage of total).

	High Forest		Coppice with Standards		Total Forest
	Coniferous	Broadleaved	Coniferous	Broadleaved	
Austria	3.0	0.8	0.0	0.1	4.0
Belgium	0.3	0.2	0.0	0.2	0.6
Denmark	0.0	0.0	0.0	0.0	0.5
Finland	19.0	1.7	0.0	0.0	20.7
France	3.9	2.5	0.7	5.7	13.5
Germany	7.0	3.5	0.0	0.1	10.8
Greece	1.0	0.3	0.0	1.2	2.6
Ireland	0.4	0.0	0.0	0.0	0.4
Italy	1.8	1.2	0.1	3.8	6.9
Lux.	0.3	0.4	0.0	0.1	0.9
Neth.	0.2	0.1	0.0	0.0	0.3
Portugal	1.4	1.0	0.0	0.4	2.8
Spain	4.1	2.2	0.0	2.4	8.6
Sweden	23.6	1.5	0.0	0.0	25.1
UK	1.6	0.6	0.0	0.0	2.3
Total EU	67.7	16.1	0.8	14.1	100.0

Source: Stanners and Bourdeau (eds), 1995.

Note: The data relate to various years, mostly during the 1980s.

The environmental consequences of land use systems often depend on the intensity of production. Much of the wildlife and landscape value of the European countryside depends on low intensity farming and might be threatened by both increased and decreased production pressures. According to Bignall and McCracken (1996), low intensity farming systems "have tended to adapt their management techniques to the natural environmental constraints of the region rather than adapting the environment (and in many cases the livestock breeds and crop varieties) to meet a standardised, often industrialised, production system".

Table 13.3 gives some of the typical characteristics of low intensity livestock and crop systems. Table 13.4 shows their overall distribution in a number of EU countries. On average, about two fifths of the agricultural area is under low intensity systems. It may seem remarkable that so much remains given the rapid pace of technological change in European agriculture during the twentieth century, and an explanation for this must rest partly on the constraints of soils and climate and partly on socio-economic factors such as distance from markets and forms of ownership.

Table 13.3: Typical characteristics of low intensity systems.

Low nutrients inputs and low output per hectare	
Livestock systems	Crop systems
• low nutrient input, predominantly organic	• low nutrient input, predominantly organic
• low stocking density	• low yield per hectare
• low agrochemical input	• low agrochemical input (usually no growth regulators)
• little investment in land drainage	• absence of irrigation
• relatively high percentage of semi-natural vegetation	• little investment in land drainage
• relatively high species composition of sward	• crops and varieties suited to specific regional conditions
• low degree of mechanisation	• use of fallow in the crop rotation
• often hardier, regional breeds of stock	• diverse rotations
• survival of long established management practices, e.g. transhumane, hay-making	• more traditional crop varieties and more 'traditional' harvesting methods
• reliance on natural suckling	• low degree of mechanisation
• limited use of concentrate feeds	• tree crops, tall rather than dwarf — not irrigated

Source: Beaufoy *et al.* (1994).

There is a rough correlation between low productivity systems and certain traditional forms of collective ownership. This would be expected because ancient assignments of property rights, especially those which govern the use of land in common, for example for grazing, transhumane or water management, often developed to deal collectively with difficult environmental circumstances. These forms of ownership have thus in the past supported farming systems which have created and sustained sensitive ecosystems. They have also tended to inhibit the introduction of many agricultural innovations. Yet in many places they have been swept away during this century by the advance of agricultural productivity. The high transactions costs involved in changing the assignment of rights (enclosing a common, for example) can only be met from a switch to high producing and intensive farming systems. Accordingly, periods of intense agricultural prosperity are likely to coincide with radical changes in the assignment of property rights which can be funded from the proceeds of intensification. Conversely, it is in areas of major environmental constraints and low productivity that there is less incentive to rationalise property rights and where the conservatism in ownership patterns and structures may act as an additional barrier to the displacement of traditional farming systems.

Table 13.4: Share of agricultural area under low-intensity farming systems: selected EU member states (%)

France	25.0
Greece	16.3
Italy	31.3
Portugal	60.0
Ireland	35.4
Spain	81.7
UK	10.9
Total	39.4

Source: Bignall and McCracken (1996)

These rights would include the historical remnants of ancient systems of land holding, many of them to do with rights of passage — such as the transhumance routes remaining in southern Europe (Whitby, 1994) and the public rights of way in the UK (Curry, 1994) — with areas of common grazing where the rights of individual graziers are more or less precisely specified and with the complex systems of communal management of such areas as the forests and mountain pastures in northern Italy (Porru, 1992) and northern Portugal (Portela, 1994). These different styles and types of ownership constrain the development of more intensive forms of agriculture and are therefore more likely to occur where there are other reasons favouring low intensity. However, such areas may also be more selectively subject to land abandonment, and this has been the fate of many commons in mountainous, upland and wetland areas of southern Europe.

Rural environmental problems: The responsibility of the CAP

During the past 20–30 years a series of environmental problems have arisen in relation to European agriculture, problems which can broadly be ascribed to the intensification and concentration of production. Intensification has involved a movement away from farming as a balanced self-sustaining cycle, in which there was extensive re-use of animal and plant wastes and little import of additional nutrients, towards an industrial model of resource throughput in which the quantity of bought-in inputs is increased in order to increase yields. With intensification, agricultural pollution becomes a major problem because farmers use increasing quantities of potent inputs (particularly pesticides), because nutrients are applied surplus to the requirements of crops and animals, and because increasing quantities of potentially polluting farm wastes are produced.

For example, the modern arable systems which are now common in most of Europe are based on heavy applications of fertiliser (17 million tonnes of plant nutrients in fertiliser are used annually in the EU, Stanners and Bourdeau (1995)) and the use of chemical pesticides to a large extent (Brouwer (1995) presents a recent survey). Many of these chemicals find their way into water courses where they may promote eutrophication and the elimination of sensitive aquatic species and into the groundwater, thus also contaminating human water supply systems. An additional problem with irrigated farming in southern Europe is salinization linked to the over-use of aquifers and salt water incursion along the Mediterranean coastline. The intensive livestock production, which is found particularly across the Low Countries, and in Brittany, Galicia, Lombardy and North Rhine-Westphalia, generates large quantities of liquid slurry which is a potent pollutant both if it enters water courses and in the noxious fumes it emits, but also when spread on land leads to nutrient leaching and gaseous emissions. The volume of animal production provides a direct indication of the extent of production of global warming gases from agriculture. In particular, the population of ruminant livestock (90 million cattle and 110 million sheep and goats) and its manures are responsible for the addition of 10 million tonnes of methane, or 12% of annual EU emissions of this very powerful greenhouse gas, to the atmosphere annually. Farm machinery also adds to agriculture's contribution to this global pollution problem. EU agriculture now has more than seven million tractors (compared with an EU total of 144 million cars).

Alongside the intensification of production has gone the concentration (and specialisation) of production as some regions have emerged with better advantages compared to others (through their superior soils or climate, better access to markets, more rationalised farm structures, or their better trained and supported farmers). The regions where production has been concentrated have lost wildlife, semi-natural habitats and traditional landscape features. Conversely, other uncompetitive regions have suffered the withdrawal of capital and labour from agriculture. At its extreme this has led to land abandonment. In southern Europe many regions with difficult terrain or poor soils have suffered extensively from the interlinked social problem of rural depopulation and a series of environmental problems arising from the withdrawal of cultivation. These problems include the abandonment of terraces and water management systems, increased incidence of soil erosion, the invasion of scrub, increased risks of forest fires and major floods, and the reversion of countryside to wilderness.

Many of these environmental impacts have been laid at the feet of the CAP (Shoard, 1980, Body, 1982, Bowler, 1985, Baldock and Long, 1987); and it must be admitted that the CAP has been in place during what must have been the most widespread and rapid transformation of the rural environment in the whole of European history. To ascribe all responsibility for this transformation and its myriad changes to the CAP would, however, be naive.

There have been associated changes in rural social and economic structure and in technology with which the CAP has interacted but which would have had profound consequences without the CAP. To pin blame on the CAP for negative environmental trends assumes a judgement of what would have happened in its absence or under a different policy regime. In fact, we have the same difficulties here, of identifying an alternative scenario, as discussed in Chapter 11 which considered the impact of the CAP on consumers. Strictly speaking we do not have this counter-factual evidence. In any case, given that many of the environmental benefits from rural land management depend upon the continuity of certain practices, it is likely that without the CAP there would have been environmental losses as well as gains. The environmental critics of the CAP, however, have tended to come from north-western Europe and to have concentrated on the apparent losses inflicted by what they see as an over-supported agriculture. What firm judgements then can be made about the environmental consequences of the CAP at the levels of support that have historically prevailed compared with a hypothetically much reduced level of support? Three broad areas can be identified: the level and efficiency of input use and the consequences for agricultural pollution; the rationalisation of farm size and structure and the consequences for rural landscapes and habitats; and the maintenance and encouragement of farming in marginal areas. Below we consider each of these in turn.

It is generally agreed that high product prices paid under the CAP have encouraged a greater use of bought-in inputs than would otherwise have been the case. This has led to a less efficient use and hence a greater polluting surplus of chemical inputs (inorganic fertilisers and plant protection products); greater use of purchased feed and thus an encouragement to overstocking; and greater reliance on bought-in fertiliser, leading to even bigger surpluses of organic manures to be disposed of. The empirical evidence confirms that high price supports under the CAP have been associated with big increases in the use of pesticides, inorganic fertilisers and surpluses of animal manures, though there are considerable variations between farms and regions (Brouwer, 1995; Brouwer *et al.*, 1995). It is generally assumed that lower agricultural supports should lead to environmental improvements, either by encouraging more efficient use of inputs or a shift to more extensive systems. The effects may not be marked, though, because of low price elasticities of demand for fertilisers and plant protection products. Unfortunately, little concrete can be learned from the experience of the two countries, New Zealand and Sweden, which have considerably reduced their farm price supports. In the case of Sweden (Vail *et al.*, 1994) the changes initiated in 1989, were not so radical and, although they were strongly influenced by environmental considerations, they had little time to demonstrate their consequences before the country entered the EU and adopted the CAP in 1995. New Zealand, though, has been farming at world prices since its major reforms in the mid-80s (Sandrey and Reynolds, 1990). Its experience clearly indicates a positive correlation between support levels and the intensity of input use (OECD, 1994, p.118). Not only has there been a considerable reduction of inputs overall, but also the area of intensively farmed land has been reduced. However, it is probably still too early to judge the full environmental consequences and care is needed in extrapolating any lessons since, by European standards, New Zealand has always been a predominantly low-input farming country.

A long-term objective of the CAP has been the improvement of the structure of farming. This has involved grants and technical aid to improve the age structure of farmers, to modernise farms and to rationalise the size and structure of farm holdings. In a number of countries, these aids have been incorporated into national programmes that have orchestrated major investment in agricultural infrastructure (such as land drainage and new buildings) and the consolidation and reparcelling of land holdings. In countries such as the Netherlands and France, such programmes have had a dramatic environmental impact regionally by removing many traditional landscape features and micro-habitats (hedgerows, trees, small woodlands, wet areas) (Boisson and Buller, 1996; van der Bijl and Oosterveld, 1996). In other countries, similar consequences may have occurred but less due to CAP structural policies than to price supports. Allanson *et al.*'s (1994) study of structural change provides "casual support" (in their words) for the argument that higher prices in the UK have provided an incentive to amalgamate farm holdings, and Munton and Marsden (1991) have shown that a change in the occupancy of all or part of a farm is one of the major factors in landscape change from agricultural intensification.

Structural polices and price support may also have helped sustain farming in marginal regions, with undoubted environmental benefits, but in some cases losses too. The main structural policy to support agriculture in marginal regions has been the Less Favoured Areas (LFAs) designation under EC Directive 75/268. LFAs contain a fifth of EU holdings and covered some 40% of the agricultural area of the EC (10) (CEC, 1990), including most of the land under low intensity systems (Baldock *et al.,* 1993). Potentially therefore they are of major significance. Although it may have helped to sustain some farmers in low intensity farming systems, the form of support, by adding to the basic headage payments for livestock, has provided a further incentive to raise stocking densities, which may be detrimental to nature conservation. A similar or even greater effect may have occurred through the beef and sheep premium which, while likewise supporting farming in marginal areas, may also have encouraged overstocking and local overgrazing and hence damage to swards and soils (Baldock *et al.,* 1993; DOE, 1990).

Efforts to reform the CAP from an environmental perspective have been aimed both to overcome the negative externalities associated with production supports and to incorporate positive environmental aims into the objectives of the CAP. Negative externalities have been attacked indirectly through measures such as the Nitrates Directive (91/676). The use of pesticides may be reduced through the implementation of Directive 91/414 which provides for the registration of products with the intent of protecting both the environment and the human population (Brouwer, 1995). Negative impacts have also been contained to some extent through the implementation of Directive 85/337 requiring the production of environmental assessments before certain types of development, such as intensive livestock units, are allowed to proceed. Positive environmental impacts have also been encouraged indirectly through the implementation of the Directives concerning habitats and birds. The direct changes have been to the CAP itself. The reform of price support and the shift to area payments promises to alter the negative externalities, while the new agri-environment regulation has made conservation an aim of the CAP, and it is to this that we now turn.

The roots of European agri-environmental policy

The first agri-environment measure at the European level is generally taken to be Article 19 of Council Regulation 797/85 on 'Improving the Efficiency of Agricultural Structures', which authorised member states to introduce "special national schemes in environmentally sensitive areas" to subsidise farming practices favourable to the environment. This amendment to the EC's Structures Directive was promoted by the British government and initially it found little favour with other member states or the European Commission. Indeed, the new Article 19 merely permitted governments to introduce environmental incentive schemes without providing any aid from the Community budget.

However, two years later, in Regulation 1760/87, it was agreed that ESA payment schemes could be eligible for up to 25% reimbursement from the European Agricultural Guidance and Guarantee Fund (FEOGA). This development must be seen in the context of the mounting budgetary crisis of the CAP, caused through overproduction. It marked the initial acceptance that supporting farmers to conserve the countryside might also help, albeit in a modest way, to curb overproduction. The 1987 change brought Article 19 into line with other elements of Regulation 797/85 concerned with extensification and set-aside, introduced at the initiative of the German government, to help reduce market surpluses.

Three years later, the Commission undertook a review of all three measures and came up with proposals for "reinforcing the relationship between agriculture and the environment" (European Commission, 1990) which led to a proposed regulation "on the introduction and the maintenance of agricultural production methods compatible with the requirements of the protection of the environment and the maintenance of the countryside" (European Commission, 1990). This eventually became the core of the agri-environment regulation that accompanied the Mac Sharry reforms of the CAP which greatly extended and broadened agri-environment policy and massively increased its funding (European Commission, 1992).

A number of considerations seem to have influenced the Agricultural Directorate (DGVI) of the Commission. The first of these was the need to give some substance to the formal commitments made in various policy documents to integrate environmental consideration into agricultural policy. The 1985 Green Paper Perspectives for the CAP (European Commission, 1985) had recognised that the role of agriculture is not only to produce food but also to manage the countryside and conserve the environment. With its 1988 statements on The Future of Rural Society (European Commission, 1988a) and on Environment and Agriculture (European Commission, 1988b) the Commission conceded the need to adapt agriculture to these other requirements. At the very least this suggested adjustments to the extensification and voluntary set-aside schemes to make them environmentally beneficial.

A second and related consideration was the need to respond to specific problems emerging from the implementation of EU environmental policy. Agricultural pollution of water sources came to public recognition in many parts of the EC in the late 1980s. The proposed Nitrates Directive envisaged specific restrictions on agriculture to reduce leaching from farm land. The DGVI was concerned over the consequences for farmers' livelihoods.

A third consideration was the very limited scope and impact of the three existing measures. In part this was seen to be due to inadequacies in their design and in the incentives available. In addition, the Commission felt that they were oriented too narrowly, reflecting specific national concerns rather than Community-wide problems, and that in general they suffered from a northern European bias which made them of little relevance to the rural problems of southern Europe, where indeed they had not been taken up. Any expanded scheme would have to address French and southern European concerns particularly over desertification (i.e. dereliction of the countryside through land abandonment). Ever since the instigation of LFA policy in 1975, the DGVI had shown itself sensitive to this more socially and agrarian oriented definition of the rural environment problem. But the Commission is also driven by an integrationist logic (witness its recurrent concerns with harmonisation, convergence and cohesion) and while it was evident that some member states (notably the UK) were pushing rural environmental concerns as part of their general opposition to the CAP, the Commission was keen to see the development of agri-environment policy within a strong, Community-wide framework.

A final consideration for the DGVI was the market situation facing farmers. In devising an agri-environment regulation, that consideration could not be put to one side. A necessary condition of any initiative was that it should help to ease overproduction. On the one hand, this meant that the Commission wished to see Article 19 schemes of a kind which contributed not only to conservation but also to the reduction of surpluses; on the other hand, it impelled the Commission to look at a considerably expanded agri-environment programme that might make a significant contribution to the control of surpluses. Insofar as the new agri-environmental schemes would provide an additional source of income for farmers in an era of price restraint, this would complement the new compensatory payments being introduced within the Mac Sharry package.

The agri-environment regulation 2078/92

This regulation as one of the 'accompanying measures' to the principal CAP reform was agreed in May 1992. It is not surprising that one of the most contentious issues during the debate in the Agricultural Council was over the decision to provide the FEOGA funding through the Guarantee, rather than the Guidance, Fund. Since the Guarantee Fund is the mechanism for supporting the CAP market measures, such as export refunds and intervention purchase, this signalled the incorporation of the agri-environment measures within the core of the CAP. More significantly perhaps, the Guarantee Fund is not subject to the same budgetary restrictions as the Guidance Fund. Whereas the Commission had dispensed 10 million ecu co-financing agri-environmental payments in 1990, it was expected that the budget would reach 1.3 billion by 1997 under the new CAP (Delpeuch, 1992; European Commission, 1992).

Member states are able to reclaim 50% of the eligible cost from the Community budget or 75% of the cost in Objective 1 regions (i.e. currently most of the Mediterranean countries, Ireland, the eastern Länder of Germany and the highlands of Scotland). It is obligatory on member states to implement a national agri-environment programme and to include within it all the individual categories of measures listed in Article 2, unless there is a clear reason why these should not apply.

The types of voluntary incentive which member states may introduce under the regulation are set out in Table 13.5. Measures (a) to (c) are all concerned with reducing the intensity of agriculture. Taken as a group, they cover both the crop and livestock sectors and make explicit reference to the conversion of arable land to extensive grassland. Measure (a) is concerned directly with reducing inputs of fertilisers and pesticides and could be seen as a means of reducing pollution arising from agriculture, as well as promoting extensification.

Table 13.5: Aid measures under regulation 2078/92, article 2.

Subject to positive effects on the environment and the countryside, the scheme may include aid for farmers who undertake:

a) to reduce substantially their use of fertilisers and/or plant protection products, or to keep the reductions already made, or to introduce or continue with organic farming methods

b) to change, by means other than those referred to in (a), to more extensive forms of crop, including forage, production, or to maintain extensive production methods introduced in the past, or to convert arable land into extensive grassland

c) to reduce the proportion of sheep and cattle per forage area

d) to use other farming practices compatible with the requirements of protection of the environment and natural resources, as well as maintenance of the countryside and the landscape, or to rear animals of local breeds in danger of extinction

e) to ensure the upkeep of abandoned farmland or woodlands

f) to set aside farmland for at least 20 years with a view to its use for purposes connected with the environment, in particular for the establishment of biotope reserves or natural parks or for the protection of hydrological systems

g) to manage land for public access and leisure activities.

Source: European Commission, 1992.

The second set of measures, (d) to (g), in contrast, provide incentives to farmers to undertake activities which maintain or enhance the countryside. They cover landscape, nature conservation and public access to the countryside, as well as a rather vague reference to "protection of the environment and natural resources" in measure (d). These measures are rather broadly framed and cover a potentially large range of different schemes, an impression confirmed by the plethora of proposals transmitted to the Commission since 1992 (most of them, though, concerned with pastoral agriculture).

Measure (d) is a development of the wording in Article 19, later Article 21. It also includes the option of providing farmers with incentives to rear breeds of livestock "in danger of extinction". The intention is to preserve a genetic inheritance in the EU which is in danger of being eroded as certain breeds disappear.

Measure (e), paying farmers for maintaining abandoned farmland or woodland, is a significant addition and permits aids of the kind put forward by the French government under the old Article 19 scheme but rejected because they fell outside the scope of the scheme which was confined exclusively to existing farmland. There is continued sensitivity over whether, in allowing stock to be reintroduced into areas from which they have been withdrawn, the new measure encourages increases in production. Measure (g) was introduced on a proposal from the British government. It reflects a concern to be able to pay farmers and other landowners for making land available for walking and other forms of public access.

The measures in Article 2 should be presented within "multi-annual zonal programmes" with a duration of at least five years. The programmes are to "reflect the diversity of environmental priorities". This underlines the intention that member states should generate schemes which are sensitive to local circumstances, rather than simply introducing standard national schemes. The reference to environmental priorities presumably refers to EU measures such as the nitrates, birds and habitats directives. Zonal programmes are meant to cover areas reasonably homogeneous in environmental terms and to be tailored to local conditions. There is, though, the option of establishing a "general regulatory framework" instead of a series of independent zonal programmes.

Evaluation

The timetable for implementation was extended further than originally intended. Member states were to have submitted their proposals by the end of July 1993 for approval by the Commission on the advice of the "STAR" Committee composed of member state representatives. Some programmes were ready on time but many trickled in after the deadline, several had to be revised following negotiations with the Commission, and the new member states also joined the process. In consequence, the approval process continued into 1995. By May, the Commission had approved at least 140 programmes for co-financing and more were to follow (de Putter, 1995). Initially, 2.16 billion ecu had been earmarked for the agri-environment regulation over five years. However, as national schemes were proposed it was necessary to revise expenditure requirements upwards. A more recent estimate suggests that 3.16 billion ecu is a more likely total for the period up to 1997 (de Putter, 1995). A preliminary review of the programmes put forward by the member states raises a number of strategic issues.

The predecessor of Regulation 2078/92 was Regulation 797/85 which, under Article 19 had provided for the establishment of environmentally sensitive areas. That had limited impact and was only taken up by northern member states. Regulation 2078/92, with its much greater resources, its range of conservation options, and its obligatory requirements on member states, is having a much wider impact, no doubt picking up on the initial sluggish response to 797/85. In consequence, some states are introducing agri-environment policies for the first time; others have been using such policies for a number of years and have adapted and expanded them to the present Regulation. 2078/92 does provide for the first time a common European framework for national policies in the agri-environment field.

Regulation 2078/92 sets certain precedents for agricultural policy which may have long-term consequences. It has established the principle that farmers, for both environmental and production control benefits, should be paid to de-intensify production and to manage the countryside. The regulation thus legitimises non-productivist agriculture, particularly low intensity pastoral farming. In certain regions and farming systems, it also brings small-scale and/or part-time farmers within the scope of agricultural support policies. These are potentially major shifts which challenge the hegemony of organisations that represent large-scale productivist agriculture.

The regulation introduces *subsidiarity* into agricultural policy to a substantial extent. Inevitably, this is leading to considerable variation in national and regional responses to the regulation. To the extent that this reflects the varied nature of European rural environments and the social values attached to them, this is a desirable outcome. But it also seems to reflect variations in resources and capacities regionally and nationally. For example, within Germany, it is the richer southern Länder which have put through the biggest agri-environment programmes.

The differential capability of different regions to co-finance the agri-environmental measures is off-set by the much greater levels of EU assistance available for Objective 1 regions. However, in such regions, the limited organisational and administrative capacity may be a more significant barrier to full involvement in the agri-environment regulation.

Member states vary in the extent to which they have devolved responsibility sub-nationally for preparing schemes under the agri-environment regulation. In Germany, responsibility rests with the individual Länder, within an agreed national framework. In the UK, slightly different schemes have been introduced for England, Wales, Scotland and Northern Ireland. In preparing zonal programmes member states have fallen back on administrative units rather than seeking to tailor their schemes to coherent geographical areas.

There is a steady advance of environmentally sensitive areas. Article 19 covered a range of agricultural practices in specific 'sensitive' and important areas for landscape or wildlife. This focus on specific zones has tended to continue even under regulation 2078/92: in Spain's national and regional parks;

special habitats in Portugal; 'operations locales' in France; British ESAs and Natural Heritage Areas in Ireland.

All types of incentive have been used, but their distribution is uneven. For example, the provision for public access has mainly been used by the UK. Maintenance of abandoned land has, not surprisingly, been a priority for Mediterranean countries. There has been strong interest in endangered breeds from Germany, Spain and Portugal. Incentives for organic farming are available to both existing producers and new entrants in Germany, the Netherlands, Denmark and Spain, but in the UK it is restricted only to those converting to organic farming, and in France the national scheme covers only new entrants, but certain regional programmes are offering payments to existing organic producers.

The variety of national and regional responses has already thrown up some anomalies. The very different levels and types of support for organic farmers in different regions may well raise questions of unfair competition and market distortion. Other variations raise questions about the equity and effectiveness of the payments being made. For example, the Bavarian extensification programme pays farmers up to a stocking density of 2.5 livestock units per hectare; this compares with lower ceilings elsewhere including in other German Länder. Certain variations of this kind might be desirable in relation to different local environmental conditions, but the Bavarian ceiling might be considered unacceptably high in relation to any environmental benefits it might possibly achieve. The French approach has been to adopt a single national ceiling of 1.4 livestock units per hectare, but the problem here may be a lack of sensitivity towards localised conservation needs; for example, the ceiling is too low to support the management of important Alpine meadows in the French Jura. This national 'prime a l'herbe' or 'grassland premium' will absorb almost two-thirds of the total budget for French agri-environment measures. An EU official has commented:

> Dreamed up at the very moment when CAP reform was adopted, this premium is designed to encourage livestock farmers to maintain their extensive pastures in the face of competition from very generous area payments for maize silage under the arable regime but given the very low level of the premium and the fairly wide conditions for eligibility (1.4 livestock units/ha and 70 kg of nitrogen per hectare) its real usefulness has been questioned – will it really dissuade farmers in western France from cultivating their pasture land, or encourage the continued grazing of distant meadows in mountainous areas?
>
> (Delpeuch, 1994)

It is safe to predict that, not only in France, but in other countries too, tensions between environmental and income support objectives will be a recurring theme in the implementation of the regulation.

Although the regulation includes a diversity of conservation options, it adopts a common means to achieve them, namely payments to farmers and their voluntary involvement in agri-environment schemes. Implicitly or explicitly therefore it promotes a particular model for resolving the tensions between agriculture and the environment and of the property rights that should regulate the matter. There are already signs that this nascent European model is challenging other approaches, including those traditionally pursued at the national level. In certain circumstances, it may undermine the polluter pays principle. Also it challenges national systems where, in the past, the maintenance of low intensity farming systems was ensured by restrictions on property rights (for example, through zoning or land use planning restrictions). In addition, it may challenge the rationality of traditional approaches to the resolution of agriculture-environment tensions pursued through the sharp geographical segregation of agricultural and environmental functions (farm land versus natural areas).

Incentives are being deployed on a large scale to address environmental issues, which have been defined partly at a European level and partly by local administrations, particularly agriculture ministries. The standard procedure of offering farmers a form of management agreement for a fixed period may not be the most appropriate response to some of these issues and there are already questions about whether some schemes will generate any significant environmental benefit. Other questions are also emerging. What will happen after the management agreements expire? How far are national and regional administrations becoming committed to incentive payments as a permanent feature of rural policy?

Monitoring and evaluation of this policy will prove complex because of the extremely diverse agendas of individual member states as well as the familiar problems of monitoring the behaviour of some millions of participants in these voluntary schemes with complex management packages. Not only is there considerable variability in the packages of measures that member states have adopted but they also have very diverse administrative structures internally with the consequence that consistency of policy implementation across the EU will be very difficult to achieve or judge.

Conclusions

In terms of expenditure, the agri-environment regulation remains a minor element in the CAP. By 1995, it accounted for nearly 4% of the total CAP budget. There are conflicts with the major commodity regimes, not least when farmers are eligible for larger payments, say, for arable set-aside than for entering environmental schemes. In addition, these and other commodity supports are making a vastly greater financial injection into farm businesses than are agri-environment measures and, by encouraging additional investment in farm machinery and other impacts, maybe causing even more environmental damage. The organisations which represent large-scale productivist agriculture continue to be a powerful force in directing the CAP and there is a widespread perception that the regulation is an adornment to the CAP rather than a more far-reaching attempt to integrate environmental concerns into the heart of the policy.

Nevertheless, significant changes have occurred to the CAP, not only through its incorporation of environmental objectives, but also in the proliferation of regional schemes that it now embodies. Moreover, it would now seem to be in the process of what may prove to be a long drawn out transformation from being essentially an agri-food policy to more of a rural environment and rural development policy. This calls for clear thinking and a long term strategy to influence these changes. In the medium term much attention will be devoted to monitoring the impact and improving the effectiveness of new instruments such as the agri-environment regulation. However, with the prospect of significant sums going to these new fields, there are larger issues at stake which demand attention to basic institutional and accountability questions. There is a pressing need to explore the institutional mechanisms that would help satisfy the following criteria in the deployment of agri-environment resources: that payments are targeted to ensure cost-effectiveness; that the level and targeting of funds are responsive to public demand; that the public benefit is clear and worth the costs; and that tangible and enduring environment benefits result.

There are essentially two types of mechanism which may be used to manipulate land use change, one direct and one indirect, and both of them are currently in use. The indirect methods involve adjusting the incentives farmers and other land managers face, mainly through manipulation of prices. This has been the main mechanism of the CAP and has proved difficult to use precisely. Thus the setting of too high a price level for agricultural products and the inability (or lack of political will) to bring it down in line with gains in productivity must be seen as a major cause of agri-environmental problems in the EU. This mechanism is easier to apply at the level of the Union than direct mechanisms because prices may be translated into members' currency units with apparent ease. Similar mechanisms have also dominated the agri-environmental field where subsidies have been the preferred method of inducing change in farming practices.

The introduction of area payments, combined with the failure of farm gate prices to fall as expected under the impact of decoupling, leaves open the possibility of an investment boom in machinery and other investment goods. This is likely to the extent that investment is still mainly funded from income and that farmers have a back-log of under investment to make good. In fact depreciation has been greater than gross fixed capital formation for some years in the UK and has only just overtaken it again as investment has grown in the past two years. The relevant magnitudes are displayed in Table 13.5.

If this hypothesis holds, then the future of farm investment depends on the extent to which world prices are maintained. If surpluses re-appear, as some believe, then investment might fall back. If not, then a more sustained period of high investment may result. That would threaten whatever gains have already been made through environment policies.

Direct methods include all administrative fiats and regulations which require or preclude particular actions. These mechanisms bring about changes in property rights. For example Regulation 337/85 obliged member states to conduct environmental assessment of various proposed developments before approving them. The introduction of that regulation thus changed the rights of would-be developers by imposing a requirement on them which would, if it had the desired effect, bring about environmentally preferred (and therefore probably more expensive and less profitable) developments. In curtailing the rights of landowners and occupiers such mechanisms reduce their potential income and curb their opportunities to make profit.

Table 13.5: Impact of CAP reform on income and investment in UK agriculture.

	1993	1994	1995
Arable Area Payments/Set-Aside (£m)	785	1,010	1,313
Net Farm Income (£m)	3,825	4,015	5,044
Depreciation (£m)	1,737	1,751	1,855
Gross Capital Formation (£m)	1,708	1,916	??

Source: MAFF, 1996.

Reflection on the preceding section and the complexity of land ownership and use shows why changes in the assignment of property rights are not popular with policy makers in the rural context. This is because any such changes will affect the wealth status of many millions of large and small proprietors who are a potent political constituency. Incentives administered through price supports and capital grants will always seem easier to apply, especially at the level of the EU which will remain a politically weak decision-making unit until it has greater democratic authority to present a more robust response to powerful lobbies.

Long term protection of the rural environment is a prerequisite for the sustainable delivery of the benefits it provides, because these benefits take long to generate or regenerate. A policy which seeks to provide such benefits must therefore accept the need to persuade farmers and landowners to relinquish some of their rights to develop land in return for commitments to long term support. The fact that many of these ecologically sensitive areas were protected historically through forms of common ownership, strongly suggests that such forms would be a promising place to start devising a secure conservation policy mechanism for the next century, that would reflect the public interest.

References

Allanson, P., Murdoch, J., Lowe, P. and Garrod, G. (1994) *The Rural Economy: an Evolutionary Perspective* Centre for Rural Economy Working Paper 1.

Baldock, D., Beaufoy, G., Haigh, N., Hewett, J., Wilkinson, D. and Wenning, M. (1992) *The Integration of Environmental Protection Requirements into the Definition and Implementation of Other EC Policies*. Institute for European Environmental Policy, London, 42pp.

Baldock, D. and Beaufoy, G. (1993) *Plough On! An Environmental Appraisal of the CAP,* a report to WWF UK. Institute for European Environmental Policy, London, 65pp.

Baldock, D., Beaufoy, G., Bennett, G. and Clark, J. (1993) *Nature Conservation and New Directions in the EC Common Agricultural Policy*. Institute for European Environmental Policy, London.

Baldock, D. and Lowe, P. (1996) The development of European Agri-environmental policy, in: Whitby, M.C. (ed) *The European Environment and CAP Reform: Policies and Prospects for Conservation*, CAB International, Wallingford.

Beaufoy, G., Baldock, D. and Clark, J. (eds) (1994) *The Nature of Farming: Low Intensity Farming Systems in Nine European Countries*. Institute for European Environmental Policy, London, 66pp.

Bignall, E. and McCracken, D. (1996) The Ecological Resources of European Farmland, in: Whitby, M.C. (ed) *The European Environment and CAP Reform: Policies and Prospects for Conservation*, CAB International, Wallingford.

Body, R. (1982) *Agriculture: the Triumph and the Shame*, Temple Smith, London, Manchester.

Boisson, J.-M. and Buller, H. (1996) France, in: Whitby, M.C. (ed) *The European Environment and CAP Reform: Policies and Prospects for Conservation*, CAB International, Wallingford.

Bowler, I.R. (1985) *Agriculture under the CAP* Manchester University Press, Manchester.

Brouwer, F. (1995) Pesticides in the European Union. In: Oskam, A.J. and Vijftigschild, R.A.N. (ed), *Policy Measures to Control Environmental Impacts from Agriculture*. Agricultural Economics Research Institute, Wageningen.

Brouwer, F.M., Godeschalk, F.E., Hellegers, P.J.G.J. and Kelholt, H. J. (1995) *Mineral Balances at Farm Level in the European Union*. Agricultural Economics Research institute (LEI-DLO).

Commission of the European Community (1990) *Farms in Mountainous and Less-Favoured Areas of the Community*. Luxembourg Office of the EC.

Curry, N. (1994) *Countryside Recreation, Access and Land Use Planning*, E. & F. Spon, London.

de Putter, J. (1995) *The Greening of Europe's Agricultural Policy: the Agri-environmental Regulation of the Mac Sharry Reform*. Ministry of Agriculture, Nature Management and Fisheries, The Hague, The Netherlands.

Delpeuch, B. (1992) PAC et environment. *Perspectives* 1 (June) 17–19.

Delpeuch, B. (1994) Ireland's Agri-Environmental Programme in the European Context. In: Maloney, M. (ed) *Agriculture and the Environment: proceedings of a Conference on the integration of EC environmental objectives with agricultural policy,* held in the Royal Dublin Society from March 9–11, 1994, Dublin: Royal Dublin Society.

Department of the Environment (1990) *This Common Inheritance: A Summary of the White Paper on the Environment*. HMSO, London.

European Commission (1985) *Perspectives for the Common Agricultural Policy*. Com (85) 333.

European Commission (1988a) *The Future of Rural Society*. Com (88) 501.

European Commission (1988b) *Environment and Agriculture*. Com (88) 338.

European Commission (1990) *Proposal for a Council Regulation (EEC) on the introduction and the maintenance of agricultural production methods compatible with the requirements of the protection of the environment and the maintenance of the countryside*. Com (90) 366.

European Commission (1992) *Council Regulation on Agricultural Methods Compatible with the Requirements of the Protection of the Environment and the Maintenance of the Countryside*. EC 2078/92.

Lowe, P.D., Cox, G., MacEwen, M., O'Riordan, T. and Winter, P.(1986) *Countryside Conflicts: the Politics of Farming, Forestry and Conservation*. Gower, Aldershot, UK.

MAFF (1996) *Agriculture in the United Kingdom 1995*, HMSO, London.

Munton, R.J.C. and Marsden, T.K. (1991) Occupancy change and the farmed landscape: analysis of farm level trends, *Environment and Planning*, 23, 499-510.

OECD (1994) *Agricultural Policy Reform*, OECD, Paris.

Peterken, G.F. (1996) *Natural Woodland: ecology and conservation in Northern temperate regions*, Cambridge University Press, Cambridge, xiii, 522.

Porru, P. (1992) Agrarian Land Law in Italy, in: Rosso Grossmann, M. and Brussard, W. (eds) *Agrarian Land Law in the Western World: essays about agrarian policy and regulation in twelve countries of the western world*, CAB International, Wallingford.

Portela, E. (1994) Manuring in Barroso. In: Van der Ploeg, J.D. and Long, A. (eds) *Born from Within* Van Gorcum, Assen, the Netherlands.

Sandrey, R. and Reynolds, R. (eds) (1990) *Farming without subsidies: New Zealand's Recent Experience*. Upper Hutt, New Zealand, GP Books.

Saunders, C.M. (1995) *Subsidies without Farming*. Department of Economics and Marketing, Lincoln University, New Zealand, pp. 15.

Saunders, C.M. and Whitby, M.C. (1996) *Estimating the Full Cost of Environmental Policies*. Working Paper, Centre for Rural Economy, forthcoming.

Shoard, M. (1980) *The Theft of the Countryside*, Temple Smith, London.

Stanners, D. and Bourdeau, P. (1995) *Europe's Environment: the Dobris Assessment (with Statistical Compendium)*, Earthscan, London.

Vail, D., Hasund, K.P. and Drake, L. (1994) *The Greening of Agricultural Policy in Industrial Societies: Swedish Reforms in a Comparative Perspective*, Cornell University Press, Ithaca.

van der Bijl, G. and Oosterveld, E. (1996) The Netherlands. In: Whitby, M.C. (ed) (1996) *The European Environment and CAP Reform: Policies and Prospects for Conservation*. CAB International, Wallingford.

Whitby, M.C. (1994) Transactions Cost and Property Rights: the Omitted Variables? Albisu L-M and Romero, C (eds) *Environmental and Land Use Issues: an economic perspective*, 34th Seminar of the EAAE. Zaragoza, Wissenschaftsverlag Vauk Kiel.

Whitby, M.C. (ed) (1994) *Incentives for Countryside Management: the Case of Environmentally Sensitive Areas*, CAB International, Wallingford.

Whitby, M.C. (ed) (1996) *The Environment and CAP Reform: Policies and Prospects for Conservation*. CAB International, Wallingford.

PART IV

The CAP and the World

Chapter 14

The CAP and Central and Eastern Europe

Allan Buckwell and Stefan Tangermann

Introduction

Until 1989, the main focus of discussions on further enlargement of the EC was on Turkey and the Mediterranean islands of Cyprus and Malta. At that stage there was little serious talk even of the Nordic enlargement. The momentous events of the autumn of 1989 changed all this. Within two years, the Marxist-Leninist political constructions of central and eastern Europe had disappeared: the iron curtain was drawn back, the Berlin Wall fragmented into souvenirs, Germany was reunited, the satellite countries of central Europe regained their independence, Czechoslovakia experienced her velvet divorce, the Soviet Union vanished, and bloody civil wars started in the former Yugoslavia and several of the republics of the former Soviet Union. These events fundamentally changed the geopolitics of Europe.

It very quickly became apparent that a high political priority for the new democracies of central Europe was to integrate into the institutions of western Europe, especially the European Union and NATO. They wanted quickly to cement themselves into the security net and the economic strength of western Europe. These feelings were reciprocated as western Europe wanted to do all it could to replace the fear (and expense) of having an antagonistic, monolithic, super-power eastern neighbour with a group of independent, friendly, democratic, freely-trading market economies. In these circumstances there has never seemed to be any question that membership of the EU would be the goal as seen from both sides.[1]

As in the creation of the EC itself, whilst the primary goal of eastern enlargement is political, the instruments are economic. By creating, ultimately, a single market in which all goods, services, people and capital can move freely (via the 1986 Single European Act) and an economic and monetary union in which there is a single currency and monetary policy and co-ordinated macroeconomic policy (via the 1992 Maastricht Treaty), the intention is to ensure the greatest integration of the European economies and maximum, interdependent, economic growth which provides the best guarantee that ancient animosities (and border and ethnic disputes) are settled amicably through economic trade-offs rather than by force of arms. There has been little, or no, explicit public debate about the desirability of integrating central Europe into the EU; as with many such 'big picture' political matters, decisions are taken without explicit calculations of the economic costs and benefits. Politicians and public alike have seemingly concluded that this development is a necessity and thus they devote their energies to thinking about when and how it might be done.

The first formal steps of this process were the setting up of programmes of technical and economic assistance by the institutions of the EU, principally the PHARE[2] programme and the activities of the European Bank for Reconstruction and Development (EBRD). This was quickly followed by the start of negotiations for association agreements between central European states and the EU. By June 1993, at the Copenhagen Summit, the European Council agreed that "the associated countries that so desire shall become members of the European Union". Between 1993 and 1995 interim versions and then the full association agreements came into force and the political discussions about the eastern enlargement gathered pace and intensity. Throughout these steps four countries, known as the Visegrad-4, (Poland, the Czech and Slovak Republics and Hungary) were ahead of the two Balkan countries, Romania and Bulgaria. By 1994 it was clear that, amongst the fragments of the former Yugoslavia, only Slovenia had sufficient claim for political and economic stability, and also that the three Baltic states, Estonia, Latvia and Lithuania would be treated differently from all other republics of the former Soviet Union. This collection of ten states who are candidates for EU entry are variously known as the Central and Eastern European Countries (CEECs, pronounced 'seeks'), or Pays d'Europe Centrale et Orientale (PECOs),[3] the latter term will be used here.[4] No timetable has been laid down for the eastern enlargement, nor is one likely to be. No decisions have been made about whether the countries will enter *en bloc* or singly. Instead, only general conditions are stated concerning political stability and institutions, the need to guarantee democracy, the rule of law, human rights and respect for and protection of minorities, the existence of a functioning market economy as well as the capacity to cope with competitive pressure and market forces within the EU.

It is reasonable to ask why these matters deserve attention in a book on the CAP. The answer is based on the simple juxtaposing of two facts: (made initially by Baldwin, 1995) that whilst the EU spends four fifths of its budget on agriculture and poor regions, the PECOs are significantly poorer and more agricultural than the EU. This has fuelled the fear that eastern enlargement is beyond the resources of the EU, and will necessitate either radical changes to EU policies or long delay before the enlargement can take place. Both of these avenues are unpalatable to some significant interests. In all enlargements of the Community to date the principle has been that the acceding countries must accept the *acquis commaunitaire*, that is they must adapt to the EC/EU by adopting all Community legislation; it is not for the EU to adjust its policies to suit new members. But equally, such strong expectations of membership have been created that it is a political imperative that concrete progress on this enlargement must take place in the final years of the 1990s. To this end, the Essen European Council in December 1994 requested the Commission to bring forward, by the end of 1995, ideas for accommodating the enlargement to the policies of the EU.

In fact, the Commission had already commenced this process. In spring 1994, DGVI sponsored a study by Nallet and Van Stolk (1994) to examine the problems and options of the PECOs adopting the CAP. Their report was very pessimistic about the recovery in PECO agriculture and urged the EU immediately to help them set up a system of co-ordinated support for agriculture along the lines of the CAP. This was itself a controversial idea and stimulated DGI (the External Affairs Directorate) to sponsor four more studies of the problems of PECO accession to the CAP and options for dealing with these problems. These reports (by Buckwell *et al.*, Tangermann and Josling, Tarditi and Marsh, and Mahé *et al.*, all published in 1995) expressed a range of views about the capacity and likelihood of agricultural recovery in the PECOs. However, all agreed that because of potential difficulties with commitments made in the Uruguay Round of the GATT and with the EU budget, to make enlargement feasible it would be advisable to continue to reform the CAP along the lines started in the 1992 Mac Sharry reforms.

These matters are both complicated and controversial. To establish a sound factual basis for discussions of the agricultural problems of eastern enlargement, DGVI itself conducted and published a series of detailed country studies of the PECO-10 (European Commission, 1995). These summarised the progress in agriculture and agricultural policy over the period 1989 to 1995 and projected expected market balances in the PECO-10 to the end of the century based on the assumption that there was no further change in policy. No attempt was made to draw conclusions concerning the difficulty of adopting the CAP. This is to be the subject of a further report.

In this chapter we first review the nature and evolution of PECO agriculture, and then the economic relations between PECOs and the EU in the first six years post-reform. This is followed by a summary of the analysis of the problems of PECO accession to the CAP and options for dealing with these problems. Our conclusion is that the CAP of 1995 is not the ideal form of agricultural policy for a European Union of 25 member states, 475 million people and 17.5 million farmers.

The PECOs, their agriculture and policy[5]

The PECOs

Absorbing the ten PECOs into the EU is a significant challenge. Table 14.1 indicates the size of the task. Whilst in proportionate terms the increases in population by 106 million (29%) and agricultural area by 61m ha (44%) are not dramatically bigger than the 1973 growth of the EC from six to nine, the absolute number of countries involved makes this enlargement of a different order than all previous ones. Also the gap in income levels and the increase in the agricultural population pose much larger problems than previously.

Comparing real standards of living is extraordinarily difficult. Simple comparisons of GDP and GDP per capita show the PECO-10 to represent only 3% of EU-15 GDP and income per head to be just 11% of the EU average. These are palpably misleading. Visitors to the PECOs can see that while material standards of living — for example housing standards, infrastructure and public services such as health and education — are at lower levels than western Europe, and on some measures considerably so, the gap does not seem as wide as the GDP figures indicate. Correcting exchange rates for purchasing power parity (PPP) is one way of trying to make more meaningful comparisons. The PPP rates quoted in Table 14.1 bring the CEFTA+ countries up to about one-third of EU per capita income levels and the Balkan-2 to one-fifth of EU levels.[6] However, even these gaps will take three to six decades of PECO growth at twice the EU rate (6% per annum rather than 3% per annum) to eliminate.

The record of economic growth in the PECOs since reform has been erratic. The initial effect of abandoning central planning and liberalising prices, production and trade, which occurred during the period 1990 to 1992, was that economic activity fell. This was an indication of the distortions in these economies. The pattern of resource allocation matched neither the real costs of production nor the preferences of consumers. To reallocate resources meant a temporary dislocation of production. For the CEFTA+ and Balkan-2 countries the worst fall in GDP was in 1991, averaging about 11%. The contraction in the Baltic states was a year later and much larger. By 1994 all PECOs except Latvia had recorded positive growth, and early indications were that 1995 should bring faster growth in most of the PECOs than in the EU-15. By the end of 1995, Polish GDP will be just 3.5% below the level in 1989, whereas for the other CEFTA countries, the Balkan-2 and the Baltics, GDP will have fallen 17%, 23% and 47% below the pre-reform level, respectively. By comparison, in the six years 1990 to 1995, which included what was considered the worst post-war recession in W Europe, there was a cumulative *growth* in the EU-15 economy of over 11%!

Table 14.1: PECO-10 in comparison with EU-15.

	Population	Agric area		Agric production		Agric employment		GDP	GDP pc	
	(mio)	(mio ha)	(%total)	(bio ECU)	(% GDP)	(000)	(% tot.empl.)	(bio ECU)	(ECU)	(ECU PPP)
Poland	38.5	18.6	59	4.684	6.3	3661	25.6	73.4	1907	4838
Hungary	10.3	6.1	66	2.068	6.4	392	10.1	32.5	3150	5967
Czech Rep.	10.3	4.3	54	0.871	3.3	271	5.6	26.7	2586	7507
Slovak Rep.	5.3	2.4	49	0.512	5.8	178	8.4	8.7	1643	6367
Slovenia	1.9	0.9	43	0.250	4.9	90	10.7	9.8	5018	7697
CEFTA+	66.4	32.3	58	8.349	5.5	4592	22.1	151.1	2277	5635
Romania	22.7	14.7	62	4.500	20.2	3537	35.2	21.8	961	2941
Bulgaria	8.5	6.2	55	1.131	10.0	694	21.2	9.4	1110	3754
Balkan	31.2	20.9	60	5.631	18.0	4231	32.9	31.2	1001	3163
Lithuania	3.8	3.5	54	0.259	11.0	399	22.4	2.3	627	n.a.
Latvia	2.6	2.5	39	0.232	10.6	299	18.4	2.2	850	n.a.
Estonia	1.6	1.4	31	0.266	10.4	89	8.2	1.5	938	n.a.
Baltics	7.9	7.4	43	0.757	10.7	717	19.4	6.0	757	n.a.
PECO-10	105.5	60.6	56	14.7	7.8	9540	26.7	188.3	1786	n.a.
EU-15	369.7	138.1	43	208.8	2.5	8190	5.7	5905.1	15972	15879
PECO/EU	29%	44%						3%	11%	

Source: European Commission (1995).

PECO agriculture during transition

Within this decline in general economic activity, agriculture too suffered a severe recession (see Table 14.2[7]). In most countries (except the Baltic states) for the first few years of transition, agricultural output declined less than the rest of the economy. However this pattern has not persisted and since 1992/93 for most countries whilst the decline in economic activity abated and then reversed, this happened later and by less for agricultural output. Thus in 1994 for all countries, except Slovenia, Romania and Lithunania, the index of agricultural output

Table 14.2: Indices of gross domestic product (GDP) and gross agricultural product (GAP) for the PECOs and the EU-15, 1989=100.

		1990	1991	1992	1993	1994	1995
Poland	GDP	88.4	82.2	84.3	87.6	91.9	96.5
	GAP	94.5	93.0	82.9	84.5	78.6	
Hungary	GDP	96.7	85.2	81.5	79.7	81.2	81.5
	GAP	95.3	89.4	71.6	64.7	65.6	
Czech Republic	GDP	98.8	84.8	79.3	78.6	80.7	84.1
	GAP	97.7	89.0	78.3	76.4	72.2	
Slovak Republic	GDP	97.5	83.5	78.6	75.4	79.0	82.6
	GAP	92.8	85.9	74.0	68.4	74.6	
Slovenia	GDP	95.3	87.6	82.9	83.9	88.1	92.5
	GAP	104.2	101.1	90.5	98.0	118.2	
CEFTA+	GDP	93.8	84.7	83.4	84.3	87.6	91.0
	GAP						
Romania	GDP	94.4	82.2	73.9	74.8	76.6	78.6
	GAP	97.1	97.9	84.9	95.7	101.0	
Bulgaria	GDP	90.9	80.3	75.6	72.4	72.6	73.3
	GAP	94.0	93.7	82.5	67.5	70.2	
Balkan-2	GDP	93.0	81.3	74.1	73.8	75.2	76.9
	GAP						
Lithuania	GDP	96.7	84.0	55.5	40.4	41.2	43.3
	GAP	91.1	87.2	66.4	61.1	47.7	
Latvia	GDP	102.9	92.2	60.0	51.1	50.0	52.5
	GAP						
Estonia	GDP	93.5	85.9	73.6	67.6	70.3	73.8
	GAP	86.9	83.4	67.9	62.7	56.4	
Baltics	GDP	99.0	88.1	60.9	49.9	50.5	53.0
	GAP						
PECO	GDP	93.8	84.3	81.4	81.5	84.3	87.4
	GAP						
EU-15	GDP	102.9	104.5	105.6	105.1	108.0	111.5

Source: Calculated from European Commission (1995) Tables 2 and 5.

(1989=100) was lower than the comparable index of GDP. In short, apart from Slovenia and Romania, economic recovery is generally not being led by agriculture, rather the other way round, agriculture is the trailing sector.

The reasons for the agricultural recession are quite clear although there is no definitive analysis partitioning the relative contributions of the three main economic elements: the collapse in demand, the changes in relative prices and the structural changes. In addition to these, nature was not very kind to PECO farmers. In 1992, and for some in 1993 also, there was a fairly severe drought which particularly hit the crop sector.

Consumption changes

The reforms caused a sharp contraction in income and this led to a fall in consumption of most foods, especially meat. For some countries who were significant agricultural exporters to COMECON and especially the Former Soviet Union (FSU), the contraction of foreign demand compounded the domestic market collapse. Because consumption levels of carbohydrate, protein and fats were high in the PECOs pre-reform, it is far from clear that previous levels of gross consumption will be regained as economic growth resumes. There are signs that the pattern of demand is changing and consumers wish a greater range of more processed and better presented foods which, pending the restructuring and re-equipping of the PECO food industry, are increasingly being imported from western Europe.

Changes in relative prices

There have been several significant switches in relative prices confronting PECO farmers. First, agricultural output prices have risen by considerably less than prices of variable inputs (fertilisers, plant protection and animal health products, energy and feeds). Second, farm prices have increased by less than retail food prices. Third, the costs of variable inputs and credit have risen by more than wages. The net effect of these changes has been a decline in ‘profitability’ of agriculture which, quite rationally led to changes in factor intensity (a decline in capital intensity) and a contraction in output. The changes are a reflection of both the distortions embedded in the planned economy,[8] and also the uneven pace of liberalisation and privatisation. Thus whilst foreign trade was initially almost completely liberalised, and partially privatised, allowing prices of tradable inputs (fertilisers, fuels, feeds) to rise immediately and fully to international levels, there was a smaller degree of liberalisation of consumer food prices. In many, but not all, countries, governments still try to ‘stabilise’ the price of some staple foods. The extreme cases are Romania where there remain strong elements of centrally planned food prices and Bulgaria where the government retains mechanisms to control price margins for basic foods. It is also true to say that privatisation has progressed more slowly in the food processing industry and there is little doubt that large inefficiencies and in some cases excess profits are being generated in the ‘downstream’ sector as a result of the lack of competition in the food chain.[9] As a result, marketing and processing margins have increased. This factor, together with the elimination of large subsidies that had been injected into the food chain pre-reform, explains why farm prices have increased less than retail food prices, in spite of all attempts at ‘stabilising’ food prices.

Farm restructuring

The third factor contributing to the fall in output is the dislocation caused by farm restructuring. This is probably the least uniform of the three factors. Poland and Slovenia were largely successful in resisting collectivisation. In the former 75% of land remained in small private farms and in the latter nearer 90% of the land was outside the 'socially owned' sector. In these two countries therefore there has been almost no restructuring and thus no dislocative effect. In all the other countries there had been a much greater degree of Soviet style collectivisation to create varying mixes of collective farms (Kolhozes) where land was collectively 'farmed'[10] and state farms (Sovhozes) based on state-owned land. In all countries there was also the phenomenon of small private plots which often produced significant amounts of the total output. Farm privatisation or restructuring has involved a complex mix of restitution of land to former owners, distribution of land to current 'farmers', compensation of former land owners, fragmentation or liquidation of former collective farms, relabelling of state farms to private co-operatives and privatisation of state farms. Each country has its own battery of new laws defining land reform, land tenure, co-operatives and commerce. The result has been a complex variety of new farming structures whose characteristics, performance and endurance will take time to describe and evaluate.[11]

There are some generalisations which emerge from this restructuring. The number of 'farms' has increased and the average size has decreased as the largest collectives and state farms have been subdivided. At the other end of the scale there has been some tendency for small private farms to emerge from the nucleus of private plots as land is restituted, and as some individuals acquire farming rights over restituted land. Thus the former bimodal structure is being replaced by a more conventional range of sizes and structures of farms: potentially with productivity gains at both ends as the over-grown collectives are divided and the over-fragmented plots coalesce into more viably sized units (Buckwell and Davidova, 1993). The range of farm structures has increased. There remain state farms, and some state co-operatives, but there are now also private co-operatives, farming companies, joint stock companies, farming associations, partnerships, and individual private family farms. Throughout the region the process of legally restoring property rights has not progressed far enough for an active land (sale) market to emerge. Land leasing is becoming increasingly common, but pending the settlement of property ownership, leases tend to be short and insecure.

There were significant differences in the size and nature of the farms created by these processes between countries. In the Czech and Slovak republics and Hungary, the restructuring has mostly taken the form of compensating former land owners and relabelling the collectives into private co-operatives. This is the least disruptive of all. Sometimes very large units were split into more manageable farms. The main changes to these farms is that the decisions are now made on the farm (usually by the same management as in the former regime) and the farm has to survive in a more rigorous climate where neither input supplies nor product markets are guaranteed, and where debts are not automatically cleared. The Balkan and Baltic countries have generally taken a more radical approach to restructuring. In Romania and the Baltics, land has been distributed to current farm workers and co-op members. This is causing two problems: the danger of creating an over-fragmented farm structure; and disruptive disputes as former owners claim they have been deprived of their land. Bulgaria avoided the latter by rigorously trying to restitute all land to former owners or their heirs. However they combined this with the politically charged decision to liquidate all collective farms. As a result neither restitution nor liquidation has been completed and the interim result is an unsatisfactory mix of very small private farms and new co-operatives whose legal status and land tenurial arrangements are not settled.

The contribution of these structural changes to the fall in agricultural output, and more importantly for the long term productivity of agriculture is hard to discern. Whilst in the short term the CEFTA countries may not have suffered from drastic restructuring, it remains to be seen how competitive and responsive the producer co-operatives turn out to be. In all countries, the incomplete process of defining land ownership rights and the lack of active land markets undoubtedly hampers agricultural development as farmers are thereby deprived of their best source of collateral for securing loans.

Agricultural policy in the PECOs

The very process of economic transformation from central control of all prices and quantities to a decentralised market system initially and inevitably results in liberalised markets.[12] Dissatisfaction with the outcome was soon apparent. The shifts in relative prices described above, which were mostly to the detriment of farmers, soon encouraged PECO governments to take an active role in agricultural markets using both domestic and border instruments. This was accentuated as the PECOs were very aware that the EU itself departs a long way from the 'market solution'. Furthermore, there is a great sensitivity in the PECOs to the lack of access to EU food markets and the local effects of subsidised exports coming from the EU. It was very predictable that PECO governments would not resist for long the temptation to try and stabilise and control agricultural markets.

Swinnen (1995) shows that the nature and extent of these interventions can be explained by a fairly simple model of the political economy in the PECOs. Countries (for example Hungary and the Czech Republic) with the highest GDP/head, the smallest dependence on agriculture and the smallest proportion of consumer expenditure spent on food, soon introduced measures to protect their farmers. Poorer countries such as Bulgaria with a greater dependence on agriculture and higher importance of food in total expenditures, tried to keep food price increases down and have, in the process, taxed their farmers.

Within this broad picture the development of agricultural policy has been complex; it has varied from year to year, between countries and commodities. In the Visegrad-4 there has been a conscious effort to model price support systems on the CAP. This was done both for defensive reasons and because it was seen as making constructive steps towards harmonising policy with EU membership in mind. Thus PECOs started using variable levies, intervention purchasing and export subsidies. However for all the superficial similarities with the CAP, there are great differences too. These countries had neither the budgetary nor administrative resources to implement the scale of support programmes developed under the CAP. Thus relatively small quantities of produce were taken off the market by the intervention agencies and intervention prices were considerably below those in the EU (for example between 30% and 45% of EU intervention prices for wheat and beef, and between 28% and 65% of EU milk prices — see European Commission (1995, Table 10). Similarly, tariffs or variable levies were generally lower than in the EU, and the use of subsidised exports was much more restrained. Given the, sometimes dramatic, fall in production there has been a great sensitivity to the issue of food security in the PECOs. Understandably, especially for governments long used to having strict quantitative controls on production and stocks this led several of them to try and purchase sufficient grain for basic bread requirements and sometimes to try and impose quantitative controls on trade (through export quotas and sometimes prohibition of exports). The resulting depression of domestic market prices then stimulated calls for offsetting subsidies on inputs, most commonly credit subsidies.

Despite this drift towards agricultural protectionism there are several constraints limiting such a move. First, most PECOs have a serious problem with their public deficit. The calls on public finances to support public services and also to provide a safety net for the rapidly increasing pool of unemployed, outstripped the ability of PECOs to expand their tax base. It takes time to create the culture and administration to capture taxes from newly emerging private economic activity. As a result, budgetary spending on agricultural support has declined in real terms for most countries during transition. Second, given that PECO citizens on average allocate 30% (Poland) to 60% (Romania) of total expenditure to food, governments are much more sensitive to the effects of farm policy on food prices than in the EU (where average food expenditure is 22% of total expenditure).[13] Third, most of the PECOs are already members of the World Trade Organisation (WTO) and thus signatories to the Uruguay Round Agreement on Agriculture (URA) which limits the extent to which they can protect their domestic market, restrict imports and subsidise exports.[14] The precise impact of these URA restrictions depends on the schedules negotiated by individual PECOs. A very wide range of commitments were made; the Poles seem to have allowed themselves considerable latitude to increase protection to close to EU levels, whereas, for example, the Romanians appear to have very little scope to increase domestic supports or subsidise exports.[15] A fourth constraint on some of the PECOs has been the conditions applied by international organisations (the IMF and World Bank) as part of structural adjustment and sectoral loans. As with the URA constraints, this has not been applied evenly across countries.

The net effect of the different agricultural structures and different policies in place in the PECOs is that there is a wide range of farm product prices between them. There is also a wide gap between their prices and those in the EU. These are exemplified in Table 14.3 which has been extracted from the Commission (1995) report. The figures should be taken only as indicative. There are immense difficulties in getting comparable figures with respect to commodity definition, season, location, stage in the marketing chain, and exchange rates. It can be seen that apart from Slovenian wheat and pork, and Polish pork,[16] all PECO farm gate prices are below those in the EU. The gap varies enormously by commodity and country: e.g. for wheat from 40% of the EU price in Bulgaria to 90% for Latvia, and for milk from 21% of the EU price in Lithuania to 92% for Slovenia.

Summarising, the accidental process of history has thrown up a massive challenge to Europe. Ten countries with very different economic and agricultural structures see their vocation as EU members. They are experiencing great differences in their success in navigating the process of economic transition, and in the process they have developed different agricultural policies. How can these differences be narrowed and a smooth adjustment towards membership of the EU and its Common Agricultural Policy be achieved? The first step has been to forge formal free trade agreements with the EU to which we now turn.

Trade relations between PECOs and the EU

When the process of domestic reforms began in the PECOs, their political and economic relations with other countries underwent fundamental reform as well. As a half-way house on the road towards membership, association agreements (somewhat emphatically called 'Europe agreements') have been concluded between the PECOs and the EU. As a part of these agreements, agricultural trading relations have been redefined.

Pre-reform trade relations

Under the old regime of central planning, foreign trade of the PECOs was, of course, centrally planned as well. Moreover, for external trade of the PECOs the world was split in two rather separate blocs. Trade with other centrally planned countries in the Soviet-ruled empire was conducted under the auspices of the Council of Mutual Economic Assistance (CMEA or COMECON). For all of the PECOs, trade with other members of COMECON tended to be the largest part of their foreign trade. Within COMECON, prices for traded products were, like prices on domestic markets, essentially politically-determined accounting prices, not directly related to world market prices. A specific unit of account, the transfer rouble, was used for settling trade balances, and this quasi-monetary unit was not convertible into other currencies. As far as agricultural products were concerned, large parts of PECO exports went to other COMECON members, above all the Soviet Union.

Table 14.3: Selected PECO, EU and world commodity prices in 1994.

	WHEAT			MAIZE			MILK		BEEF		PORK		POULTRY	
	ECU/t	% EU	% world	ECU/t	% EU	% world	ECU/t	% EU	ECU/t	% EU	ECU/t	% EU	ECU/t	% EU
Poland	98	73	104				103	33	1240	40	1320	103	1179	88
Hungary	75	56	80	72	52	97	220	70	1630	52	1260	98	1038	77
Czech R	88	66	94	100	72	135	172	54	1850	59	1200	94	910	68
Slovak R	84	63	89	93	67	126	164	52	1580	50	1130	88	987	74
Slovenia	175	131	186	123	89	166	292	92	2510	80	1710	134	1090	81
Romania	81	60	86	75	54	101	179	57						
Bulgaria	54	40	57	71	51	96	114	36	750	24	680	53	590	44
Lithuania	60	45	64				66	21	680	22	1040	81		
Latvia	121	90	129				83	26	560	18	980	77		
Estonia	75	56	80				83	26	360	12	550	43		
EU	134		143	138		186	316		3130		1280		1340	
World	94	70		74	54									

Wheat, maize and milk prices are farm gate prices. The world wheat and maize prices are notional farm gate prices by deducting 10ECU/t from the fob export price (Argentine and US Gulf price, respectively). EU beef and port prices are wholesale prices; CEEC meat prices are farm gate prices converted from liveweight.

Trading relations of the PECOs with western countries were also governed by special regimes which generally differed from those applied to trade among western countries. The Visegrad countries and Romania were members of the GATT. However, as centrally planned countries, where market prices and hence GATT-regulated trade measures such as tariffs did not have much meaning, the PECOs had specific GATT commitments. Rather than having to observe tariff bindings and other related GATT rules the PECOs which were GATT members had committed themselves to import a given total value of products from other GATT members, and that import value had to grow by an agreed minimum percentage every year. Western GATT countries, on the other hand, could and did apply quantitative restrictions to imports from these COMECON countries. The only exception from this regime among the PECOs was Hungary which had a more liberal trading regime even before the process of reforms began in 1989. In addition to these general rules for GATT members among the COMECON countries there were many bilateral trading arrangements with western countries, relating both to trade in particular products and to the settlement of trading accounts. Trade relations of the PECOs with the EU, including agricultural trade, were also governed by that same overall regime.

As a result, trade between the PECOs and the EC was not very dynamic before the reforms started. During most of the 1980s, the PECOs had a slightly positive balance in trade with the EC, both in general and for agricultural products. As the reforms occurred, trade with the EC began to increase rapidly, and the balance turned negative for the PECOs. In agricultural trade of the PECOs, the intensity of their links with the EC pre-reform differed significantly among the individual PECOs. In the first half of the 1980s, nearly half of Poland's agricultural exports went to the EC. Czechoslovakia, Hungary and Romania each exported around one quarter of their agricultural products to the EC. On the other hand, Bulgaria shipped less than 5% to the EC. Again, the major changes in agricultural trade relations with the EC which were to come about as a result of the reforms were indicated by a large upward jump in the EC share of total PECO agricultural exports. In the case of Bulgaria, for example, the EC share in total agricultural exports jumped from 5% in the first half of the 1980s to around one quarter in 1989–91.[17]

The growing intensity of PECO trade with the EU as reforms began was in part a result of the special trading arrangements between the PECOs and the EU, to be discussed in detail below. However, it also reflected the breakdown of the COMECON soon after the reform process began. Trade among the former COMECON members suddenly had to be conducted at true world market prices, and paid for in convertible currencies. Moreover, with the steep economic decline in all former COMECON countries, above all the former Soviet Union, import demand dwindled. As a result, the PECOs lost large parts of their former export markets among their eastern neighbours, for both agricultural and other products. Against this background it came as no surprise that their trade was redirected towards western countries.

Trade relations since the reforms

When the reform process began in the PECOs, the EC was quick in responding by revising its trade regime *vis-à-vis* the PECOs. As a first step, the EC lifted its previous quantitative restrictions against imports from Hungary and Poland in 1990, and beginning in 1991 it also began to eliminate them step by step *vis-à-vis* Bulgaria, Czechoslovakia and Romania. The second step was to provide preferential market access to the PECOs. The EC included most of the PECOs in its Generalised System of Preferences (GSP) extended to developing countries, beginning with Hungary and Poland in 1990, and followed by other PECOs in 1991.[18] As far as agricultural products were concerned the GSP covered mainly tropical products which were of no interest to the PECOs. In order to help the PECOs, the EU therefore also included some types of pork and poultry meat in its GSP, with a 50% levy reduction on limited quantities. A further step taken by the EU was the conclusion of trade and co-operation agreements with some of the PECOs; but these did not have much economic effect.

The most important phase in EU trade relations with the PECOs so far was reached when association agreements were concluded. As explicitly stated in the preambles of these agreements, this is the first direct step towards eventual membership of the PECOs in the EU. The association agreements contain elaborate trade preferences, including agricultural products.[19] In order to allow these trading arrangements to take effect before the time consuming process of ratification of the overall association agreements was brought to an end, the trade parts of the agreements were allowed to enter into force on the basis of interim agreements. For Hungary, Poland and the then Republic of Czechoslovakia this commenced on March 1, 1992, for Romania on May 1, 1993, and for Bulgaria on December 31, 1993. The full association agreements have entered into force, after ratification by all national parliaments, on February 1, 1994, with Hungary and Poland, and on February 2, 1995, with the remaining PECO-6 (OECD, 1994).[20]

The trade provisions under the association agreements are, in principle and in GATT terms, supposed to establish free trade arrangements between the PECOs and the EU. They are mutual but asymmetric, in the sense that both sides eliminate trade barriers, though the PECOs have more time for eliminating their barriers against imports from the EU. For most industrial products, trade will eventually be completely free. In agriculture, however, neither the PECOs nor in particular the EU had the courage to open markets up completely. Indeed, when it came to negotiate the agricultural concessions there were a number of occasions when the overall negotiations were close to breakdown because some EU member states felt that too much access to EU agricultural markets might be provided for the PECOs. In the end, most trade preferences in agriculture granted by the EU were strictly limited to given, and generally rather small, quantities. The agreements were to be phased in over five years, with the tariff preferences and tariff quotas increasing in steps. In the event this transition period was accelerated for most of the PECO-6. Whilst the agreements do not seem generous towards the PECOs, it must be noted that there has not yet been any other group of third countries to which the EU has granted similarly far-reaching trade preferences in agriculture, including reductions of variable levies on a wide array of core CAP products.

Agricultural trade preferences granted to the PECOs under the association agreements, equivalent in nature for all PECOs, come in several different forms. However, with the exception of very few products, all preferences are strictly limited to maximum quantities. Both tariff/levy reductions and quotas are defined at a highly disaggregated product level. For each of the PECO-6 some 250 to 400 individual products are listed. Quotas set in the negotiations are generally based on actual PECO exports to the EU (including the former GDR) in a past reference period (1988–90 for CSFR, Hungary and Poland, 1989–91 for Bulgaria, and 1988–89 for Romania). Products not exported to the EU in the past, such as cereals, many dairy products, sugar and many sugar products, are not included in the preferential arrangements. In effect this means that products where the PECOs pre-reform could not export to the EU, because EU import barriers were prohibitive or for any other reason, do not benefit from preferential treatment. Products where market conditions have changed, either in the PECOs or in the EU, after the reforms and where the PECOs now have a comparative advantage and therefore could profitably export to the EU, have often not been granted preferential access to EU markets.

The quantitative significance of the agricultural trade preferences granted under the association agreements differs significantly among the PECO-6, for a number of reasons. First, as stated above, the share of their total agricultural exports going to the EU varies between 27% for the Czech and Slovak Republics and Romania, and 59% for Poland (see Table 14.4, row 1). Second, the product coverage of preferences also differs. For Hungary, nearly 66% of all agricultural exports to the EU are in products which receive some form of preferential treatment. For Poland, on the other hand, only 45% of its agricultural exports to the EU are covered by preferences (Table 14.4, row 2). If agricultural exports which enter the EU free of tariffs (and therefore could not receive preferential treatment) are excluded from the base, the product coverage is somewhat larger, as high as 77% for Hungary, but still as low as 51% for Poland (Table 14.4, row 3). The low percentages for Poland are in part due to the fact that the relatively large Polish exports of fish and live animals to the EU have been covered only to a small extent. Relative to overall exports to the EU (including non-agricultural exports), the agricultural preferences are only minimal, covering between 1.2% of total export value for Romania and 4.4% for Poland (Table 14.4, row 4).

The quantities allowed to be exported to the EU under preferential conditions increase over time. If the PECOs can make full use of the preferential quotas, this allows for a growing value of agricultural exports to the EU. Given the varying product composition of preferences and the different rates of quota growth under the different types of preferences, the potential for export growth stemming from the expansion of quotas over time varies considerably among the individual PECO-6. For example, if Poland were to make full use of the growing quotas, this would add, in the fifth year of the agreement (i.e. when the maximum of quotas is reached), nearly 12% to its base period agricultural exports to the EU. The Czech and Slovak Republics taken together, however, could expand their agricultural exports to the EU by more than 49% (Table 14.4, row 5).

As far as economic benefits to the PECOs are concerned, however, the possibility of exporting more to the EU may in the long run be less valuable than the possibility of receiving higher prices as a consequence of lower levies and tariffs to be paid on exports to the EU under preferential conditions. The reasoning behind this statement is simple, but important to understand. Growing quantities exported to the EU tend to cause some sort of opportunity cost. These extra quantities have to be produced over and above existing production volumes. In this case they cause resource costs, unless they were produced with otherwise completely unemployed resources (which is unlikely, in particular as far as variable inputs are concerned).[21] Alternatively, extra exports to the EU could originate from diverting exports to the EU which would otherwise have gone to different destinations. In this case the opportunity cost is the loss in revenues from exports to those other destinations. Finally, extra exports to the EU could also originate from lower domestic consumption, in which case there is an opportunity cost from consumption foregone in the PECOs. In any case it would be wrong to consider the revenues from extra exports to the EU as a result of growing preferential quotas as a full addition to economic welfare in the PECOs.[22]

On the other hand, higher prices received by the PECOs as a consequence of lower EU levies and tariffs are a cost free increase in PECO welfare, as no resources have to be employed in order to gain this extra export revenue. A rough estimate of the magnitude of this potential price advantage can be gained by assessing the preference margin on exports concerned, i.e. the size of the levy or tariff reduction. This preference margin is the upper bound of the potential price advantage, as the maximum price gain achievable is less where EU market prices are depressed as a result of growing exports from the PECOs (though price depression is likely to be very small, see below). The size of that preference margin again varies considerably among the individual PECO-6.[23] In the fifth year of the agreements (i.e. when the highest quotas and the maximum levy reductions have been reached), the potential price advantage resulting from preference margins might be as much as nearly 25% of base period agricultural exports to the EU for Romania, but as little as 5% for Bulgaria (Table 14.4, row 6). The different size of these potential price gains results again from the varying product composition of preferences and from the varying levels of EU levies and tariffs on different products.

These *potential* price gains, however, accrue to the PECOs in reality only if they can manage to collect the preference margin. Alternatively it is also possible that importing companies in the EU get hold of the preference margin. What happens in practice very much depends on which administrative procedures are used to implement the levy and tariff reductions. In particular, where preferences are limited to given quotas, as is the case for most products, the approach used for allocating licences under preferential quotas largely determines the distribution of the preference margin. After all, as is the case with all quantitative restrictions, rents tend to accrue to those agents who hold the licences. Unfortunately, for the PECOs, the EU decided to issue licences itself, and only to EU based trading companies. PECO exporters often cannot even find out whether the EU company they are dealing with has a licence to import under preferential conditions. As a result, preference margins tend to flow largely to importing companies in the EU, rather than to exporters from the PECOs. Even in cases where licences are not issued, but quotas are filled on a first-come-first-serve basis as in the case of fruit and vegetables, preference margins tend to flow

Table 14.4: Quantitative implications of agricultural preferences under association agreements on PECO-6 exports.

	Poland	Czech+ Slovak Republic	Hungary	Romania	Bulgaria
1. Share of agricultural export to EU in total agricultural exports	58.7%	27.1%	39.9%	26.7%	35.0%
2. Share of preference products in total agricultural export to EU	45.0%	51.5%	65.6%	54.7%	62.0%
3. Share of pref. prod. in total ag. exp. to EU exc. tariff free exports	51.0%	57.0%	77.0%	60.0%	67.0%
4. Share of pref. prod. in total exports	4.4%	1.4%	6.0%	1.2%	3.1%
Potential growth of exports in fifth year of agreement as percentage of base period agricultural exports to EU:					
5. through full use of quotas	11.8%	49.3%	29.9%	46.6%	12.8%
6. through full benefit of pref. margin	7.0%	22.3%	13.9%	24.8%	5.1%
7. through full use of quotas and benefit of pref. margin	18.8%	71.6%	43.8%	71.4%	17.9%

Base period exports set equal to 100% are 1991 exports for Poland, CSFR and Hungary, and 1992.

Source: Overberg, 1995, Tracy, 1994.

to EU importers. This is because the EU does not publish information on the extent of quota use at any particular point in time. As a consequence, price negotiations between EU importers and PECO exporters are based on the worst-case assumption that the full duty has to be paid. The PECO exporter then gets the price he would also have received if no preferences had existed at all, and the preference margin accrues fully to the EU importer. Empirical surveys in the PECOs have indeed indicated that quota rents rarely flow to PECO exporters.[24]

In other words, impressive as the agricultural preferences granted under the association agreements may appear at the first glance, the economic benefits which they are likely to confer to the PECOs are small in quantitative terms. The PECOs have therefore repeatedly requested that the EU should improve preferential treatment of their agricultural exports to the EU.[25] In addition to revising the procedures for implementing the current preferences, the PECOs are requesting an increase in quota volumes for preferential treatment. In the EU, many farmers and ministers of agriculture are strongly opposed to such requests, fearing that larger preferential access for the PECOs to EU markets could depress EU market prices. However, what is generally overlooked in these discussions is the small size of preferred PECO exports relative to total market size of the products concerned in the EU market. In particular, the growth of PECO-6 exports allowed for by the expansion of quotas under the agreements adds only minimal amounts to total market supplies in the EU. Out of the 53 major product groups concerned, only three add more than 1% to total supplies on the relevant EU markets (goose meat 28%, hops 5%, onions 3%). In only 13 of these 53 product groups, would expansion of PECO-6 quotas under current agreements add more than 10% to total EU imports from third countries. Assuming reasonable price elasticities, only three out of these 53 markets can be expected to suffer a price decrease by more than 3% as a result of growing preferential exports from the PECO-6.[26] Hence, some further expansion of preferential quotas for PECOs could hardly be assumed to put much pressure on EU market prices.

The changing balance of PECO trade with the EU

As stated above, since the beginning of the reform process the EU has rapidly become an increasingly important trading partner for the PECOs. In recent years, the new trading arrangements under the association agreements may have contributed to that development. Of course, one hope attached to the agreements was that they might support the PECOs in their efforts to raise the growing amounts of foreign exchange so urgently needed to foster the process of modernising their economies. In agricultural trade, however, their balance of trade with the EU has deteriorated significantly since the late 1980s. In 1988, 1989 and 1990, the PECO-6 on aggregate had a positive balance of agricultural trade with the EC of slightly less than 1 billion ecu. In 1991 the surplus began to dwindle, in 1992 it was down to 0.3 billion ecu, and in 1993 it had turned into a deficit of nearly 0.5 billion ecu. Behind this rapid turnaround were both a decrease in PECO exports and an increase in PECO imports of agricultural and food products in trade with the EU. A number of factors have contributed to these developments.

The decline in PECO agricultural exports had much to do with problems on the supply side. As described above, the dramatic adjustments going on in agriculture and the food industry have resulted, initially, in a significant drop of agricultural output in most PECOs, and this has reduced export availability. In addition, new marketing channels and procedures had to be established in many cases. Liquidity problems made it difficult for processors and traders to continue their businesses, and many other difficulties of a similar nature got into the way of flourishing agricultural exports. What was particularly sad was that even many of the quotas for preferential export to the EU could not be filled. On aggregate across all preference products for the Visegrad countries, only around one half of what could have been exported preferentially within quotas was actually shipped in 1993.[27] This disappointing failure of the agricultural preferences under the association agreements, however, was not only due to supply side problems in the PECOs. It also had to do with the lack of commodity fit of preferences granted, and with the unfavourable mechanisms used by the EU for administering the quotas. A few interesting cases illustrating the nature of difficulties encountered are reported in Table 14.5.

Table 14.5: Use of preferential quotas under association agreements in 1993, selected PECOs and products.

	Licences issued[1] (%)	Actual exports[1] (%)
HUNGARY		
Duck meat	100	337
Chicken parts	88	116
Cheese	46	14
POLAND		
Duck meat	100	325
Whole chicken	0	11
Sausages	54	43

Source: Overberg (1995).
[1] As a percentage of quota volume available.

Where established trading relations could continue to be used and market conditions were favourable, actual deliveries exceeded the preferential quotas by far, as in the case of duck meat from Hungary and Poland. In other cases, actual deliveries were higher than licences issued under preferential quotas, indicating that the process of issuing licences did not operate satisfactorily and that exports within preferential quotas were not more attractive than exports under ordinary levies, presumably because quota rents anyhow did not accrue to exporters from the PECOs (examples of this are chicken parts from Hungary and whole chicken from Poland). In still other cases, actual exports were less than preferential licences issued, presumably because there were supply side problems in the PECOs and EU importers could not find the supplies they had hoped for (e.g. cheese from Hungary and sausages from Poland).

While agricultural exports from the PECOs to the EU were weak for such reasons, agricultural and food imports of the PECOs from the EU have grown significantly after the reforms. This development has often been attributed to export subsidies granted by the EU, and to the lack of matching protective import barriers in the PECOs. To some extent this explanation is correct.

However, a look at the commodity structure of EU agricultural and food exports to the PECOs indicates that other factors have been at work as well. As shown in Fig.14.1, EU exports to the PECO-6 of products where no or only minimal

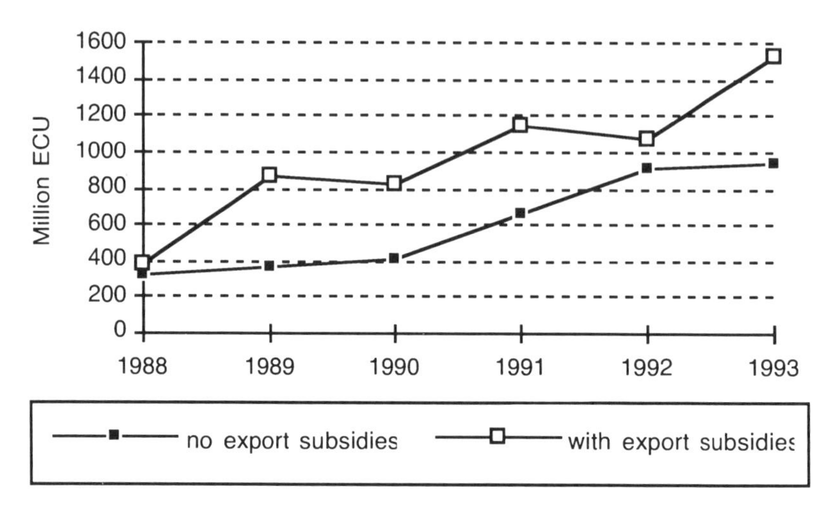

Fig.14.1: EU exports of agricultural and food products to the PECO-6 by product category: With and without export subsidies.

Source: Tangermann and Josling (1994).

export subsidies are paid have also grown significantly in recent years, and now have a large share in total agricultural and food exports to the PECOs. Examples are fish, flowers, coffee, oilseeds and their products. Moreover, as shown in Fig.14.2, a large part of both the value of exports and its growth in recent years consists of lightly and highly processed products. The large increase in EU exports of subsidised products and raw materials in 1993 was mainly due to cereals, explained by the low cereals crop in the PECOs due to drought in that year. In other words, the growing PECO imports of agricultural and food products from the EU reflect, in addition to EU export subsidies, also shifting consumer preferences in the PECOs, towards products such as tropical commodities (which in part are transhipped through the EU) and western style processed foods. As the food industry in the PECOs recovers from its adjustment problems there may be scope for much of the latter type of imports to be replaced by domestic production in the PECOs.

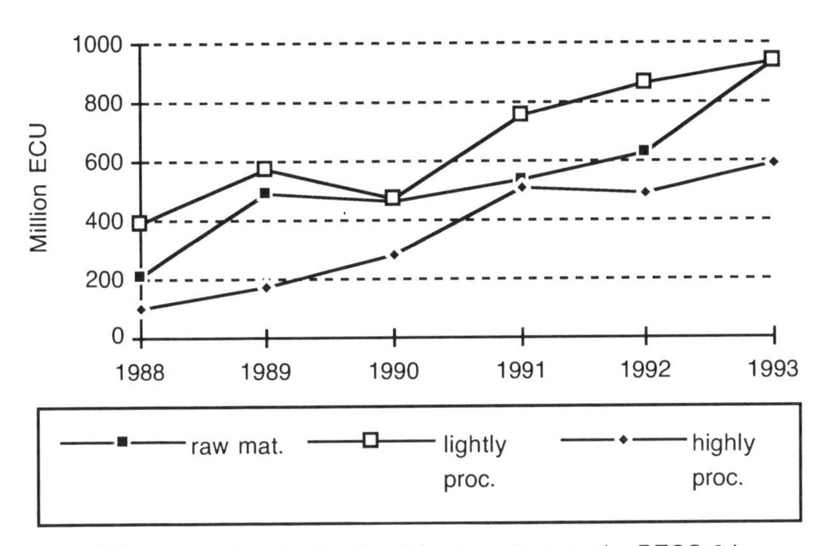

Fig.14.2: EU exports of agricultural and food products to the PECO-6 by product category: Different degrees of processing.

Source: Tangermann and Josling (1994).

PECO accession to the EU and adoption of the CAP

Timetable and issues

As discussed above, accession to the EU is essentially a political matter and its timing will thus be determined by political considerations. The process of agreeing the terms of entry requires extremely detailed bilateral negotiations. Because agriculture is the most 'developed' area of EU policy, it can be expected that the agricultural part of the entry negotiations will be amongst the most time consuming and troublesome. It is generally accepted that entry negotiations will not start in earnest until after the 1996 Inter-Governmental Conference (IGC) has been completed. There is no way of knowing how long the IGC will take; one year seems a minimum. Once entry negotiations commence, it is unlikely that they could be completed within four years. The least time this has taken in all enlargements to date is three years for the EFTA countries. For countries with a large, and potentially expanding, agricultural sector and many sensitive commodities it could take even longer; the Iberian accession provides an example. Thus the most optimistic outcome suggests that the first PECO to join the EU might enter on 1 January 2000.

Whilst accession requires a treaty between each individual new entrant and the EU members, it is anticipated that the same general principles will apply to all the PECOs. This is likely to be an important point for agriculture as the EU has yet to decide what will be the principles behind its approach to agricultural policy for the much-enlarged EU. These are big issues and they will have to take into account the wider international context of agricultural policy at the time of eastern enlargement. The 2000 'mini-round' of further agricultural negotiations in the WTO, foreseen in the URA, will have commenced. Given the commitments made in the Uruguay Round, it must be presumed that these negotiations will revolve around the nature, size and timing of further cuts in domestic agricultural supports, import tariffs and subsidised exports. Thus the EU will have to decide its approach to agricultural policy to accommodate both further enlargement and the WTO.

In the meantime, the immediate analytical task is to assess whether the CAP could be extended to embrace the PECOs. If it does seem possible, what changes or decisions would be required to do so? Or if there appear to be insurmountable problems, what are the broad alternatives? The analyses published to date on these questions identify two major problem areas. Can the enlarged EU live within its Uruguay Round commitments? Would the budgetary cost of the CAP in an enlarged EU be unbearable? Not surprisingly, it is very difficult to provide definitive answers to these questions. Analysis requires many assumptions about the developments in the EU-15 and world markets pre-enlargement, the way the CAP is applied to the PECOs — particularly the application of supply controls and compensation payments — and the likely reaction of farmers, the food industry, and consumers in the PECOs. The next section summarises the arguments as seen in 1995. It comprises mostly qualitative arguments supplemented with some figures derived from the four DGI sponsored studies cited above. It is impossible to be precise, the analyses cited were done in a short space of time[28] and their base data referred to the EU-12 and the PECO-6.

Problems and options for agricultural accession

To analyse the consequences of extending the CAP to the PECOs it is first necessary to define the supports given under the CAP and how they are applied to the PECOs. It is simplest to assume that the CAP at the time of entry is no different than the CAP of 1995. That is, all the intervention prices, aids, payments, tariffs and supply control measures are unchanged from their 1995 levels, export subsidies continue to operate in the same way, and that there will be no significant further reforms of the so-called unreformed sectors (wine, sugar, fruit and vegetables). Applying the CAP to the PECOs means that they instantly abandon any such measures they were independently operating pre-accession and fully adopt the EU prices and support mechanisms. For some countries such a switch in policy might come as a profound shock. In the past there has often been a long transition period to allow prices to be aligned gradually. The 'no transition' assumption can be defended on practical and analytical grounds. In reality, since the advent of the single market, it is not possible to have different administered prices in different parts of the EU. Such differentials would distort trade, necessitating border taxes; but border taxes cannot be levied if there are no frontiers. This might be regarded as too purist an argument. It might take time to abolish all other border controls thus some could be retained for the duration of agricultural transition. This becomes a political judgement; the precedent of the EFTA enlargement on 1 January 1995 was that no transition period was allowed. Analytically, it is simplest to assume instant adjustment. It avoids having to specify the length of the transition period and details of expectation formation and the dynamic behaviour of economic agents. The analysis is thus a comparative static appraisal of effects before and after adoption of the CAP.

The adoption of the CAP involves a great deal more than the mechanical application of pre-determined intervention prices and aids per hectare and per animal. The CAP is the outcome of a complex and sensitive negotiating process in which member states obtain the best conditions for their own farmers. This especially applies to supply controls, but it also applies to many of the aids and payment rates. For the supply controls, the analytical approach implicitly or explicitly taken is to project the market situation in the PECOs as if there are no such controls and use this to determine the magnitude of controls necessary to avoid any consequential problems. This applies to milk and sugar quotas, the limits on the numbers of beef and sheep eligible for livestock payments, and the base areas for cereal, oilseeds and protein set-asides. Such calculations can give some feel of whether and to what extent supply controls will be necessary. They may also give useful indications of the principles on which the base areas and livestock numbers can be determined. However, they will not give any indication of the administrative problems and costs of imposing and implementing these controls.

The issue of compensation payments poses similar difficulties. As we saw in previous chapters, these payments were introduced in the 1992 Mac Sharry reforms to compensate the growers of cereals, oilseeds and protein crops for the 30% cuts in support prices which took place over the three years 1993 to 1995. For larger farmers (defined as growing more than 92 tonnes of grain) the payments are conditional on setting aside a proportion of their base area of these crops. Although the payments were agreed as a rate per tonne, they are paid on a hectarage basis by multiplying by a regional average yield. Will PECO farmers be eligible for these payments? Based on their origin and official name — compensation payments — the answer is 'no'. PECO farmers will enjoy a price rise on accession to the CAP, not a fall, thus compensation cannot be justified.

However, this logically rigorous approach will be hard to apply. It does not seem plausible that in a stable redistributive policy, the poorer majority of farmers from the newest member states would be denied payments made to the richer farmers of the older members. At the very least the basis of the payments and their financing would have to change. But in any case, by the time these issues become relevant, the definition and original justification of the arable payments will be ten years old. Assuming the payments are still in place, their justification must surely shift from politically necessary compensation to farmers for accepting a once-for-all price cut, to some kind of direct income support payment based on social, regional or environmental criteria. It is impossible to argue that indefinite compensation is justified for a once-off change in policy.[29] Whether the switch in justification for the payments occurs by stealth or as a result of explicit discussion and negotiation is perhaps less important than the consequences which will flow from the change.

First, if the payments are acknowledged to be social, regional or environmental supports, it will be impossible to claim that farmers in the PECOs are ineligible. On any objective measure of these criteria PECOs will be seen as having a strong case for receiving payments. Second, if this change in the basis of payments occurs, it will be obvious that the different member states will have a very different interest in payments under these criteria. Conditions are so variable around the EU that it will be difficult to agree universal payment rates which make sense. Thus the change will necessitate greater member state differentiation of the payments. Third, a glance at the numbers of farmers in, prospectively, the 25 member states and their relative income levels suggests that there could be large flows of funds towards the PECOs as a result of these payments. The change could thus be very expensive for the net budget contributors in the western part of the EU.

In turn, these consequences of a reinterpretation of the compensation payments will create a strong pressure to ensure that they are demonstrably decoupled from resource allocation decisions. This is particularly in the interests of the poorer member states. They will be concerned that the richer countries will distort markets by offering more generous support for their own farmers. This implies that payments should be divorced from current input and output levels and prices.[30] Furthermore, it is highly likely that the net contributor members will argue, on the basis of subsidiarity, that it is logical that nationally differentiated payments at nationally determined rates should be largely, if not wholly, nationally financed. These are difficult issues to analyse empirically. The debate on such changes in the CAP is still at a very early stage and the outcome will make a big difference to the impact of the CAP in the PECOs.

Analysing the application of supply controls and compensation payment in the PECOs is bedevilled by the problem of judgements about the political nature of these aspects of the CAP. The problems of analysing the effects of market supports mostly concern assumptions about technicalities and economic behaviour. These are no easier. It was shown in Table 14.3 that there is currently a large gap between producer prices for the major products between the PECOs and the EU. Whilst it is assumed that EU prices will change little in real terms by the time of accession, some of this gap may be closed by inflation in the EU and continued creeping protectionism in the PECOs prior to entry (the next section analyses pre-accession policy options in the PECOs). However it is presumed that EU prices will still be above those in the PECOs and both will be above world market levels at the time of entry. Qualitatively, the effects of entry are straightforward. The higher producer prices will stimulate production. Consumption may be depressed by the higher prices, but this may be offset, for some products, by a stimulus from growth in incomes. The net effect on commodity trade balances is an empirical matter. Analysing the consumption response is comparatively straightforward: there already exist price and income elasticities of demand estimated from household panel data.[31] The bigger analytical problems are on the supply side. Price elasticities cannot meaningfully be estimated for PECOs where there are too few post-reform observations for econometric work. However this may be a less serious problem than the assumptions about the rates of structural and technical change. The processes at work here are not well understood at the theoretical level and empirical research is non-existent. Analysts are left with little choice but to define broad scenarios of optimistic or pessimistic rates of technical progress and to make corresponding broad assumptions about the rate of yield growth and conversion efficiency improvements. Conclusions from such analyses are therefore highly conditional on the assumptions made.

The analyses made so far have identified principally two kinds of problems emerging from PECO adoption of the CAP: meeting the commitments made under the Uruguay Round and living within the budgetary guidelines for FEOGA. These are spelled out in detail in the four DGI reports and will be summarised here.

The Uruguay Round commitments

In outline, as PECOs join the EU they will have to harmonise their tariff bindings with the EU's, and they will aggregate their commitments on domestic supports, minimum access, and value and volume of subsidised exports. The URA schedules of the PECOs required rather more judgement and political negotiation than those for market economies which were based on the factual situation in agreed base periods. The problem for the PECOs was that the trade flows and prices in the 1986 to 1990 base period were so far removed from international values that it was difficult objectively to measure tariff equivalents and export subsidies. The result was that the individual PECOs took very different approaches. The Poles, farsightedly, successfully negotiated tariff bindings which were close to the EU levels and defined their domestic supports and export subsidy values in US dollars (thereby protecting them against inflation and depreciation of the Zloty). Other countries were not so focused on EU membership and defined their AMS and export values in their own currencies, and generally had lower tariff bindings and mostly quite small subsidised export volumes.

It is agreed by most analysts that for the EU-15 the most restrictive of the URA commitments is the limitation on the volume of subsidised exports. There also seems to be a consensus that given likely market developments until the end of the Century, the EU can live within even this constraint. In the event that stocks of grains, dairy produce or beef seem to be rebuilding rapidly enough to threaten to cause a problem, there are sufficient production control instruments already available to manage the situation. However, given the pressure of continued technical progress, it is expected that the subsidised export volume constraint will become progressively more of a problem during the first decade of the 21st century. A further reduction in the allowable volume of subsidised exports in the 'mini-round' of agricultural negotiations under the WTO would of course worsen the problem. Now consider the effect of admitting the PECOs into the CAP.

PECOs generally have not utilised export subsidies to a large extent: either they have not exported much or they have been able to sell produce at world market prices. However as they have introduced more protection this has stimulated a recovery in production, and more exports must be subsidised if they are to be exported. By 1993, the Visegrad-4 were already exporting in excess of their combined URA subsidised export limit for grain, beef, pork, milk powder and sugar.[32] This does not necessarily require any action as it is only if these exports are subsidised that a problem arises. According to the European Commission (1995) the net export position of most products from the PECO-10 is expected to grow by the time of accession.[33] Thus if at the time of accession the PECOs are expected to be net exporters at significantly greater quantities than the present levels which already exceed their URA limits, as soon as they adopt the higher EU prices which necessitate export subsidies, there is a great problem.

This problem is not insoluble, but it will require either the EU-15 to reduce its own subsidised exports to make way for the PECOs; or the PECOs, as part of their entry conditions, may have to impose strict production controls which limit their own exports to whatever URA limits they have in their individual schedules; or the EU will have to try and negotiate higher export subsidy volumes with other WTO members. None of these solutions is very easy.

There could also be problems in merging the domestic support limits, the so-called Aggregate Measures of Support (AMS). Whilst the EU-15 has considerable 'slack' in its own AMS, several of the PECOs will bring very small, or in the case of Romania, no AMS. When these countries adopt the CAP commodity regimes this will immediately raise their measured domestic support. The enlarged EU could also find itself challenged under article 13 (the 'due restraint' clause of the URA) if protection can be demonstrated to have risen for specific commodities beyond their level in 1992. This will certainly happen for many commodities in the PECOs. How this will be interpreted when they are part of a larger entity (the EU) remains to be tested.

Merging the tariff bindings will not be a problem if higher PECO bindings are reduced to EU levels. If the tariff bindings are higher in the EU than the PECOs, there could be objections raised by other WTO members if PECOs attempt to raise their previously bound tariffs. These issues are further complicated as applied tariffs are usually below the bindings and raising the maximum binding does not necessarily raise the tariff applied. However, because tariff levels are negotiable, these difficulties may be less of an obstacle to the accession than the export subsidy volumes and the AMS. In the last resort the EU-15 could offer to make small reductions in their tariffs to make room for raising certain PECO tariffs.

The budgetary costs of PECO adoption of the CAP

There are as many estimates of this as there are analysts who have tried to answer the question. This is not surprising. There are so many assumptions which have to be made that objective scientists can make estimates which are easily different by an order of magnitude. The main explanations for the wide range of estimates of the FEOGA guarantee cost of eastern enlargement, which lie between 2.8 and 42 billion ecu,[34] reside in the lack of agreement about the assumptions with respect to: the year, the PECOs included, the commodities analysed, the elements of CAP included (especially supply controls and compensation payments) and the EU support levels, world prices, the supply response and technical progress in the PECOs, the economic growth rates and other demand shifters in the PECOs, and the extent to which URA constraints are incorporated in the analysis. In addition, the base data about PECO agriculture has been subject to error. Figures are confidently produced by many international agencies which show surprising differences in basic information on crop areas and yields and livestock numbers. Price information is even more variable. Of course, if analysts used the same base data and normalised their assumptions on the variables listed above then they would produce estimates which were of a similar magnitude.

The PECOs themselves will of course contribute to the EU budget, but given their economic weight they would not expect to be net contributors. Whatever the FEOGA costs of enlargement turn out to be, and even if they can be contained within the agricultural guideline, it cannot be assumed that the additional funds can easily be raised. If contributions have to be increased by, for example, raising the GDP key, individual member states, who have to vote such a change through their parliament, could obstruct the extra funds. This makes it very difficult to judge what constitutes an excessive cost: it is whatever the least expansionist EU member state decides it should be.

The budget costs are thus a real issue, although there are a number of factors which suggest that too much emphasis has been placed on the controversy about the FEOGA costs of enlargement. First, the enlargement is highly unlikely to involve all PECOs joining the EU on the same day. Therefore, the additional FEOGA costs of the PECOs will only build to the 'true' total cost over a period of years. But FEOGA guarantee costs are only part of the picture. It is certain that the PECOs will try and negotiate substantial help from the social, regional, structural and cohesion funds. Given their poorly developed infrastructure and low levels of income they will clearly be eligible for assistance from these funds. The amounts involved here are more difficult to estimate as they have a larger political element. On some estimates, e.g. Baldwin (1995) they could be substantial. A third consideration is that many argue that the CAP will have to be reformed because the costs of extending it to the PECOs will be prohibitive, yet these analysts do not demonstrate whether a reformed CAP would be significantly cheaper. This is not a trivial point, but it raises even greater analytical problems as it requires specification of feasible alternatives to the present CAP and their costs. Fourth, in the heat of the debate about the budget costs of an enlarged CAP it is all too easy to lose from sight that there will be benefits from enlargement too. These have not been estimated, and yet without some idea of what they could be it seems unbalanced to put so much weight on one aspects of the costs of enlargement.

Summarising, analysis to date suggests that if, at the turn of the century, the PECOs were to adopt the CAP as it was in the mid-1990s, the impact of the higher prices and other supports would cause the expanded EU to exceed its Uruguay Round commitments and also raise budgetary problems. Could these problems be avoided, or if not, what steps could be taken to reduce or eliminate them?

Options which have been discussed, and rejected, are for the PECOs to join the EU but not the CAP or to create a different CAP for the PECOs. Both options conflict with the Treaty of Rome and the Single European Act. Article 38 of the Rome Treaty requires that "the common market shall extend to agriculture and trade in agricultural products" and that this "must be accompanied by the establishment of a common agricultural policy amongst the member states". In principle, if all existing and the new members agree to change these treaty requirements they could do so, but this seems a cumbersome route to take. Both of these options would presumably lead to a situation in which prices and aids were different between the EU-15 and the new members from the PECOs.

To prevent these differentials creating distortions to trade it would be necessary to reinvent the internal border taxes and subsidies (compensatory amounts) which were abolished in 1993. Furthermore, if the PECOs were either excluded from the benefits of the CAP or only enjoyed part of the benefits they would presumably not be required to contribute so much to the financing of the policy. Thus two of the principles of the CAP would be breached, common pricing and financial solidarity. A further objection to the no-CAP or two-CAP options is that they would increase regressivity in EU policies, conflicting with one of the objectives of EU action which is to help convergence of the European economies. This would come about because the richer areas (EU-15) would be applying an agricultural policy with higher prices and support payments and the lower income areas would have lower prices and lower or even no income payments at all.

Another way of trying to avoid the problems of PECO adoption of the CAP is to restrict membership to those PECOs which caused fewest such problems. This stepwise enlargement approach is likely to occur in any case. Clearly some PECOs will achieve the necessary conditions of macroeconomic and political stability before others. It may enable the EU to delay confronting the problems of its overprotective agricultural policy, but it will not remove the pressure altogether. Similarly, countries could be admitted but required to adopt a long transition period in which prices and support levels are slowly harmonised. This would require the reintroduction of accessionary compensatory amounts (ACAs) — taxes which correct for temporary price differences in the 'common' market. These were abolished with the completion of the single market in 1993 and were not applied for the three EFTA countries who joined in 1995. Their reintroduction could be done if the political will existed to allow it; the complete abolition of borders between the PECOs and the rest of the EU-15 is likely to be a sensitive issue for a long time anyway.

If none of these avoidance measures is acceptable or practical the EU is left with only two choices for overcoming the problems of enlargement: to utilise supply controls more fully and rigorously or to reduce prices to world market levels. Both the URA problems and the budget problems arise because of the stimulation to production given by prices above world market levels. This can therefore be tackled by attacking the effect — the overproduction, or the cause — the high prices. The 1992 Mac Sharry reform went as far as politically possible to reduce prices, but because they were not able to go far enough, a massive extension of supply controls, the set-aside scheme had to be introduced. It seems unavoidable that as the prospect of eastern enlargement approaches, this debate has to be reopened to find a way to continue and ultimately complete the reforms started in 1992.

In the meantime the PECOs are left in a state of uncertainty about the CAP they will eventually join. What are their options in the intervening years? This is the subject of the final section.

Pre-accession options for the PECOs

Until PECO accession to the EU actually occurs, at a point in time still difficult to predict, the PECOs have to design, implement and pay for their own agricultural policies, ranging from structural policies to market and trade policies. As far as strategic choices in view of future accession to the CAP are concerned, the basic direction of market and trade policies in the PECOs relative to levels of support and protection prevailing in the EU is a particularly important decision to be taken in the years to come. The choices to be made in this regard are difficult as PECO market policies to date are still significantly different from the CAP, and there are several options as to how to approach the process of eventual harmonisation with the CAP.

Given the current price gap between the PECOs and the EU, and assuming the need to harmonise PECO prices with those under the CAP when it comes to EU accession (i.e. disregarding the option of a long transition period after EU accession), one strategic choice to be made by the PECOs is which time path to choose for bringing PECO prices in line with CAP prices. Three major options can be considered, schematically shown in Fig.14.3, where the assumption has been made that the year 2000 is a potential date of entry into the EU. A first option is to harmonise PECO prices rapidly with the CAP. A second option would be to embark on a process which would gradually align PECO prices with those likely to exist in the CAP around the year 2000. A third option is to keep PECO prices at their current lower level until accession to the EU actually occurs. Obviously these three options have different implications, for both the PECOs and the EU.

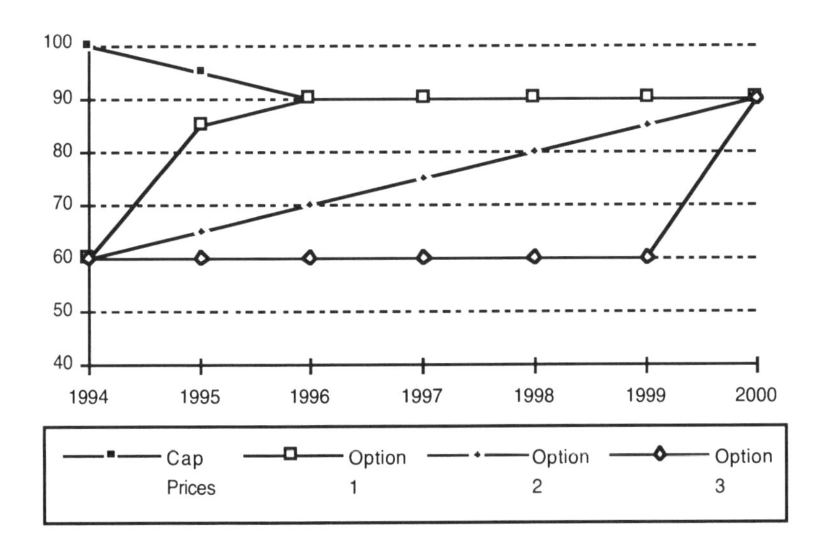

Fig.14.3: PECO pre-accession options.

From the point of view of the PECOs, the strategy of a rapid price alignment would have a number of serious disadvantages, even for many PECO farmers. While it may be possible to raise support prices for crops relatively soon, prices for livestock products are more difficult to raise, not the least because of the heavy budgetary costs which would be involved. Livestock producers in the PECOs might then be hard hit by rising feed prices, and their international competitiveness would be undermined. For the food industry in the PECOs, rapidly rising raw material prices would make life even more difficult, and recovery of that industry might be seriously retarded. Consumers in the PECOs, who still spend a much larger part of their total expenditure on food compared to the EU, would be frustrated as their hoped-for improvement of living standards would be further delayed. It is for such good reasons that the option of a rapid alignment with CAP prices appears not to be seriously considered in the PECOs.

The second option of a gradual alignment with CAP prices, may appear more attractive for the PECOs. However, even that strategy may run against a number of serious constraints. Once productivity and output levels in PECO agriculture recover from the current adjustment problems, price incentives resulting from higher levels of support and protection would tend to add to a growing level of production. The result might well be rising levels of surplus production which, at higher levels of price support, can be disposed of only with growing export subsidies. The required levels of budgetary outlays could be very significant, and could far over-stretch the ability of the PECO macro-economies to finance growing budget deficits.[35] In this context it is important to remember that until the day of accession to the EU the PECOs have to finance their agricultural policy expenditure out of their own domestic budgets. Another important constraint is the GATT/WTO. Under the URA the PECOs, too, have to honour the GATT disciplines they have accepted in their national schedules of commitments. As quantitative estimates have shown, a policy of gradual alignment of PECO prices with the CAP (as well as a policy of rapid price alignment) would tend to have market and policy implications which are inconsistent with the specific GATT commitments accepted by the PECOs, in particular in the area of export subsidies.[36]

Hence, from the point of view of the PECOs the preferable strategy is probably option 3, i.e. to keep support and protection in agriculture low until accession to the EU actually occurs. This strategy is, also, the one which is most likely to result in an agriculture and food industry which is internationally competitive. Moreover, it may well happen that the EU continues to reform the CAP, with the result that CAP support prices are further reduced by the time the PECOs enter the EU. It would then be difficult for the PECOs to engage in downward price adjustments after they had first raised their prices towards what they wrongly assumed to be CAP prices at the time of their accession.

Conclusions

The Common Agricultural Policy has proved for almost three decades to be an extremely adaptable set of instruments. It has survived many challenges:

monetary and international commodity market shocks, four enlargements, several budgetary crises and the Uruguay Round negotiations. It would be a brave person who predicted that it could not survive the challenge of eastern enlargement. This chapter has examined the nature of this challenge, outlining the very different features of the next round of EU members and the evolution of their relationship with the EU in the first five years of transition. It is argued that the biggest problems of extending the CAP to the PECOs will be respecting the Uruguay Round commitments of the enlarged EU and financing the application of the CAP in ten much poorer countries. A large part of the genesis of these problems is that the EU maintains agricultural support prices above world market levels.

If the strategy is to try and avoid these problems, three options are to offer membership of the EU without adopting the CAP; to offer a scaled down version of the CAP for the PECOs; or to admit only those countries whose agriculture could be absorbed without creating these problems. None of these was thought to be acceptable. If the alternative strategy of trying to deal directly with the problems is taken there are two more options: to use rigorous supply controls in both the PECOs and the EU-15 to ensure that production remained within the limits imposed by the URA and the budget, or to bring EU prices closer to world market levels thereby removing the need to subsidise exports and artificially to restrain output. From the point of view of economic efficiency there is little doubt of the preferred option.

Faced with these difficult challenges, it is easy to identify two extreme approaches the EU could adopt. To use equestrian metaphors the initially least painful approach is to jump each fence encountered with the tried old horse making minimal adjustments necessary to scrape over. The other extreme, which may be more disruptive in the short run, is to anticipate the difficulty of the course and select a different and more appropriate horse. The decision making process and institutions in the EU strongly predispose to the first approach. Most of the radical changes in the CAP over the last thirty years have been of this kind; examples are the introduction of the agrimoney system in 1969 and its subsequent modifications, milk quotas in 1984, maximum guaranteed quantities in 1988, and set-aside and arable payments in 1992. In each case the changes were introduced in the face of a crisis induced either by budgetary or international pressures. By the mid-1990s neither of these pressures were present. Stocks were at very low levels, international commodity prices were at a high, and so the CAP was operating well within its budget ceiling. As the URA was only just being implemented and the US was preoccupied with its own 1995 Farm Bill, international pressure had abated. The case for considering further reform of the CAP requires arguments concerning possible events in five to ten years time. The case for reform would thus have to be very strong and clear to generate the political momentum necessary. This requires a demonstration that the present CAP, even fully adjusted within its present mechanisms, could not cope with the twin challenge of eastern enlargement and a further round of WTO induced support cuts.[37] Much more work requires to be done to make this case convincingly.

References

Baldwin, R. (1995) *Towards an integrated Europe*, CEPR, London.

Buckwell, A.E. and Davidova, S.M. (1993) Potential implications for productivity of land reform in Bulgaria, *Food Policy*, December, 493–506.

Buckwell, A.E. (1994) *The restructuring of Agriculture in central and eastern European Countries: Progress, Problems and Policies*, paper presented to the OECD ad hoc group on East–West economic relations in Agriculture, AGR/EW/E6(94)25.

Buckwell, A.E., Haynes, J., Davidova, S.M., Courboin, V. and Kwiecinski, A. (1995) *Feasibility of an agricultural strategy to prepare countries of central and eastern Europe for EU accession*, Study prepared for DGI of the European Commission, Brussels.

Buckwell, A.E. and Davidova, S.M. (1995) *Are the PECOs in favour of CAP reform?* Paper presented at the Conference on Agriculture and Trade in Transition Economies: Policy Design and Implementation, Prague 27–28 July.

European Commission (1995) *Agricultural Situation and Prospects in the central and eastern European Countries: Summary Report*, Directorate General for Agriculture (DGVI) Brussels.

Ivanova, N., Lingard, J., Buckwell, A.E. and Burrell, A. (1995) Impact of Changes in Agricultural Policy on the Agro-food chain in Bulgaria. *European Review of Agricultural Economics*, forthcoming.

Kjeldahl, R (1995) *Direct Income Payments to Farmers: uses, implications and an empirical investigation of labour supply response in a sample of Danish farm households*, unpublished PhD thesis, Wye College, University of London.

Mahé, L-P., Cordier, J., Guyomard, H. and Roe, T. (1995) *L'Agriculture et L'Elargissement de l'Union Europeene aux pays d'Europe Centrale et Orientale: transition en vue de l'intégration ou intégration pour la transition?* Study prepared for DGI of the European Commission, Brussels.

Ministry of Agriculture Fisheries and Food (1995) *European Agriculture: the case for radical reform*, MAFF, London.

Nallet, H. and Van Stolk, A. (1994) *Relations between the European Union and central and eastern European Countries in matters concerning agriculture and food production*. Report to the European Commission DGVI, Brussels.

Overberg, B. (1995) *Die Auswirkungen der Europa-Abkommen mit den mitteleuropäischen Ländern auf den Agrarhandel*. Doctoral dissertation. Göttingen.

OECD (annual reports) *Agricultural Policies, Markets and Trade. Monitoring and Outlook in the central and eastern European Countries, the New Independent States, Mongolia and China*. Paris.

OECD (1994) *Review of Agricultural Policies: Hungary*. Centre for co-operation with the economies in transition, Paris.

OECD (1995) *Review of Agricultural Policies: Poland*. Centre for co-operation with the economies in transition, Paris.

Phimister, E. (1994) *Farm Household Production under CAP Reform, the impact of Borrowing Restraints*. Working paper, Unite Politiques Agricole et Modelisation, INRA, Rennes.

Swinnen, J.F.H. (1995) *Endogenous price and trade policy developments in central European Agriculture*. Paper presented at the Conference on Agriculture and Trade in Transition Economies: Policy Design and Implementation, Prague 27–28 July.

Swinnen, J.F.H. *et al.* (1995) *Restructuring central and eastern European Agriculture*. Forthcoming report of a COST network on privatisation and restructuring of CEEC agriculture, Leuven.

Tangermann, S. (1993) Some Economic Effects of Preferential Trading Arrangements between the European Community and central Europe. *Journal of Economic Integration*, **8** 152–174.

Tangermann, S. and Josling, T.E. (1994) *Pre-Accession Agricultural Policies for central Europe and the European Union*. Study prepared for DG I of the European Commission. Brussels 1994.

Tarditi, S. and Marsh, J. (1995) *Agricultural Strategies for the enlargement of the European Union to central and eastern European countries*. Study prepared for DGI of the European Commission, Brussels.

Tracy, M. (1994) *East-West European Trade — The Impact of Agreements*. APS Agricultural Policy Studies, La Hutte, Belgium.

---

[1]Membership of NATO of course is a very different issue. Not all EU members are full members of NATO and it is an issue which extends beyond western Europe.

[2]The acronym refers to Poland, Hungary, Assistance for Reconstruction of the Economy. The programme was launched in 1990, and was rapidly expanded over the next four years to incorporate all ten countries in central and eastern Europe plus Albania. The corresponding programme for the Commonwealth of Independent States is referred to as TACIS (Technical Assistance for the CIS).

[3]The list of candidate states from this region will doubtless continue to grow. Negotiations for an association agreement with Albania commence in 1996, and it seems reasonable to assume that following a political settlement in the rest of Yugoslavia, others will follow.

[4]The Visegrad countries set up, in December 1992, a Central European Free Trade Area, CEFTA. This is an alternative designation for these countries. The abbreviation CEFTA+ is used to indicate the addition of Slovenia to this group as she had completed a free trade agreement with three of the CEFTA members in 1994 and was expected to complete the process with Poland soon after.

[5]This section draws heavily on the Summary Report of the Agricultural Situation and Prospects in the central and eastern European Countries prepared by the European Commission (1995).

[6]A worrying feature of these PPP corrections is that they all imply that PECO currencies are significantly *under*valued. This does not fit well with the observation that most PECOs have significant government and trade deficits and faster inflation than that in strong currency countries, without their exchange rates in recent years having depreciated fully in proportion to their rates of inflation, which *ceteris paribus* would lead one to suspect *over*valued currencies.

[7]It is important to note that such is the heterogeneity of the PECOs, that there are few generalisations which apply to them all. In the case of the post-reform development of Gross Agricultural Output, Slovenia was the exception suffering

only a 10% fall in output in 1992 and enjoying by 1994 output almost 20% above the pre-reform level.

[8]The most graphic of which was the not-uncommon practice of feeding bread to pigs because it was 'cheaper' than grain!

[9]See Ivanova *et al.* (1995) for a detailed analysis of the very large economic transfers which take place in the Bulgarian grain and meat chains as a result of the incomplete liberalisation and lack of competition in the food chain.

[10]Technically, the land on collective farms was still privately owned in the sense that titles were never formally taken away from the owners. However, effectively the owners were deprived of almost all rights of ownership; they could not transact the land nor determine the use to which it was put.

[11]See Swinnen *et al.* (1995) for a review of the agricultural structures emerging in central Europe and Buckwell (1994) for a discussion of the characteristics and potential performance of the structures emerging.

[12]That is markets in which prices are not determined by government but by the interaction between market agents. However this does not mean that they are the classic 'free market' prices of a perfectly competitive system. The reason is quite simply that market price and quantity liberalisation can be done almost overnight, but the demonopolisation and privatisation of economic agents takes years, leaving in the meantime markets characterised by a high degree of concentration with market power in few hands.

[13]Although this does not explain why most of these governments tolerate a high degree of inefficiency and profiteering in the downstream sector which significantly raises retail food prices.

[14]The CEFTA+ countries and Romania are all members of the WTO. Bulgaria and the Baltic states are in the process of joining.

[15]See Buckwell *et al.* (1995) for a review of the Uruguay Round Schedules of the Visegrad-4 and Romania.

[16]Polish pork prices in 1994 were considerably above trend, reflecting high feed grain prices due to drought conditions in 1993.

[17]For these figures and more detail on PECO agricultural trade with the EU, see Overberg (1995).

[18]Romania had already been included in the EU's GSP since the late 1970s. However it had remained excluded from many individual tariff preferences under that system. In 1991, all GSP benefits were extended to Romania as well.

[19]For a detailed analysis of the agricultural preferences granted under the association agreements, see Overberg (1995). See also Tracy *et al.* (1994) and Tangermann (1993).

[20]In the following, the term PECO-6 will be used to refer to the six PECOs whose association agreements had entered into force at the time of writing (September 1995), i.e. Bulgaria, the Czech and Slovak republics, Hungary, Poland and Romania. The EU has negotiated and initialed association agreements also with Slovenia and the Baltic states. These agreements are rather similar to the Europe agreements concluded earlier. However, these agreements have not yet entered into force. Negotiations on an association agreement with Albania are expected to be completed during 1996. With the republics of the former Soviet

Union the EU still has only co-operation agreements which do not contain any preferential trade provisions.
[21]In the short run these resource costs of extra production for larger exports to the EU may, though, be low. As PECO agricultural output has slumped so much in recent years there are now idle resources, in particular land and labour.
[22]This argument should not be taken to suggest that there is no reason for the EU to expand the quotas.
[23]The estimates of preference margins reported here are based on actual levies and tariffs collected by the EU on average between July 1990 and March 1991. EU levies and tariffs in this period reflect well the conditions in the second half of the 1980s and the early 1990s. See Overberg (1995).
[24]See Overberg (1995).
[25]In addition, negotiations were being held on adjusting the agricultural preferences to the new trade regime introduced by the EU as a result of the GATT agreement on agriculture and to the EU enlargement to include some EFTA countries.
[26]See Overberg (1995).
[27]See Overberg (1995).
[28]Five months to be exact.
[29]See MAFF (1995) for a strong defence of this argument.
[30]Strict definitions of decoupled payments are contained in the Uruguay Round agreement on agriculture in specifying domestic supports which are exempt from reduction commitments. Whether payments can ever be totally divorced from resource allocation decisions is doubtful. Kjeldahl (1995) shows that on-farm labour use is sensitive to such payments, and Phimister (1994) shows that credit market failures provide another route through which apparently decoupled payments affect resource use and thus product markets.
[31]More sophisticated analysis would pay greater attention to the relationships between farm gate and retail prices. This brings substantial problems. There can be no presumption that processing and distribution efficiency and margins will be constant or unaffected by accession. The anticipation and fact of EU accession may stimulate inward investment into the PECO food industry, and it will certainly have to become more competitive in the single EU market. These changes will have complex effects on the demand for PECO agricultural products.
[32]See Fig.4.1 of Buckwell *et al.* (1995).
[33]The Commission's summary report on the PECO-10 (Tables 15 to 23) projects that even in the absence of further policy change there will be significant growth in net exports of grain (x6), oilseeds (x5), potatoes (x8), milk (x4), poultry (x7) and beef (x5) from their 1994 levels.
[34]See Table 4.6 of Buckwell *et al.* (1995).
[35] For a quantitative estimate of the market and budget implications of this option, see Tangermann and Josling (1994).

---

[36] See Tangermann and Josling (1994).
[37]Another ingredient in this case is that if further CAP reform is not implemented before the next enlargement it may be even more difficult to do it afterwards. This depends on the likely attitude towards the CAP in the prospective members, see Buckwell and Davidova (1995).

Chapter 15

The CAP and the Developing World

John Lingard and Lionel Hubbard

Introduction

The effects of the protectionist agricultural policies of rich countries on economic development in poorer countries (or Less Developed Countries — LDCs) have been an issue of growing international concern for some time. Many have written of the global imbalance of food supplies and the moral obscenity of developed country surpluses compared with enduring hunger and chronic and acute famines in parts of the developing world, particularly in sub-Saharan Africa. The CAP has often been accused of indirectly contributing to the so-called world food problem and the long-term negative effects of the CAP are only partially appeased, not totally offset, by periodic short-term EU shipments of food aid.

Recently world food production has been increasing faster than population growth, with per capita production up by 5% in the 1980s, but worryingly the rate of increase is now slowing down. However, with more than 700 million people in the developing countries having limited access to sufficient food to lead healthy and productive lives and with over 180 million children classified as being chronically underweight, disease, hunger, poverty and malnutrition are still widespread (Pinstrup-Andersen 1994). Real food prices have been generally declining and were at historic lows until 1995 but may now rise with market liberalisation following the Uruguay Round. A counterbalancing force could however result from increasing grain exports from eastern Europe and the Former Soviet Union (FSU) as their livestock numbers fall and their agricultural sectors are stimulated by structural adjustment (see Chapter 14). Whatever the outcome, the likelihood is a continuation of generally depressed international food prices.

The effects of the CAP and the EU's agricultural trade on both the level and stability of world agricultural prices came under close scrutiny during the Uruguay Round of the GATT negotiations. However, the implications of the CAP, and its reform, for the LDCs are not immediately obvious. The distorting effects of European protectionism have undoubtedly hit agricultural exports of some LDCs, particularly in commodities like sugar, beef, and cereals; but the impacts are neither universally harmful nor uniform. Some groups of LDCs (food importers) and some groups within LDCs (food consumers) may actually benefit from the trade effects of the EU's agricultural policies. There are different implications for African food importers compared with Latin American food exporters, and it is not possible, a priori, to determine which particular changes to the CAP would unambiguously safeguard LDCs' interests. The export markets for LDCs, particularly in temperate products, have been reduced by the EU's external tariff, but world food prices have been lowered by subsidised Community exports, which could have aided the food importing low-income countries of Africa and Asia. In general, LDCs are net food importers and in 1989/91 imported 120 million tonnes of cereals whilst, at the same time, exporting 32 million tonnes.

Agricultural trade between the EU and LDCs is also influenced by a set of preferential trading relationships, mostly under the auspices of the Lomé Convention. We refer briefly to these towards the end of the chapter. However, in the main the trade concessions granted have not involved products covered by the CAP. A much more important matter for many LDCs has been the way the CAP cereals policy depresses and destabilises world cereal prices and this is the issue highlighted in the present chapter .

The CAP and the world market

The Uruguay Round of international trade negotiations under the GATT aimed to liberalise trade in agriculture. Domestic support policies will as a consequence gradually be made more responsive to international market signals, thereby reducing the trade-distorting influences that inevitably followed when the US and the EU, in particular, competed with each other to subsidise exports of their surplus production. The Punta del Este Declaration at the outset of the Uruguay round recognised the need to accord special and differential treatment to developing countries, and ways were sought to take into account the possible negative effects of agricultural trade reform on net food-importing LDCs .

In order to maintain prices paid to European farmers above levels determined in world markets, the CAP used a system of variable import levies and variable export subsidies, combined with the provision for intervention buying, as described in Chapter 1. The CAP effectively insulates the EU domestic agricultural sector from world market forces and trends, in order to maintain internal Community objectives of farm income support, stable prices and secure food supplies. The consequences of this policy are well known — a rapid growth in production, increased self-sufficiency ratios, surpluses, subsidised exports onto world markets, income transfers from consumers and taxpayers to producers, financial transfers between member states and a continuous budgetary burden. Of particular note in the UK has been the doubling of wheat production since 1971 (the UK became a net cereal exporter in 1981) and the substantial increases in self-sufficiency in dairy products and sugar, which have reduced dependence upon imports. Caribbean countries have seen their traditional export markets for sugar reduced as domestically produced sugar-beet is substituted for imported cane sugar within Europe. Botswana's beef exports have been similarly hit. The EU has emerged as a major exporter of agricultural produce and is now the second largest agricultural exporter after the US. EU incursions into export markets, plus a reduced EU import market for traditional suppliers, including LDCs, have been a growing source of tension in the international trade arena.

The question 'what would be the situation if EU policies were wholly or partly removed?' is a complex one already confronted in Chapters 8, 11 and 13. To answer it one must first assume alternative EU arrangements to protect domestic agriculture, and then model the new world agricultural and food trade flows that would ensue. Europe is so large that any change in its level and methods of protection will have complex repercussions on the whole of the world trading system. Cross-linkages between world agricultural prices are diverse and the market responses and associated substitution mechanisms via demand and supply are not immediately obvious. Exercises of this type calculate the changes in trade balance and welfare for a variety of trade liberalisation scenarios. Burniaux and Waelbroeck (1988) use a general equilibrium model to calculate that, with no agricultural protection in Europe (the free trade option), output would fall by 16.8% and world agricultural prices would rise substantially, with grain prices increasing by 13.4%. Colman (1984) summarised some earlier results and suggests that the overall benefits to the LDCs as a whole, from the reduction of agricultural trade barriers, would be quite small. World grain price rises of between 10 to 15% are now thought likely following the GATT-induced market liberalisation in developed countries (Pinstrup-Andersen, 1994). The CAP exercises a major constraint upon the agricultural export earnings of LDCs, but offsetting this, many of the poorest, which are becoming increasingly dependent upon food imports, would be adversely affected by policy changes which caused international food prices to rise substantially. However, higher world food prices should be transmitted internally into LDC markets and force their governments to pay more attention to domestic agricultural production and rural development.

It is generally accepted that, in addition to the effects noted above, agricultural policies, such as the CAP, force price instability onto world markets to the detriment of less developed countries. It is held that stabilising EU farm prices has increased price instability on world markets. The issue of instability and variability of world markets is an important one, and for illustrative purposes the following two sections examine, in some detail, the world market for cereals.

The world cereals market

Recent trends in the world cereal market are shown in Table 15.1 for wheat and coarse grains. For the other major cereal, rice, international trade accounts for only 4% of world production — in 1991 less than 13 million tonnes out of a total world production of 520 million tonnes. Trade is mostly in quality rice (e.g. basmati), the world market is thin and volatile and most major rice-eating nations strive for self-sufficiency (IRRI Rice Facts, 1994). As the table shows, the US is still the major player in the world cereals market in terms of production, stockholding and exports but since the mid-1980s the EU has become the second most important source of grain exports.

Despite the headline-making problems in starving Africa and recent droughts in the US, China and western Europe we have entered an era of grain surpluses, buyers' markets and high storage costs of surplus grain stocks. Production has expanded by 40% since 1974, more than matching effective world consumption. A growing number of grain exporters now face a shrinking set of importers and the long-term trend of prices, as we shall see, is generally downwards. Until 1995 large surpluses existed in both the EU and the US with world stocks of cereals of 350 million tonnes or roughly 20% of annual consumption. In addition potential surpluses exist in the form of fertile cereals land, temporarily retired or set aside from present production.

Table 15.1: World production and trade of wheat and coarse grains.

	1974/75	1984/85	1994/95
Total World Production (m tonnes)	1014	1334	1420
(as % of total)			
United States	20	23	24
China	11	14	16
European Union	11	11	11
(Former) Soviet Union	18	12	11
Closing Stocks (m tonnes)	52	255	278
(as % of total)			
United States	45	34	27
European Union	20	7	3
Total Trade* (m tonnes)	123	205	176
Exports: (as % of total)			
United States	51	47	46
European Union	7	12	15
Canada	11	11	14
Imports: (as % of total)			
Japan	15	13	16
Latin America	10	11	15
Africa	9	8	10
China	5	6	9
European Union	18	3	2

Source: FAO. * excluding intra-EU trade.

There has been intense competition by exporters to dispose of surplus stocks in world markets, involving a variety of concessional credit programmes, export subsidies and food aid, but the capacity of many food-deficit areas to import is constrained by lack of foreign exchange. As ever, the issue of food aid is clouded with political and logistical questions. In the future, world production is expected to continue to grow and export competition to increase. Surpluses will increase and prices will tend to move further downwards. Lack of per capita income growth in many LDCs will curtail consumption increases for grain, both for direct human usage and as livestock feed. The world food problem is not one of food availability but uneven distribution globally and lack of effective demand and access to food by many poor people and many poor nations. With the costly process of stockholding there will be increased pressure to apply production controls (crop retirement or set-aside schemes) in developed countries, but income support to farmers is likely, as in the past, often to override both cost and trade considerations albeit now within the constraints of the GATT and its successor, the World Trade Organisation (WTO).

Cereal supply is crucial to solving the malnutrition problems in LDCs. In general, agricultural production growth in LDCs is failing to keep pace with growth in demand arising out of increases in population and income. This is especially the case in Africa. The developing countries' share of the world market in foodstuffs, which is predominantly cereals, was estimated to be 45% in 1981 and imports and food aid continue to rise. Imports of grain into sub-Saharan agriculture were over 12 million tonnes in 1985, only 10% below the previous year's record level. The EU has emerged as the major agricultural supplier, supplying the region with one-third of its import needs since 1980, whilst the US has held only a 15% share. In the foreseeable future there will be a continuation of the trend for LDCs to increase their imports of temperate zone food, implying that international trade will remain an important instrument as these countries attempt to stabilise domestic consumption levels. Globally, food aid shipments were running at 15 million tonnes of cereals in 1992/93, up from 9 million tonnes a decade earlier, with eastern Europe and the FSU being new recipients of this form of assistance. In future as the Uruguay Round of the GATT and the WTO is implemented, food aid may be expected to be reduced as food surpluses in developed countries diminish and disappear. Food aid shipments of cereals fell to 9.8 million tonnes in 1994–95 and future commitments have recently been drastically reduced (FAO, 1995).

Of increasing concern is the increased variability in world cereal production and prices in recent years. The causes of this phenomenon are only partially understood, but are linked to individual crop yield variability associated with the adoption of new seed varieties, fertiliser-intensive technologies, a reduction in offsetting patterns between different crops and regions and periodic oil price shocks. Hazell (1985) calculated that world cereal production grew at 2.7% per annum between 1960/61 and 1982/83 but that there was a widening band of variability around the trend underlying this increase. Total cereal production rose by 305 million tonnes, of which wheat and maize each contributed 33% of the increase, rice 12% and barley 18%. Comparing the two periods 1960/61 to 1970/71 and 1971/72 to 1982/83 the variability of production increased markedly in the second period particularly in the wheat/barley regions of Africa, South America, Oceania, Canada and the USSR. Rice regions are less variable since much of the crop is irrigated.

It is not easy to account for the increased production variability. Similar technologies, similar weather, reduced offsetting yield effects between different cereals, oil price effects, the expansion of crops to marginal areas, fewer drought-resistant crops and a narrowing genetic base all play their part, along with the pricing policy for these commodities. However, a continued high level of variability in world cereal production and hence prices seems likely. Inevitably, stockpiling to cope with this will be a costly operation, and the developed countries may be unwilling to hold and fund world grain stocks in the future to ensure food security for the less developed countries.

Instability in the world wheat market

Taking the fob (free on board) price of US exports of Hard Red Winter wheat as indicative of price on the world market, Fig.15.1 shows the annual average world price of wheat over the period 1960–93. In nominal dollar terms (that is, unadjusted for inflation) the world price showed remarkable uniformity from 1960 to 1971, at around US$60 per ton. The commodity crisis of the early 1970s resulted in a tripling of this price by 1973. Since then it has undergone considerable fluctuation, and in 1993 was US$ 143 per ton.

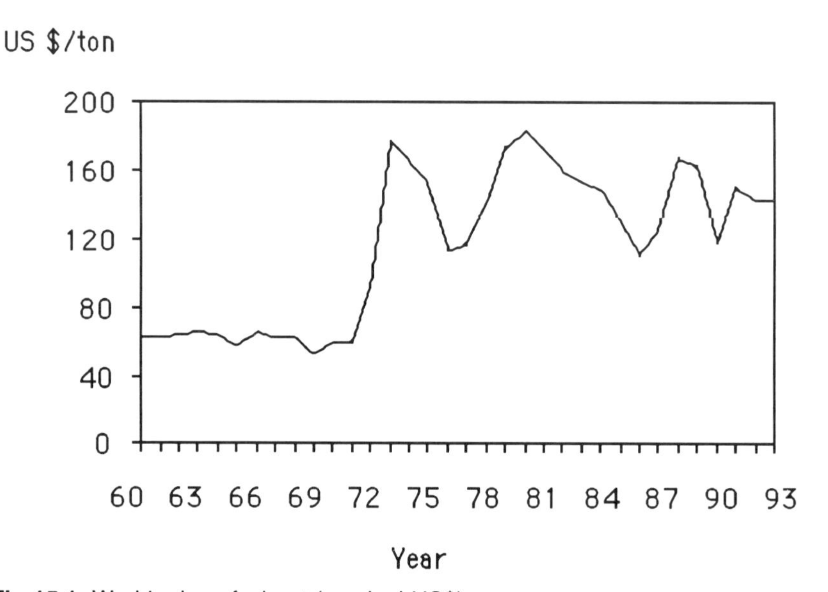

Fig.15.1: World price of wheat (nominal US$).

Source: International Wheat Council

In Fig.15.2 these nominal world market prices are expressed in 1993 values, using a GDP index for the industrial countries (IMF, 1995). In real terms the world market price had been falling steadily prior to the commodity crisis and, indeed, has been falling since. If the abnormal prices of the mid-1970s are discounted, world market price exhibits a continuing downward trend. Wheat prices have fallen dramatically in real terms, from around US$ 350 per ton in the early 1960s (in 1993 values) to around US$ 150 per ton in the early 1990s.

The mean price, again expressed in 1993 values, over 1960–93 is US$ 280 per ton, but this average conceals price variability during the period. As a measure of the extent of this variability it is useful to calculate the coefficient of variation (c.o.v.). This statistic relates the standard deviation to the mean, and is expressed on a percentage basis. The higher the c.o.v. the greater the degree of fluctuation of price about its mean level. The c.o.v. associated with the mean of US$ 280 tonne is 38%.

In an attempt to isolate the effects of the commodity price boom of 1972–75 and to examine more closely what has been happening to world prices over time, let us divide the 34-year period into three separate periods, representing the pre-commodity crisis period (1960–71), the commodity crisis period itself (1972–75) and the post-commodity crisis period (1976–93). The mean and c.o.v. for each of these periods are shown in Table15.2. The three periods exhibit markedly different summary statistics. For reasons that are well documented, world price increased dramatically in the early 1970s, but perhaps of greater surprise is the equally dramatic fall in price in the last of the three periods. At US$ 217/ton the average price of wheat since 1976 is only 70% of the average price during the 1960s. The c.o.v. rises from 17% through 25% to 32% over the three periods, suggesting increased price variability. On examination of Fig.15.2 it can be seen in both the first and third periods that the downward trend in price is quite smooth. Over the period 1960–71 the world price of wheat, in real terms, fell at an average rate of 4% a year, and indeed since 1976 has continued to fall at this same average rate.

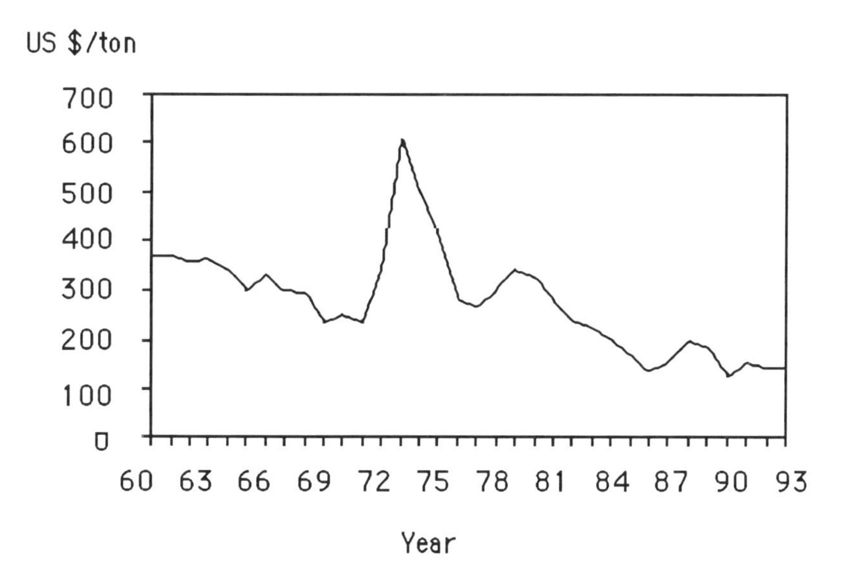

Fig.15.2: World price of wheat (1993 US$).

Table 15.2: World price of wheat.

Period	Mean (US$/ton; 1993 prices)	c.o.v. (%)*
1960–71	312	17
1972–75	464	25
1976–93	217	32
1960–93	280	38

Source: Compiled from International Wheat Council data.

* The coefficient of variation (c.o.v.) is calculated as the standard deviation as a percentage of the mean.

Let us now turn to look at the EU's external trade in wheat. Fig.15.3 shows the EU's exports and imports (intra-EU trade is excluded) of wheat and wheat flour since the establishment of the common market for cereals in the late 1960s. Exports have increased from a low of 3.1 million tons in 1970 to a high of 22.7 million tons in 1992, whilst imports have fallen from a high of 7 million tons in 1972 to a low of 1.2 million tons in both 1991 and 1992. The difference between the two sets of bar charts in Fig.15.3 represents the Community's net external trade in wheat and wheat flour. The Community was a net importer in 1970, 1972, 1976 and 1977, but in more recent times exports have substantially exceeded imports, with net exports peaking at 21.5 million tons in 1992.

Summary statistics for these trade data are presented in Table 15.3, with the overall period split into three. In the first period, 1968–75, exports of wheat and wheat flour from the Community averaged 5.9 million tons with a c.o.v. of 27% Net exports over the period averaged only 0.9 million tonnes per annum, but were subject to significantly greater variability, as is indicated by the c.o.v. of 179%. In the second period, 1976–84, whilst gross exports nearly doubled, net exports increased by more than seven-fold. Again, variability of net exports during this period was high, though less than over the previous period. In the most recent period, 1985–93, gross exports have again almost doubled to an annual average of around 19 million tons, and net exports have more than doubled to around 17 million tons. With the EU emerging as a major and consistent exporter of wheat and wheat flour, the year-to-year fluctuations associated with these annual averages has fallen markedly.

Thus, our simple analysis indicates that, over the last 30 years, the real price of wheat on the world market has been on a downward trend and has exhibited somewhat greater instability, whilst over the last 25 years net exports from the EU have risen dramatically. However, attributing any cause and effect between these occurrences requires a fuller analysis. US exports have risen dramatically as well, and they still account for a large share of the world market. Other developed country exporters have also contributed to the destabilisation of world markets, and it is difficult to isolate the EU influences alone.

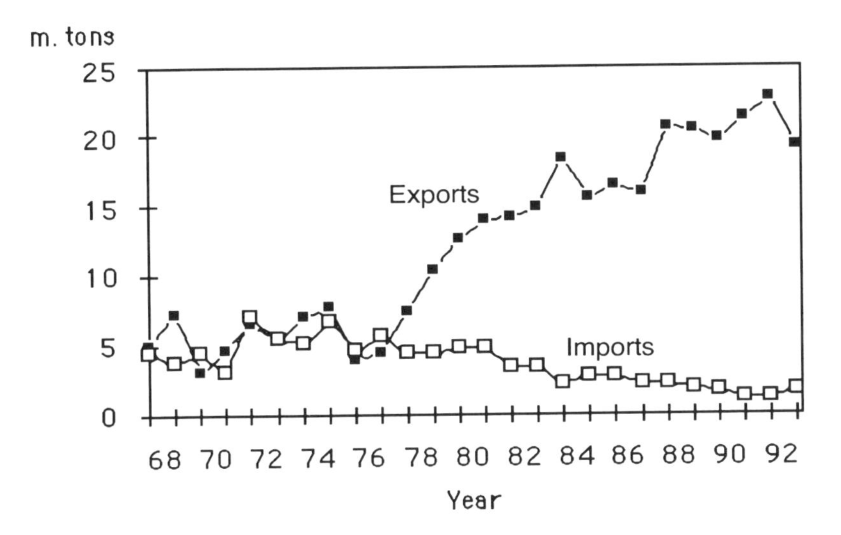

Fig.15.3: EU trade in wheat and wheat flour.

Source: International Wheat Council. Data incorporate the enlargement of the Community over the period, including from 1990 the former German D.R. Excludes intra-trade.

Table 15.3: Annual exports of EU wheat and wheat flour.

Period	Gross Exports		Net Exports	
	Mean (m. tons)	c.o.v. (%)	Mean (m. tons)	c.o.v. (%)
1968–75	5.9	27	0.9	179
1976–84	11.1	45	6.9	83
1985–93	19.1	13	17.1	18
1968–93	12.3	52	8.6	91

Source: Compiled from International Wheat Council data.

Nevertheless, the impact on the world price of EU exports of wheat is generally regarded as being two-fold. First, any increase in the EU's excess supply (net exports) can be expected to depress the world price. Second, as a consequence of the workings of the cereals regime within the CAP, variability of world price can be expected to increase. There have been a number of studies illustrating these effects. For example, Sarris and Freebairn (1983) used their own model of the world wheat market to show that if the EU were to dismantle the CAP and adopt the unlikely position of free trade, the world wheat price would increase by 9% (an example of the price-depressing effect of the CAP). Moreover, the c.o.v. of world price would decrease by 20% (an example of the instability-increasing effect of the CAP). If the entire world followed suit and moved to free trade, Sarris and Freebairn estimated the world price would rise by 11% and the c.o.v. would fall by 35% (an illustration of the effects on the world market of other countries' agricultural policies). Using these results, Sarris and Freebairn concluded that the CAP was responsible for 84% of the price depression in the world market and for 56% of increased price variability, as compared with a position of global free trade. Comparable results to those of Sarris and Freebairn have been obtained from other studies.

So much for the effect of the CAP on the level and stability of world prices. Some authors have attempted to show the impact on world markets of the elimination of all forms of agricultural support in the developed world. The work of Sarris and Freebairn (1983) has been cited; other examples include Tyers and Anderson (1986), the OECD (1987) and Anderson and Tyers (1993). Whilst the scope of these studies is largely beyond the remit of this chapter, the results highlight the problems faced by LDCs as a direct consequence of agricultural protection in developed countries. Anderson and Tyers more recent work contradicts their previous findings and suggests that not only food-exporting LDCs but also food-importing, poor countries could gain through the liberalisation of agricultural protection and market insulation policies. The subsequent increasing of food prices in LDCs is likely to boost agricultural productivity, whilst at the same time lowering food prices in industrialised developed countries, which could slow down their productivity growth. The possible gain for a food-importing nation from trade liberalisation depends upon how it transmits and responds to the rising food prices. The negative effect of developed country agricultural protection on the net economic welfare of LDCs as a whole could be as high as $17 billion per year in 1985 dollars.

Agricultural protection reduces the volume of world agricultural trade and causes world prices to be lower and less stable. Of this global impact, the EU must bear a major share of the blame. However, in future the EU must reduce the volume and spending on export subsidies to comply with the WTO's rules following completion of the Uruguay Round of the GATT. The binding constraint on EU exports of cereals is a volume constraint requiring a 21% reduction in volumes of subsidised exports relative to the base period — this is likely seriously to curtail EU export capacity, although higher world wheat prices in 1995–96 have thus far limited the effectiveness of the constraint.

Effects of instability on less developed countries

Fluctuations in production levels and food availability frequently occur in LDCs consequent upon uncertain and variable climates and production environments, coupled with a less sophisticated traditional infrastructure. Food security (the ability of LDCs to meet target levels of consumption on a year-to-year basis) is not easy to plan. Access to assured imports at known prices would help the planning process. Reliance on imports by food-deficit LDCs requires that they have confidence that grain imports will be available in the future within some foreseen price range. In addition they must have access to the foreign exchange needed to maintain consumption given shortfalls in domestic production. The combination of domestic production instability and world price instability makes effective planning very difficult and hazardous or infeasibly expensive. In past years of world grain shortages, the richer food-importing nations, like the Soviet Union and China, have always outbid the poor.

It is now largely accepted that the CAP has achieved internal price stabilisation within the EU but at the expense of increasing instability outside the Community. The burden of adjustment to changes in global production of agricultural commodities is borne by the world market, resulting in larger price fluctuations, compared with a freer trade situation. Liberalising the CAP would substantially reduce price instability on world commodity markets. This argument also extends to export markets of LDCs. Increased price and revenue instability, brought about by the CAP, adds to the production risks of agricultural exporters of the Third World, undermining investment and future productivity (Roarty, 1985).

Production and price fluctuations in wheat in developed countries can be a potential problem for the rest of the world. Some changes in production directly reflect protectionist policy decisions of the US, the EU and Japan. Greater stability in production levels would be to the advantage of developing countries in planning their food supplies and ensuring food security. Reducing some variations in production may require developments in technology and husbandry, but eliminating policy-induced variations is an urgent priority for the international community. The CAP has been an adverse influence on the food security of LDCs. Any improvement must involve policy within the EU, adjusting so that stocks and prices are more sensitive and responsive to world market conditions.

Trade preferences

The EU operates a number of preferential trading agreements with developing countries. These offer some trade advantages over other third countries by making access to the Community market somewhat easier. Of the agreements, the most elaborate is the package of arrangements under the Lomé Convention, covering nearly 70 African, Caribbean and Pacific (ACP) countries. Under Lomé, many exports from ACP countries enter the Community free of customs duties and free of quantitative restrictions, whilst the ACP countries themselves are not required to offer reciprocal trade concessions. The Lomé Convention is "a symbol for the

Community and of North-South relations — and therein lies both its strength and its weakness, for the hopes it has generated turn to disappointment at its slightest shortcoming" (European Commission, 1985). As this statement implies, Lomé is not without its critics.

The Lomé Conventions are quite separate from the CAP, though some common ground does exist. The principle of free, or freer, access to the Community market does not extend, in general, to those commodities covered by the CAP, the main exception being fruit and vegetables (Ritson and Swinbank, 1993). There are, in addition, special trade arrangements for sugar. Sugar is a commodity which can be produced in both temperate and tropical regions, as sugar beet and cane sugar respectively, and is an important crop in both the Community and some ACP countries. Under the sugar protocol, ACP countries[1] not only have free access to the Community market for around two-thirds of their sugar exports, but also benefit from the price they receive being tied to the high support price paid to European farmers. This arrangement has been very controversial, as indeed has the whole sugar regime under the CAP. A ceiling of 1.3 million tonnes of ACP sugar is covered by the arrangement, which means, in effect, that surplus sugar production within the Community (which is around 2.5 million tonnes) is increased by an equivalent amount. This excess is disposed of on the world market, depressing world price and thereby lessening the export earnings of those developing countries not party to the protocol.

One special feature of Lomé has been a system of 'compensatory finance' known as Stabex. This is a form of insurance that operates in poor harvest years and seeks to compensate ACP countries for losses in export earnings, thereby attempting to stabilise their export income. Fifty commodities are covered by Stabex,[2] but, whilst it may go some way to providing temporary relief in times of crisis, it operates within a limited budget and cannot preclude serious fluctuations in export earnings. The commodities covered are such that they affect the CAP only to a minor extent.

LDCs and future reform of the CAP

For many years reform of the CAP has been a popular topic for debate (see Chapter 4). The CAP is frequently criticised both from within the Community and from third countries. A major political pressure for reform was a group of agricultural exporters, the Cairns Group, named after the Australian city in which they first met. The group included both developed and developing countries and threatened to block agreement on many important issues to other countries in order to keep agricultural reforms as part of the GATT negotiations. In addition, the US has periodically threatened the EU with agricultural trade wars. In response to these criticisms there have been minor changes in the workings of the CAP over the years, but the principle of high domestic support prices remained unchanged until relatively recently with the Mac Sharry reforms and GATT pressures.

Criticism from within the Community stemmed mainly from the financial cost of the CAP. Budgetary expenditure (and to a lesser extent the consumer cost) was the major concern of those seeking reform. The budgetary cost of the CAP is financed principally by the Community's taxpayers, and as this cost escalated as a consequence of increasing surpluses, so the demands for reform have grown. Consumers in the Community, as we saw in Chapter 11, also pay heavily for the CAP, but they have failed collectively to exert much of an influence in the political process. Since the burden on consumers does not increase as production increases, there is little reason to suppose their pressure will be any more effective in the future.

However, there are signs that environmental pressures against intensive agriculture are growing, particularly from the 'Green Movement' within European politics. Major reform in the early 1990s (see Chapter 13) occurred partly as a response to these internal pressures and partly to the external pressures brought to bear during the GATT process. Previously EU policy-makers were reluctant to heed the criticisms of their own taxpayers and it is unlikely that they were swayed by the claims of poor third world countries. Thus it is fair to say that pressure from the LDCs had only a minor impact in reforming the CAP and regrettably will continue to do so. In any case, many LDCs must also put their 'own house in order' with respect to the pricing policy for agricultural output and inputs. Food prices play a dual and conflicting role in the elimination of poverty and hunger; on the one hand they provide an incentive to production, thus generating employment and income; on the other they help determine the entitlement to food, the extent of hunger and malnutrition and thus real income levels. Short-sighted governments in some LDCs often keep the price of foodstuffs low for the urban proletariat, thereby undermining incentives to local farmers and building up a dependence on imported foods. Policy changes are required within many of these countries to overcome the past neglect of agriculture.

International political pressure should continue to be directed to convincing the EU that it is responsible for some of the instability in world markets. Accepting its responsibilities, it may be possible to institute a continuing food aid programme whereby aid donors provide food aid to meet grain production shortfalls which exceed a given percentage of the trend level of production in LDCs. However, the sources of food aid, namely CAP induced food surpluses, are likely to be much lower in the future as protection to EU farmers is lowered. Matthews (1985) suggested that another route to reform would be to maintain the average level of protection to EU agriculture by means of a constant tariff, so that fluctuations in world prices were reflected in EU markets.[3] Permitting the EU market to react to world price fluctuations would help dampen those fluctuations which occur. This would be a positive benefit to LDCs by lowering the probability that they could not afford food imports in periods of high food prices. However, it would mean that EU farmers could no longer depend on a guaranteed minimum price for their produce, while EU consumers would find food prices more variable. There is thus a direct trade-off between the food security of LDC populations and price stability within the EU.

In terms of future potential changes to the CAP, reform is likely to continue with significant, but gently phased, price cuts for the main agricultural commodities, including cereals combined with some physical controls on production for which EU farmers will be directly compensated. The result will be a reduction in the surpluses and thus in EU exports. Firm conclusions about reforming the CAP from an LDC perspective are difficult. LDCs are heterogeneous, and higher world food prices, as explained earlier, would have different consequences for different groups both between and within LDCs.

Nevertheless, internal and external pressures are likely to prevent any worsening of the impact of the CAP on world markets in the longer term. The dark days of the 1980s are over! Budgetary pressures alone should be sufficient to limit the growth of EU exports. However, a continued internal EU emphasis on coping with crises in a partial, commodity-by-commodity approach may well mean that international markets are subject to short-term fluctuations and uncertainties, as evidenced by the consequences of milk quotas on beef markets. If the WTO is not effectively policed and implemented, EU action may yet again destabilise world food markets. Unfortunately, living through the 'short term' is the overriding ambition of many LDCs. Given the time taken by the EU to reach decisions about inevitable policy changes and the protracted nature of international negotiations leading to reduced agricultural protection at any future GATT talks, the short-term uncertainty on world food markets is likely to continue. Notwithstanding, the world grain trade and food aid are of limited use in solving the enduring and serious problems of poverty, hunger, food security and world food distribution. All we can ask is that the EU and the CAP contribute to their solutions rather than, as in the past, adding to the problems!

References

Anderson, K. and Tyers, R. (1993) More on Welfare Gains to Developing Countries for Liberalising World Food Trade, *Journal of Agricultural Economics*, 44(2), 189–204.

Burniaux, J.M. and Waelbroeck, J. (1988) Agricultural Protection in Europe. In: Langhammer, R.J. and Rieger, H.C. (eds) *ASEAN and the EC: Trade in Tropical Agricultural Products,* ASEAN Economic Research Unit, Singapore.

Colman, D. (1984) EEC Agriculture in Conflict with Trade and Development, *Manchester Papers on Development,* No 10, 1–12.

European Commission (1985) *Europe-South Dialogue*. Directorate-General for Information.

FAO (1995) *The State of Food and Agriculture — Agricultural Trade: Entering A New Era?* FAO, Rome.

FAO. Commodity Review and Outlook, annual, Rome.

Hazell, P.B.R. (1985) Sources of Increased Variability in World Cereal Production Since the 1960s. *Journal of Agricultural Economics,* 36(2), 145–159.

IMF (1995) International Financial Statistics. Yearbook, Washington.

IRRI (1994) *IRRI Rice Facts*. Los Banos, Philippines.

International Wheat Council (various years) *World Wheat Statistics and World Grain Statistics*. Annual Publications. International Wheat Council, London.

Matthews, A. (1985) *The Common Agricultural Policy and the Less Developed Countries*. Gill and Macmillan, Dublin.

OECD (1987) *National Policies and Agricultural Trade*. Paris.

Pinstrup-Andersen, P. (1994) *World Food Trends and Future Food Security*. IFPRI, Washington.
Ritson, C. and Swinbank, A. (1993) Prospects for Exports of Fruit and Vegetables to the European Community after 1992. *FAO*, Rome, pp.163.
Roarty, M.J. (1985) The EEC Common Agricultural Policy and its Effects on Less Developed Countries. *National Westminster Bank Quarterly Review*, 2–17.
Sarris, A.H. and Freebairn, J. (1983) Endogenous Price Policies and International Wheat Prices. *American Journal of Agricultural Economics*. 65, 214–224.
Tyers, R. and Anderson, K. (1986) *Distortions in World Food Markets: a Quantitative Assessment*. Background Paper for the World Bank's World Development Report. Oxford University Press, New York.

---

[1]The ACP countries in question are: Mauritius, Fiji, Guyana, Jamaica, Swaziland, Barbados, Trinidad and Tobago, Belize, Zimbabwe, Malawi, St Christopher and Nevis, Madagascar, Tanzania, Congo, Ivory Coast, Kenya, Uganda and Surinam.
[2]The main commodities are coffee, cocoa, cotton, groundnuts, sisal, timber, oilcake, bananas, tea and palm products.
[3]Tariffication under the GATT Agreement has involved a move in this direction, but not, as explained in Chapter 5, for cereals.

Chapter 16

The CAP and North America

Tim Josling

Introduction

Agriculture has proved a painful area of commercial conflict between North America and the EU for several decades. The list of commodities over which trade skirmishes have taken place is long. It embraces wheat, corn, soybeans, wine, citrus fruits, sugar, beef, poultry and pigmeat. At times the conflicts have threatened relations in other areas, in particular in other aspects of commercial policy. Understanding of the roots of EU agricultural policies is not widespread in North America. There is little patience for arguments based on historical, cultural or social protection. The same is true of European attitudes towards US policy problems. US agriculture is seen as a poor role model by many in Europe, involving the impersonal exploitation of vast acreages by corporate farms. Thus sermons from US politicians about needed changes in EU agriculture usually fall on deaf ears.

The transatlantic rhetoric would seem to indicate deep philosophical differences between the US and the EU, showing up as divergent trade and farm policies. It is easy to make the case that the US and the EU have some fundamental differences in their respective approaches to trade in general and agricultural trade in particular. The last three US administrations have been adamant in their belief that the US is an open market, a place where the rest of the world can come and sell their goods without hindrance, whereas the other developed countries, Japan and the EU in particular, guard their markets with a mix of subtle and blatant trade barriers. The US economy itself is held to be more market-oriented and less directed by government policy. One ongoing big debate in economic and trade policy in the US in recent years has been over the desirability of moving towards more management of the economy, through industrial policy and active trade policy, to compete better with the regulated economies of Japan and Europe. The success of Japan, in particular, has been seen by many as a challenge to the notion that open markets and a 'hands-off' industrial policy are good for growth.

Of course, from outside the US, the picture looks very different. The US is indeed a relatively open economy, but with all manner of residual restrictions on commerce – even between states – and with protectionist measures emanating from every level of government. A fundamentally protectionist alliance of organised labour and an urban, Democratic Congress was for years held at bay by staunch free-trade rhetoric from the Republican Administrations and with support of their allies in the business and agriculture community. The election of a Democratic President with a personal preference for freer trade made it possible to pass the legislation needed to implement the North American Free Trade Area (NAFTA) and the GATT agreement with help from the Republican minority. The trade remedy laws are however still strongly tilted in favour of US businesses, and act to scare off potential exporters. Harassment through state taxation laws and minor court decisions adds to this view of the US economy as ambivalent if not actually hostile to trade.

North American views on Europe also depend upon the location of the viewer. Is the EU the great experiment in a single, seamless market, open to foreign firms and products, as advertised by the Commission? Or is the single market a device to develop European competitiveness at the expense of overseas firms, and a way of consolidating at the Community level a range of protective devices? It is possible to find support for almost any interpretation of the EU's liberalism. The EU is in some ways an 'open' agricultural market, importing more farm produce than any other region. But this does not ring true to the supplier of cereals, dairy products, meat and sugar who face high and sometimes prohibitive import levies, or seem relevant to those competing with the EU's open-ended export restrictions.

Though political structures differ, and the domestic politics of farm support require a large dose of 'we-they' system confrontation, the similarities between the EU and the US are more striking than the differences. The US-EU farm trade clashes are better seen as conflicts among similar systems, each of which would work better if only the other's were to be modified. The tensions are more a result of this unfortunate fact than any fundamental difference in philosophy.

Market pressures and trade tensions

The US has a strong belief in the capacity of its own agriculture to compete on level terms with other countries. Its objection is to subsidised competition from other countries. These countries seek to use world markets to avoid domestic resource adjustments and hence prevent true competitiveness from emerging. The EU is singled out as a major player in the game of subsidised exports and high domestic market prices.

The conflicts between the US and the EU in agricultural trade go back 30 years to the founding of the CAP in the early 1960s. The sharpness of the conflict has increased in the past decade. The exacerbation of tensions in the 1980s is a clear reflection of the emergence of EU agriculture as a major exporter in temperate-zone markets. The US has seen its exports to the EU diminish over the last decade, and has faced increasing competition in third country markets. EU exports have also penetrated the US domestic market to an increasing extent.

During the 1970s, US exports rose rapidly, as shown in Fig.16.1. The United States had in place the capacity to meet the surge in demand from the USSR and from the developing world. European exports lagged until 1977, but accelerated with the generally favourable trade conditions until 1981. That year represented a peak for US agricultural exports, US$43 billion. Over the next few years, US exports declined dramatically, to a low point of US$26.1 billion in 1986. EC exports also stalled a little after 1981, as the debt crisis and a sharp rise in oil prices reduced international demand, and then recovered. By 1986 EC exports were actually higher than those of the US, and have stayed at or above US levels ever since. The EU, for years touted as the "largest import market" for farm products had become the world's largest exporter (though imports remained significant as well). A recovery in US exports since 1987 has helped to restore confidence somewhat in the ability of the US to export farm products to world markets. However, the view persists that the EC achieved its position in the export league by means of subsidies rather than fair competition.

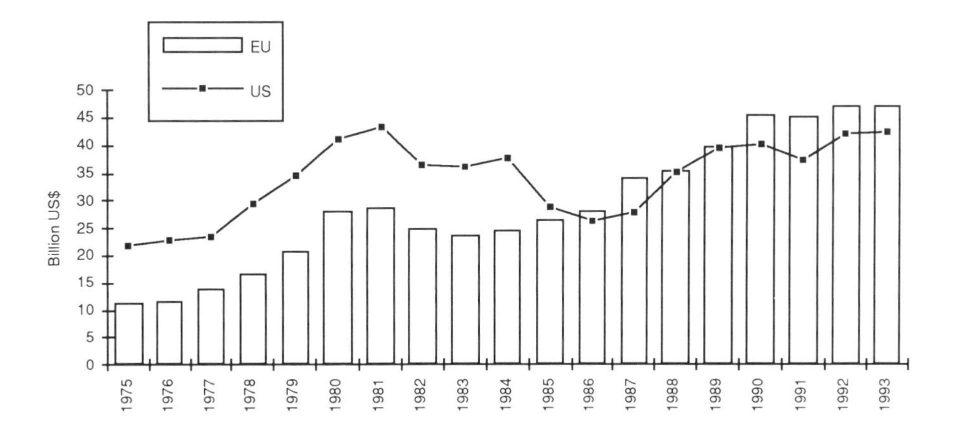

Fig.16.1: Export of agricultural goods, EU and US, 1975–1993.

The trend in overall exports sets the tone for US farm policy. The budget cost largely reflects the balance between domestic production and export demand. Escalation of budget costs over the 1980s was a major policy concern in the US as it was in the EC. The comparison of budget costs is shown in Fig.16.2. Throughout the 1970s, US farm support costs stayed low, at about US$3–5 billion, helped by buoyant export markets and firm world prices.

Support costs in the US took off in the wake of the drop in exports in 1982, coupled with the generous policy prices set under the 1981 Farm Bill. A drought-induced drop in production in 1983, coupled with the cost-shifting effect of the payment-in-kind (PIK) scheme, gave a temporary respite in expenditure in 1984, but by 1986, expenditure had reached a peak of US$25 billion, leading to the interesting phenomenon of higher US programme costs than those in the EU. More recent developments have returned this relationship to its more normal state: US spending since 1986 has had a marked downward drift, in contrast to the seemingly inexorable rise in spending on farm programmes in the EU.

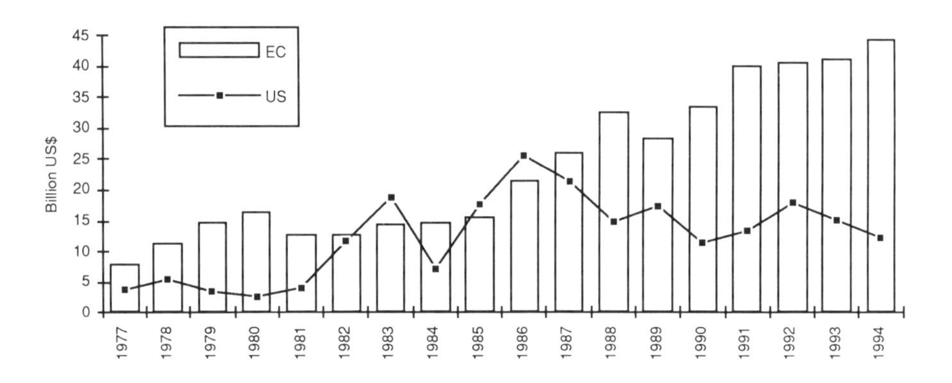

Fig.16.2: Spending on farm programmes, EU and US, 1977–1994.

Bilateral trade figures point to another aspect of the story. Fig.16.3 shows bilateral trade between the US and the EC over the period 1975-93. Over the last half of the 1970s, trade was increasing steadily in both directions. The situation changed abruptly after 1980. US exports of agricultural products to the EU-10 fell from a high of US$9.6 billion in 1980 to US$5.2 billion in 1985, recovering the next year to US$6.6 billion in part because of the enlargement of the Community in 1986 and continuing at the US$6-7 billion range through to the present. By contrast, EU agricultural exports to the US continued to rise in the 1980s, reaching US$4.1 billion in 1986 (US$3.8 billion for the EU-10) and nearly US$5 billion by 1992. Though still significant, the bilateral trade gap is now relatively minor in relation to total bilateral trade between the US and the EU.[1]

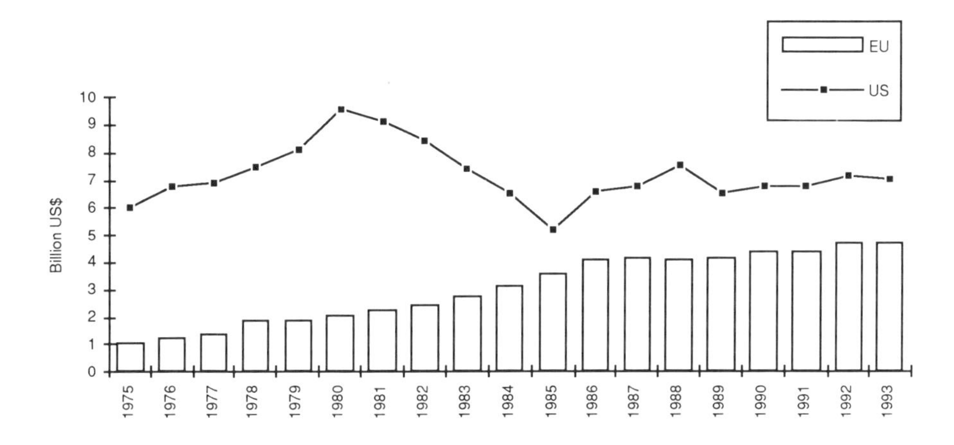

Fig.16.3: Bilateral exports of agricultural goods, EU and US, 1975–1993.

The fall in US exports to the Community in the 1980s was primarily in the 'big-ticket' items of grains, non-grain feed ingredients and oilseeds and oilseed products. Corn sales were hard hit by emerging surpluses of barley in the EC and by the continued attractiveness of non-grain feed ingredients in livestock rations. Soybean sales were hit by increased oilseed production in the EU, which grew from 3.2 million tons in 1980/81 to 11.6 million tons in 1987/88. Domestic oilseed production now accounts for nearly half of the total EU oilseed use (USDA, 1993). More recently, other export markets have been affected. Beef and variety meat sales to the EC from the US had reached over US$100 million in 1988, when a ban on imported meat produced with hormones was introduced. Cotton sales fell away after 1985, in part as a result of Spanish and Portuguese accession to the EC, and the US has lost some of its rather small share in the EU wine market in recent years.

Added to the concerns about import access has been the intensified competition in third countries. This has been particularly true in the case of wheat, as highlighted in the previous chapter. EU wheat production has more than doubled since 1962, largely as a result of yield increases (USDA, 1993). In the face of stagnant consumption, self-sufficiency was reached in the mid-1970s, and the EC emerged as a major exporter of wheat, capturing 15% of the world market by 1987, and maintaining that share for the next five years. The US share in wheat sales declined sharply in the 1980s from 48% in 1981 to below 30% in 1985 (ABARE, 1989). Whatever the many causes of this decline, the Community's generous export subsidies were held in large part to blame. This experience in the wheat market led to a minor subsidy war, with the US using an export enhancement program (EEP) to target export subsidies at markets lost to the EC. These EEP sales accounted for about 50% of total US wheat exports from 1985–1988. The struggle for markets was particularly intense in the north African countries of Morocco, Algeria, Tunisia, and Egypt, which import large quantities of wheat and wheat flour. With the aid of considerable export subsidies, the US was able to regain market shares in this region. The incident left a lasting mark on US-EC relations which perhaps exceeded its real significance.

Trade relations in the 1980s

Agricultural trade relations between the US and the EU reached a low ebb on November 5, 1992. On that day the US announced the imposition of punitive tariffs on US$300 million worth of agricultural imports from the EU. Specifically targeted was white wine, mainly from France and Italy. Two days earlier the 'last ditch' talks to settle the twin problems of EU oilseed subsidies, twice ruled illegal under the GATT, and the agricultural component of the Uruguay Round of trade negotiations had broken up without agreement. A delay in the application of the punitive tariffs allowed time for an additional set of negotiations. In the hiatus between the November election, when President Bush was defeated, and the inauguration of President Clinton, the 'lame duck' administration in the US quickly cut a deal with the EU Commission. On November 20, 1992 a settlement was announced of both the oilseed and the Uruguay Round issues, to the intense relief of the negotiators and the consternation and anger of French politicians and farmers. The 'Blair House Accord', as it became known, was also less than popular in the US, where it was widely seen as a desperate attempt to get a GATT agreement through before the Clinton administration took office.

The longest-running transatlantic trade battle is over soybean exports to the EU. The US received a concession from the Community in the Dillon Round that the tariff on soybeans and meal would be bound at zero.[2] This one action has turned out to dominate more than any other the next 25 years of trade and trade relations. To the EU it was a concession easily made, since trade was fairly small in those commodities and domestic production almost non-existent. Low tariffs on imports of oilseeds in any case favoured the domestic crushing and refining sectors. To the US it seemed like a useful concession, but the US was generally unhappy with the outcome on other products. It carried out of the Dillon Round "unsatisfied negotiating rights" to be used in later trade talks.

What followed profoundly affected US-EU trade relations. The high wheat prices set for the CAP (the background to which is explained in Chapter 2) drove feed demand towards corn – much of it imported from the US. Later, when grain prices were harmonised in the EC under the so-called silo system (moving corn prices up to meet the higher wheat prices), feed compounders switched to non-grain feeds. The pressure was particularly strong in Germany and the Netherlands, where 'green' exchange rates (Chapter 6) ensured higher cereal prices than in other parts of the EC. Compounders found that a mix of soybean meal (from US soybeans) and cassava chips (from Thailand) made an excellent substitute for corn, barley, and wheat.

Soybean sales to the EC boomed over the 1970s and continued to be high when EC corn market imports slowed down. The Dillon Round hole in the dike was letting in a flood of imports. The EC negotiated and entered into a bilateral agreement with Thailand to set a limit to cassava imports. The result was that other starchy feed ingredients such as citrus pulp, maize germ meal and distillers dried grains began to be used to replace cereals in feed rations — many imported from the US. Corn gluten feed, a by-product of the corn wet-milling process in the US, became the ingredient of choice in the 1980s.[3]

The US view is that the Dillon Round concession was negotiated in good faith, and the fact that it has turned out to be valuable is not through any fault of the US. It is one of the few things that the US has 'won' in the GATT: to let the concession slip would undermine the domestic credibility of that organisation. Soybean and non-grain feeds are among the few market-oriented sectors in world trade. An assault on this trade by the EC would represent a major set-back for the aims of liberalisation.

The conflict surfaced on three fronts. The EC proposed on several occasions to institute a fats-and-oils tax, which would have increased the price of seed-based oils. This proposal was repeatedly rejected by the Council of Ministers but never quite abandoned by the Commission. In the US it was seen as a scarcely disguised tax on soybean imports. The fact that the tax would have applied to domestic as well as imported oils cut little ice. The fats-and-oils tax to be effective would undoubtedly have had to reduce consumption of soybean oil: its intention was to benefit the market for butter and olive oil. US opposition to the tax was made abundantly clear, and this perhaps influenced the member states (the UK and West Germany) that had consistently opposed it within the EC.

The second front to the conflict was the attempt in 1985 to limit imports of corn gluten feed from the US, following the 'success' of the voluntary export restraint agreement with Thailand. This also received a rebuff from the US. Why enter voluntarily into an agreement to limit a booming export market when the open market was due to a firm commitment in GATT?[4] At least, if the negotiations were undertaken in the GATT, the US could have expected to be offered compensation. Unilateral action of this kind was ruled out of the question by Washington.

The third front of this conflict has been the action within the GATT on oilseed subsidy. This was a reaction to the growth of oilseed production in the EC promoted in part by subsidies to crushers who use domestic oilseeds (and pay at least a minimum price). Production of soybeans increased rapidly in Italy, rapeseed became a popular crop in northern Europe, and sunflower production expanded in the south. Once again, the US view was clear cut: the EC was trying to promote the production of one of the few products that it had actually allowed in from abroad. The GATT binding was being avoided by the back-door method of subsidising production. Nothing could show more clearly the true aim of the CAP to eliminate imports and close the door to the rest of the world.

The threat to US soybean exports did not go unnoticed. The United States Soybean Association lodged a complaint against the EC oilseed programme, and the US Administration successfully pushed for a GATT investigation of the case. The GATT panel ruled twice on this complaint. The first time, they found that the way in which the EC operated its oilseed policy did indeed interfere with trade, over and above the bound tariff agreed in the GATT. The EC changed its oilseed policy, in effect advancing its 'reform' proposals by applying them to oilseeds in advance of other arable sectors. This was insufficient to placate the exporters, who argued that 'hectarage payments' still encouraged oilseed production. A second panel agreed with this assessment, to the chagrin of the EC, and led to the crisis of November 1992.

The second GATT panel report became inextricably entwined in the Uruguay Round final stages. For the EU, the second panel appeared to be saying that the Community should either offer compensation to the US (and other suppliers) or change its newly reformed oilseed policy yet again. Reform would mean cutting even further the link between production and support (i.e. fully decoupling) in order to avoid the implication that exporter's rights were being impaired. But to do this for oilseeds would have implications for other products. The 'Green Box' for non-trade distorting subsidies being negotiated in the Round could hardly hold the Mac Sharry price compensation payments if some part of these payments (for oilseeds) were deemed trade distorting by a GATT panel.[5]

Another recent trade conflict that illustrates well the predominant US view of EU agricultural policies is that over beef imports. The EU is not a major market for US beef. Most exported beef goes to Japan, a market which seems to appreciate the grain-fed beef with the marbled texture produced on US feedlots. The US in turn imports considerable quantities of grass-fed manufacturing beef from Australia and Central America, together with some cooked meat from South America. The EU has moved from being a steady importer of South American and Australian beef to one of the world's largest beef exporters. Beef imports into the EU are controlled by both a tariff (of 20%, pre-Uruguay Round) and by variable levies, to protect the domestic market price. However, beef is still imported under a number of schemes for preferential access, such as levy-free quotas and tariff-free quotas. The US has had a levy-free quota into the EU market for high-quality ('Hilton') beef, accounting for some US$10 million annual sales. But much more important has been the growth in the market for 'variety meats', comprising beef offals used in making pies and sausages in Europe. This market had by the late 1980s reached US$100 million, and provided a valuable outlet for beef by-products not commonly used at home. It was this trade that was threatened by the import ban.

In December 1985, the EC decided to ban the use of anabolic hormones in livestock production. The ban followed rising consumer concern over the health effects of such hormones, spurred by the well-publicised incidents in Italy involving the use of a synthetic oestrogen, Diethylstilbestrol (DES). DES had been banned in the EC (and the US), but consumer confidence in the safety of cattle feeding and in the ability of the regulatory authorities to control such practices was shaken. The hormone ban, applicable at first to domestic production and then extended to imports in January 1989, was ostensibly a reaction to this consumer pressure. The US reacted to the loss of the trade in beef and offals by imposing a 100% tariff on an equivalent value of EC exports to the US. An EC counter-retaliation was announced, but did not come into force, pending bilateral negotiations.

The US view of the beef hormone ban saw the Commission's actions as essentially of a protectionist nature, bowing to pressure from cattle producers and being concerned with the build-up of intervention stocks of beef. No one doubted that there had been consumer concerns over the health effects of hormone use in cattle. But there was no credible scientific evidence that hormone treatment of beef cattle, under proper conditions, leaves any harmful residue in the meat sold to consumers. The US Food and Drug Administration tests artificial growth hormones before approving their use in livestock production, and the USDA has a test programme to check for chemical residues in meat. Therefore, in the US view, the role of public authorities in the EC should have been to educate consumers on the matter, and to make sure that producers followed accepted practices. Consumer views don't seem to be heeded when it comes to heavy taxes on food: why should they suddenly be so important when the issue is hormones in beef production? The answer seems clear to the transatlantic observer. Allow the issue to be settled by scientific evidence and not by uninformed public sentiment.[6]

Conflicts in the GATT Round

The start of the Uruguay Round of GATT negotiations, in September 1986, provided a new setting for US-EC discussions on agricultural policy. US reaction to the EC position in the GATT talks reflects the underlying view on the CAP, in the context of a formal negotiation, these perceptions begin to take on an added significance. This is particularly true of the Uruguay Round, where by mutual agreement domestic agricultural policies were 'on the table' for the first time.

The first step in the Uruguay Round negotiations on agriculture was to elicit from the major participants their ideas for the conduct of the talks. This prompted the US administration to table a somewhat radical proposal to eliminate over a ten-year period all trade-distorting price support for agricultural commodities. The Cairns Group (of smaller agricultural exporters, led by Australia and Canada) also proposed liberalisation, but seemed to offer a less abrupt transition to this state of grace. The EC countered with a paper which emphasised short term action to correct the depressed situation on world markets, which would then be followed by 'significant' reductions in levels of support and some rebalancing of that support.

The year 1988 saw intensive negotiations on these proposals in Geneva. The US wanted the EC to sign on to the notion of long-term trade liberalisation (the 'zero' option): once that had been agreed then shorter term issues could be addressed.[7] The EC saw no reason to move so far beyond the Punta de Este agreement (for "greater liberalisation of trade in agriculture") and doubted that the US was really serious in its proposal. The stand-off lasted throughout the year, and caused the 'failure' of the Mid-Term Review in December 1988, the ministerial meeting that was to have provided the agenda for the remainder of the Round. In the event, an agreement was put together in April 1989 which allowed negotiations to proceed.

To the US, the EC reaction to its initial proposal confirmed the view of the Community as a protectionist group. Why would the EU not come back with a counter-proposal for long-run policy reform, if the zero-option was too drastic? Would it not assist the economic development of Europe, as well as its political cohesion, to scale down or remove the element of protectionism in the CAP? Farm incomes could still be supported by a variety of production-neutral programmes, and research and extension programmes, together with development assistance and food aid, would not be touched. What would go would be the troublesome set of market policies which had proved so costly. Surely the EC could not resist the logic of the argument for agricultural market liberalisation!

The EC response was also deemed unsatisfactory in three other regards, besides not taking seriously the US proposal. First, the EC proposal had unmistakable traces of a market sharing philosophy, anathema to US competitive instincts. Even if market shares were only suggested as a way out of temporary problems in a few markets, the signs were ominous. It is widely remembered in the US that the EC went into the Kennedy Round with notions of world commodity agreements (as spelled out in the Baumgartner/Pisani Plan). The view of the EC as favouring market management over liberalised trade is pervasive. For the EC to propose such solutions in the GATT was regarded as evidence of an underlying preference for managed trade.

A second concern was the hint that the GATT Round would be used as a way of correcting the relative price distortions that had dogged the cereals, oils, and fats markets, as discussed above. If reducing protection overall could allow for some increase in protection, then the genie would be out of the bottle. The soybean and corn gluten feed markets would disappear under some broad cloak of GATT respectability, and the US farmer would once again have been duped by the clever Europeans. In part for this reason, the US has emphasised in its own submissions to the GATT the notion of 'country plans' which would have to be acceptable to trading partners. Support reduction would not be tied just to some formula and based on technical calculations: it would be tangible and transparent, and allow for the calculation of mutual advantage.

The third caution raised by the EC position in the GATT related to the notion of credit for actions taken. The trade ministers at Punta del Este decided that it would be wise to allow countries who constructively modified domestic policies in advance of any GATT agreement to claim credit for their actions. The EC had been asking for credit for its reform programme, on occasions even suggesting that the 1984 dairy policy changes be counted. Such requests for credit had two implications: they would have taken the pressure off the EC to make further changes in the CAP in advance of an agreement (and possibly not even then, if world prices remained firm); and they could have allowed the EC to shift the focus onto others, and to request that they too make a serious start on policy reform.

Neither of these implications was particularly acceptable to the US. It seemed implausible that a GATT agreement would be saleable to domestic interests if it did not include some major liberalising moves by the EC. The politics of 'equal degrees of pain' would ensure that the EC could not just live on credit for the next few years, even if an indicator could be found that would validate a claim for such credit. Nor was it likely that the US could undertake domestic reform as a way of catching-up with the EC. The US strategy was rather to 'keep the pressure on' the EC through an expanded EEP and a relaxed set-aside. This was not building up credit, it was accumulating debts. The US might still have been able to adopt a plan for a radical liberalisation of trade, which it would claim was inspired by the tactics of pressure, in conjunction with other countries. It was unlikely to be persuaded to take the first steps alone.

Developments after April 1989 did not resolve any of these issues. The US modified its own position, but not in a way that appealed to the EC negotiators. Instead of phased reductions in support, allowing a degree of policy choice by individual governments, the US began to be more explicit as to which policy instruments were to be modified. By the time of the tabling of the US Comprehensive Proposal at the end of 1989, the US was asking for conversion of the EC variable levy to a fixed tariff, the phasing out of export subsidies and the abolition of all price-related domestic measures such as intervention buying and direct payments. A more comprehensive attack on the CAP could not have been imagined. The EC, meanwhile, had embraced the notion of phased reductions in overall support (though not to the same extent as the US had suggested in 1987) and shunned the notion of singling out individual instruments of policy. The EC comprehensive proposal did offer to discuss limited movements towards the tariffication of levies, making them less variable, but only in return for agreement to 'rebalance' support among commodities. The stage was set for the confrontation that delayed the end of the GATT Uruguay Round for about two years. To see how this was finally resolved requires a discussion of the state of CAP reform and the role that transatlantic relations played in that process.

CAP reform and the North American view

The evolution of reform of the CAP is traced out in Chapter 4, and a detailed outline of the 'new', post Mac Sharry/GATT agreement, CAP is provided in Chapter 5. The issue of CAP reform has always been of great interest on the other side of the Atlantic. The US saw floods of subsidised exports, shrinking EC markets for imports and a succession of trade squabbles. Reform in the CAP would surely imply drastic changes in support prices or sharp policy shifts to remove production incentives. How far reform had proceeded before 1984 could be argued among reasonable observers, but from the predominant US viewpoint the process had been woefully inadequate.

Policy reform in the context of the CAP started with the introduction of milk quotas in March 1984, and their strengthening in December 1986; continued with the adoption of the 'stabiliser' package in February 1988; and culminated with the Mac Sharry reform package of June 1992.[8] The changes in milk policy were not of great importance to US trade interests. The US is roughly self-sufficient in dairy products. It imports speciality cheeses from the EC, while disposing of American cheese under domestic and foreign concessional programmes. The US disposes of limited amounts of skimmed milk powder (non-fat dried milk) abroad, mainly through overseas aid programmes. There is little US trade in butter, although there are occasional domestic surpluses. The introduction of quotas in the dairy sector clearly reduced EC dairy capacity and firmed up world dairy product prices, but the effect on the US was minimal.[9] It was seen as a domestic reform, aimed at cutting budget costs. The introduction of quotas, as an alternative to the politically more difficult price cuts, was not regarded as a great contribution to farsighted policy making. Though the US and many other countries use quota-like policies in controlling the dairy industry, it was seen as moving in a direction counter to that of deregulation. Since quotas once introduced are hard to remove, the net result may have been to reduce the chances for true liberalisation in the future.

The relative lack of impact of EC dairy quotas on US trade may explain why the first stages of CAP reform were not widely recognised across the Atlantic. The same cannot be said about the 'stabilisers' programme, and the introduction of set-asides and direct income payments that together made up the second tentative steps towards CAP reform in 1988. How the EC tackled its grain surplus and the growing expenditure on oilseeds payments was of direct interest to the US. The prevailing US impression of this package was that it represented a small start to correcting a large problem, and that the applause should be withheld at least until there was evidence of further similar measures. To tout it as the Commission did, as a major change in EC agricultural policy, a significant contribution to the adjustment in world agriculture, and a step which shows to others the way of the future, was to stretch credibility to the limit.

As we saw in Chapter 4, the stabilisers programme itself was a fairly modest quasi-automatic device for linking price levels to output at the Community level. Guarantee thresholds have been around for much of the 1980s, but were generally regarded as ineffective. The 1988 programme extended their use to most major products, including cereals, and made more certain the penalties for overproduction. Large-scale cereal producers now faced an additional co-responsibility levy of 3% (in addition to the 3% levy already in place) which was only to be reimbursed if output fell short of the maximum guaranteed quantity (MGQ). The levy was to be translated into a fall in the intervention price the following year. Assuming that the Commission did not negate these automatic price reductions at the annual price review, the price restraint could have had an effect on output. But set against productivity changes in the cereals sector, the fact that small farmers were exempt from the co-responsibility levies (as was grain used for feed on farm), and the generally upward movement of prices due to the green currency system, the restraint was modest. And since the cereals stabiliser only lasted for three years, there was no guarantee that any permanent change had been achieved.

Oilseed and protein crops were included in the stabilisers programme in part because of the steady rise in support costs for these products and in part to anticipate any possible shift out of cereals toward these commodities.[10] The net effect on the price of rapeseed and protein crops (field peas and beans) was expected to be minimal, although the programme may have helped to inhibit the growth of soybean production in Spain. It could be argued that the US did not show enough appreciation for this attempt to control the growth of oilseed production, but from the 'outside' the policies looked less than dramatic.

The introduction of set-asides in the early stages of CAP reform should have triggered a more sympathetic response. The US has been using set-asides as a major aspect of policy since the 1930s, and in recent years has made participation in the major price support programmes conditional upon compliance with mandatory acreage adjustment. The reaction to EC set-asides was hardly enthusiastic. It was apparent that the national schemes would not be sufficiently attractive to entice many acres out of production. The Commission's estimate of 1 million hectares (and 3.5 million tons of grain) was considered optimistic. Individual countries had varying degrees of commitment to the scheme, which was pushed by the West German government as an alternative to the stabiliser price cuts. Since national governments had to pay up to 65% of the cost, their incentives to encourage participation were relatively weak. Add to this the fact that even the most extensive set-aside programmes in the US seem to have a relatively modest impact on production, as farmers find ways of idling acres without losing output, and the attitude was cautious at best and scornful at worst.

One aspect of the EC's initial 1988 reform package that might seem to have been more warmly welcomed by the US was the element in the socio-structural measures which allowed for direct income assistance to farmers. Farmers earning significantly below the national average farm income level could receive up to 1500 ecu per year, with the member states covering up to 70% of the cost. From the Community's perspective, the introduction of a scheme of this type seemed significant. It established the principle of direct payments and allowed the political process to get used to the idea of non-price income policies. With hindsight this scheme looks to have been a relatively minor add-on to the range of structural programmes that were available.

The muted US response to what was regarded at the time in Europe as a major change in policy may seem churlish. In part this response was conditioned by the experience of years of well-meaning efforts by the Commission to make significant changes in the CAP, only to be rejected by the Council of Ministers. In part it reflected the detachment of a foreign point of view, where results count but the amount of political contortions needed to achieve the decision is not of interest. In part, it was a function of the stage of the multilateral negotiations, where to have recognised that the EC had made major changes may itself have squandered negotiating capital. The next stage of the reform process could not be so easily dismissed.

With the passage in June 1992 of the Mac Sharry package, the CAP was changed more radically and more visibly. The reforms might not have been as radical as most economists would have desired, nor as sweeping in their scope as the EC's trading partners would have hoped. But no return to the old CAP is likely in the next decade. The GATT Uruguay Round agreement locks in these policy changes and makes any backsliding difficult. As importantly, future developments in the construction of Europe will make it almost impossible to regress to the policy as it existed before 1992. The element of the new CAP which qualifies it for the status of a true reform is the sharp reduction in the price at which the cereal market is supported, combined with compensation payments to farmers which do not depend on current output levels. This change gives a vital degree of flexibility to EC policy-making, and breaks the direct link between price support levels and farm receipts.

It is this change that has important implications for the US and for other trading countries. It changes qualitatively the way in which the Community can deal with other countries. It gives back to foreign commercial policy some flexibility not seen since the inception of the CAP. It opens up the possibility of reduced conflicts between domestic and trade policies. And it moves the Community into the growing group of countries that have chosen not to rely exclusively on price-for-income-support policies.

The reconciliation of 1992

Despite vigorous denials by the European Commission, most observers consider the 1992 CAP reforms to be an outcome of pressures from trading partners in the GATT. The Community had resisted for much of the GATT Round the notion that it must change its policy instruments. The negotiations had laid down a direct challenge to the EC to abandon its support system. The variable levy was to be turned into a fixed tariff, the export subsidies were to be eliminated, and all domestic support was to be phased out with the exception of research and extension, domestic and foreign food subsidies and 'decoupled' income payments. The EC market management system would have been effectively dismantled. In addition, protection levels would have been slashed by up to 75% over a period of a decade. The EC response was to agree conditionally to modify its levy system, to be a combined tariff and supplementary levy, and to scale back the overall level of support by 30% (as measured by a Support Measurement Unit, a variant of the PSE) over five years. This would have in effect rescued the CAP, subject to a modification in the levy system. No change in internal market management would be dictated, and price levels would not have had to be reduced markedly. The level of support was already falling relative to 1986, the start of the Round. Thus the issue of whether the CAP would have to be further reformed had become a key aspect of the talks.

The issue came to a head in the summer of 1990, as discussed above, when the chairman of the agricultural negotiating group of the Uruguay Round put forth a paper for use as a basis for the completion of the negotiations. The 'de Zeeuw Draft' picked up the notions of the US (and the Cairns group) as to the framework of the agreement. The EC rejected this paper as the basis for negotiations. Despite the diplomatic 'arm twisting' by President Bush at the Houston Summit in July, the issue was still not resolved at the time of the inconclusive 'final' meeting of the trade negotiators in Brussels in December 1990. An attempt by the Swedish Minister of Agriculture, Mats Hellstrom, to broker a compromise failed to convince the EC.

Within hours of the breakdown of the talks, Commissioner Mac Sharry was circulating a document in Brussels that proposed a new policy development which was to lead eventually to the 1992 CAP reform. The support price for cereals, oilseeds and pulses was to be cut by 35%, to be offset by hectarage compensation payments to farmers. Dairy and beef prices were also to be cut, with compensation also paid to these producers. Whether or not the plan originated as a way out of the trade dilemma, the implication for the GATT Round was clear. The Community could, if it introduced such reforms, live with a requirement for a cut in support of Hellstrom proportions. The market price cuts would stimulate consumption, and along with some reduction in output from a set-aside programme this would enable the Community to meet the targets for reduction in export subsidies. In the light of this change in position the EC removed its objection to negotiations along the lines accepted by others, and talks began in earnest. The culmination of this burst of activity was the Dunkel Draft of December 1991, which incorporated the ideas of tariffication, export subsidy reduction and the definition of non-trade-distorting policies (GATT, 1991). After about a year of further negotiations, the EC and the US reached an agreement on the terms of the agricultural component of the Uruguay Round in November 1992, in the Blair House Accord. This was followed in December 1993 by the conclusion of the Round as a whole, after some last minute 'clarification' of the Blair House Accord, and the signing of the Uruguay Round Agreement at Marrakech in April 1994. In the end, agricultural trade conflicts had not brought the trade system down.

Conclusion

Significant changes in attitudes towards farm policy and trade have occurred in the developed countries in the past five years. In the United States, budget pressures have reduced price supports considerably since 1987: although further policy reform seems difficult to achieve. Canada has been experimenting with income insurance schemes that are production-neutral, but succeeded in defending its protectionist provincial marketing legislation from the influence of the NAFTA. The EC has gone ahead with the 1992 CAP reform, as discussed above, and seems likely to increase the reform efforts as the prospect of further enlargement to the east becomes closer. Though well behind the Latin American nations, most of whom have reduced market support considerably, and those countries such as New Zealand and Australia which have undergone even more radical policy adjustments, the transatlantic partners seem to be moving slowly in the direction of freer markets and more targeted farm programmes. The end point is a domestic system which does not require trade interventions to be effective. This will eliminate most of the causes of trade tensions between the US and the EC. The GATT agreement will presumably help this process of reform of domestic policies, though the GATT timetable is not in most cases forcing the pace of reforms, if it ever could have done so. The agreement will be most useful as a backstop, to prevent backsliding, and to begin a process which can be continued, and a push for those countries that are lagging in the process.

The GATT Agreement on Agriculture should also help directly to improve trade relations between the US and the EC. One part of that agreement is a 'peace clause' which exempts from certain challenges under the GATT any domestic policy actions which are in conformity with the agreement. This includes a shelter from challenge for the major cereal policies of both the US and the EC. As a part of the Blair House deal that allowed the Round to move to its final stage, the US and the EC agreed to exempt each other's direct payments programmes from reduction. The protection for these policies however is conditional on there being no increase in such payments from the 1992 level. In effect these policies are bound, with penalties for increasing protection. By constraining domestic policies within effective GATT rules, each country will incur much greater risks when running policies which offer excessive protection to agriculture. Domestic policies will run up against the constraints and adapt more quickly. In addition, disputes over differences in standards could be reduced by the new rules on the use of sanitary and phytosanitary standards as hidden trade barriers. It remains to be seen whether the new rules, stressing the need for scientific evidence to support any trade disrupting measure, will stand up to the test of adjudicating sensitive and emotional issues in agricultural and food markets.

US-EC agricultural conflicts will not of course disappear overnight, but they will take place within a clearer framework of rules and obligations. Few would put the probability of a period of perfect peace in agricultural trade relations very high. More likely is a continued period of underlying tension, with a generally improving climate of trust as the magnitude of trade interventions is reduced. There is a however still a significant risk that both the EC and the US will try to interpret the terms of the Uruguay Round Agreement in their favour, at the same time accusing the other of bad faith. If the Agreement begins to lose credibility then the prospects for a peaceful decade in farm trade would not look so bright.

References

ABARE (1989) *The 1988 EC Budget and Production Stabilisers*, Discussion Paper, No. 89. 3, Canberra.

Friedeberg, A.S. (1993) *Solved or Shelved: The Oilseed Trade Conflict in Focus*, (unpublished).

GATT (1991) *Draft Final Act* ("Dunkel Draft"), MTN.TNC/W/FA.

Moyer, H., Wayne and Josling, T. (1990) *Agricultural Policy Reform: Politics and Process in the EC and the US,* Harvester Wheatsheaf, Hemel Hemstead.

Josling, T. (1994) The Reformed CAP and the Industrial World, *European Review of Agricultural Economics,* 21, 513-527.

Josling, T. (1993) Agricultural Trade Issues in Transatlantic Trade Relations, *The World Economy*, 16(5), 553-573.

USDA (1993) *Europe: International Agriculture and Trade Report,* USDA/ERS, Washington DC.

---

[1]Economists point out the irrelevance of bilateral deficits in a multilateral trade system. Such balances do however play a significant role in shaping trade policy attitudes.

[2]The concession was actually negotiated in the Article XXIV (6) negotiations which preceded the Dillon Round proper.

[3]The wet-milling of corn is used to produce high fructose corn syrup (isoglucose), a substitute for cane and beet sugar, and ethanol, used as an additive to gasoline. The EU classifies all these starchy products as 'cereal substitutes': the high protein meals are, however, better considered 'cereal complements'. The distinction is not always so clear. At times feed ingredients high in protein have been used for their carbohydrate content.

[4] The duty on corn gluten feed had also been bound at zero in the Dillon Round.

[5]See Alfred Friedeberg (1993).

[6]The time significance of the beef controversy will be more apparent in future years. Several similar instances of trade conflicts arising from different health and safety standards are likely to emerge.
[7]The shorter term issues themselves became less urgent over the year, as the effect of the US drought caused world prices to rise.
[8]For a discussion of these reforms see Moyer and Josling (1991).
[9]The US has had its own dairy scheme in recent years, the Dairy Herd Replacement Program, which significantly (if temporarily) reduced the dairy cow numbers.
[10]One might have expected the Community to have welcomed any shift out of cereals, since that presumably was the rationale for price restraints in that sector.

Chapter 17
The GATT, the WTO and the CAP

David Harvey

"It is not the end, it is not even the beginning of the end. But it is, perhaps, the end of the beginning" Winston S. Churchill

Introduction

The last chapter has emphasised the role of GATT in conditioning and materialising the conflicts between the EU and the US on agricultural policy. In many ways, the development of these conflicts is a convenient and largely accurate reflection of the interactions between the CAP and the rest of the world. The position of the GATT in regulating these interactions is critical to both understanding the past history of the CAP and, more importantly, setting the stage for the future evolution of the policy. This chapter first outlines the history of the GATT with special reference to agriculture, including a brief outline of the elementary economics supporting the notion of trade liberalisation. We then turn to a consideration of the Uruguay Round Agreement (URA) — the GATT Final Act, 1994 — as far as agriculture is concerned and the implications of this act for the future of the CAP. The next section provides an outline analysis of the key issues which seem likely to dominate future Multilateral Trade Negotiation (MTN) agendas, though apart from the formal review of the agriculture agreement already mandated in the Final Act and due in 1999, the form and mechanisms of these negotiations are not yet clear. As is apparent from the references, this chapter owes a considerable debt to the International Agricultural Trade Research Consortium (IATRC) and to the invaluable debate and discussions generated through the Consortium.

History of the GATT and agriculture

Born at Bretton Woods, Massachusetts, in 1947, the General Agreement on Tariffs and Trade (GATT) has long been the 'poor relation' of its two sister organisations, the International Monetary Fund (the IMF, which is really a 'bank') and the International Bank for Reconstruction and Development (the

World Bank, which is really a 'fund'). The original intention was that the GATT should become the International Trade Organisation, but the necessary agreement to establish the ITO was not ratified. The GATT persisted as:

a) an agreement of 38 articles between contracting parties; of which the most important is the "most favoured nation" article (A.1) which extends to all countries any preferential trade treatment afforded to one or a group of countries. Article 3 ensures that favoured treatment of traders within countries is to be extended beyond national boundaries. Article 24 contains the exceptions, which are used to justify trade discrimination practices.
b) a process for negotiation on trade rules and trade dispute settlement; serviced by a secretariat for the negotiations and dispute settlement, and, more recently, a Trade Policy Review Mechanism, launched in September 1988.

The primary objective of the GATT has been the reduction of tariff barriers, quantitative restrictions and subsidies on trade, attempted through a series of conferences (Multilateral Trade Negotiation or MTN 'rounds') in which the participating countries either requested and offered reductions in each other's trade impediments, or (more recently) reached agreement on formulae and codes governing the use and level of such restrictions.

GATT's history may be characterised in three phases. The first phase comprised four rounds of negotiations: two in Geneva in 1947, one in Annecy and one in Torquay in 1955, each lasting about three months. This phase was largely concerned with coverage (which commodities and sectors should be included) and with freezing and 'binding' existing tariff levels. The second phase included three major rounds: the Dillon, 1959–1961, which achieved an average 8% tariff reduction on 20,000 line items on a request/offer basis; the Kennedy, 1963–1967, achieving a 35% reduction in tariffs through agreement on formulae and codes; the Tokyo, 1973–1979, again achieving a 35% reduction in tariffs on a formulae/code basis. The third phase begins and ends with the Uruguay Round, formally begun in September 1986 and ending with 111 countries meeting in Marrakesh to the sign the Final Act of the Uruguay Round on April 15, 1994.

Agriculture and the GATT

Agriculture was effectively left out of the 1947 agreement. The sector was given special status in two key articles: 11 (on import quotas) and 16 (on export subsidies), where farm programmes for both production and export subsidies were allowed, subject only to somewhat mild requirement that countries seriously disadvantaged through their use could seek to negotiate limits on such programmes. Modest revisions of Article 16 in the 1950s only succeeded in adding a requirement that such policies should not be used to generate "more than an equitable share of world trade" (Warley, 1989). Successive negotiations gave rise to an increased protection of agricultural policies under GATT, as opposed to liberalisation in other product sectors, exemplified by the 1955 waiver for certain key US farm policies and by the 1956 inability to rule on the beginnings of the Community policy. The US waiver covered exemption from GATT obligations for cotton, sugar, some dairy products and peanuts, which were protected through import restrictions under Section 22 of the US Agricultural Adjustment Act, 1933, amended 1948. In effect, this waiver granted supremacy to domestic farm policies over the GATT agreement, and established a convenient precedent for other countries, making agriculture a 'special case' (Blandford, 1990).

Furthermore, there was a manifest reluctance or inability by GATT to enforce existing rules on farm policies — the European variable import levy system clearly violated the principle of bound tariffs. Indeed, the Dillon Round, 1961–66, confirmed agriculture's special status by accepting the principles of the CAP, specifically the minimum import price and variable levy/export refund mechanisms. By 1982, the anomalous position of agriculture within the GATT had become too obvious to ignore. Given the progress made under GATT in other product trade areas, the special treatment of agriculture was beginning to threaten the coherence and integrity of the whole GATT system. However, strong world markets in many farm products following the end of the Tokyo Round had concentrated early planning discussions for the next (Uruguay) Round on rule changes and dispute settlement procedures as far as agriculture was concerned (IATRC, 1994).

By 1986, however, agriculture's place on the agenda was raised by pressures of declining world prices, growing domestic support costs of farm policies on both sides of the Atlantic, and associated increased dispute activity as export subsidies appeared to grow without limit (see Chapter 16). Growing pressure from agricultural exporters reinforced the position, especially through the 'Cairns Group', a group of 14 small and medium sized developed and developing agricultural exporters, including Australia, Argentina, Brazil, Canada, Hungary, New Zealand, named after the location of their first conference. Not only could agriculture no longer be excluded from the GATT, but major steps to bring farm (trade) policies within the ambit of GATT had become a key requirement for an agreement. Without such agreement, a substantial number of countries, led by the Cairns Group, would reject the complete package.

"By the time governments met in Punta del Este to launch the Uruguay Round, a general consensus had been reached that it was necessary to reform agricultural policies in order to achieve trade liberalisation" (IATRC, 1994). The Punta del Este Declaration in July 1986 made reform of trade in farm products a central objective of the negotiations, and furthermore made it clear that this was to be achieved by "bringing all measures affecting import access and export competition under strengthened and more operationally efficient GATT rules and disciplines" (Ingersent *et al.*, 1994). The clear intention to include all policies, including those previously declared and accepted as domestic rather than trade policies, was highlighted by the OECD Ministers in May, 1987. The objective of the Uruguay Round was to end the special status of agriculture within the GATT. This, alone, made the Round very different from its predecessors and justifies its place as a separate phase of GATT history.

The Uruguay Round 'in the large'

While agriculture became central to the Uruguay Round, the negotiations included the largest and most complex list of matters ever attempted which, reflecting the patterns and issues of world trade itself, were far broader than farming (Greenaway, 1994). In particular, the totally new areas of services, capital and intellectual property rights were included in the list, with the major negotiating groups focused on the following topics and issues:

- tariffs;
- non-tariff measures;
- natural resource based products;
- textiles and clothing (Multifibre agreement);
- agriculture (including phytosanitary regulations and plant health issues);
- tropical products;
- trade restrictions on intellectual property — TRIPs;
- trade restrictions on investment measures — TRIMs;
- dispute settlement procedures;
- safeguards (Article 19), subsidies and countervailing measures;
- GATT articles review;
- functioning of the GATT system — FOGs;
- multilateral trade negotiations (MTN): agreements and arrangements for future rounds.

If ambition is the parent of achievement, this would be the most colossally successful international negotiation in history. More realistically, almost any agreement reached on this full list would be a major success. For present purposes, however, embedding agricultural negotiations within the wider multilateral package has two important implications for the outcome. First, the dependence of the whole Round on an agricultural agreement widened the interests in farm policies beyond the traditional farm ministers and, in most states, finance ministers. Trade, commerce and industry ministers now had considerable interests in a farm policy agreement. Second, nations facing perceived losses on one (agricultural) front could seek offsetting gains on another, hence increasing the scope for a bargained solution, the classic argument in favour of 'multilateral rounds' of negotiations rather than a sequential single-issue approach.

The breaking of the hegemony of farm ministers (especially of the large developed countries) and the inclusion of agriculture in a wider negotiating arena increased the pressure on agricultural special interests. However, Zeitz (1988) noted that "with negotiations limited to agricultural specialists with more or less close ties to the domestic farm lobby and no restraining countervailing power from other domestic special interest groups, bold moves for trade reform are unlikely", though he goes on, "involvement of government at the ministerial level could make a difference [and] could go some way in restraining the influence of domestic farm lobbies on the talks" (p383).

The major agricultural arguments

That world agricultural markets were in disarray had long been recognised, even by the European Community. At the inception of the CAP in 1960, Tracy (1994) says: "the Commission's general principle was that prices in the Community would have to be stabilised above world market levels, which it regarded as being falsified by 'artificial' measures; but every effort should be made both to reduce EEC production costs and to 'normalise' world market conditions" (p361). The fate of the Commission's good intentions are now a matter of history (and outlined elsewhere in this volume), though the spirit of the principle perhaps informs interpretation of the EU's positions in the UR negotiations. The analysis of disarray in world agriculture is classically captured by Johnson (1973) in his book of the same name, at least as far as the industrialised countries were concerned. The second edition of this volume (1991) needed only to incorporate macroeconomic policy effects, especially on interest and exchange rates, and extend analysis to the developing world. Otherwise the classic arguments needed no revision.

The essential argument as far as international trade is concerned is illustrated in the box on the next page. This shows the effects of border protection by an 'exporting' country (such as the EU) on the world market. The box depicts the same market circumstances as Fig.7.3, here emphasising the trade effects. Domestic market support, resulting in greater levels of supply (and, in the case of border protection, though not with deficiency payments or 'direct' producer subsidy, lower levels of demand) either reduces imports from, or increases exports to the world market. In either case, the effects are to depress world prices, reduce export earnings from other exporters, and make world markets more volatile.

The magnitude of these effects depends, of course, on the relative size of the domestic farm sector — the bigger it is, the more damage it does to world markets. Based on such a framework, Tyers and Anderson (1986, 1992) estimated that "distortions to grain, livestock product and sugar (GLS) markets in industrial countries alone cost the world economy of the order of US$35 billion per year (in 1980 dollars), and they cost producers in developing countries — who constitute the vast majority of the world's poor people — $28 billion per year ...Moreover, the cost to industrial market economies of their own GLS policies, at almost $50 billion per year, is 80% more than those countries provide as official development assistance to developing economies"(1986, p117).

However, it is naive to suppose that the global or developing country effects of industrialised country farm policies were the major driving force for inclusion of agriculture in the UR. The major impetuses for both the EU and the US to consider abandoning the special status of agriculture within GATT were the repercussions of depressed world markets on their own domestic costs of support, as noted in the previous chapter. "Global expenditures on domestic farm programmes nearly doubled during the first five years of the 1980s. In 1986, the US and EC each spent nearly $35 billion on farm programmes" (Ronningen and Dixit, 1989, p1). In short, the major players were looking to the multilateral trade negotiations (MTN) to solve domestic problems through 'fixing-up' world

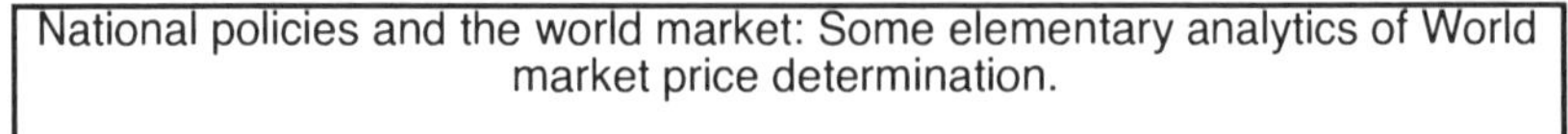

National policies and the world market: Some elementary analytics of World market price determination.

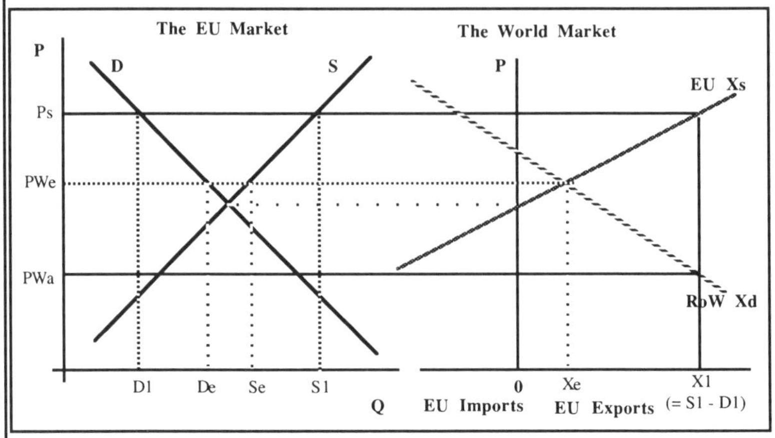

The EU Excess Supply (EU Xs) is defined as the difference between EU quantities supplied (S)and demanded (D) at each price, and thus is 0 at that price at which EU supply (S) is equal to EU demand (D). The Excess Demand curve from the rest of the world (RoW Xd) is defined similarly to the EU Xs curve, though from the market situation in the rest of the world (not shown, but implicitly on the right hand side of the diagram).

Under domestic market intervention, with market prices in the EU set at Ps, the EU is a net exporter, exports being X1 (equal to S1 - D1). For world markets to clear, RoW Xd must equal EU Xs, thus the world price becomes PWa. The equilibrium world price under free trade is, however, PWe (ignoring transport costs, which would drive a wedge between the excess demand and excess supply price), where EU Xd=RoW Xs, with the EU exporting a smaller quantity under free trade on this diagram.

It can be clearly seen that the effect of EU domestic policies on the world market (and, by the same token, the consequences of free-trade for the EU market) depend critically on the slopes (elasticities) of supply and demand in the EU and in the rest of the world. The steeper (more inelastic) is the RoW excess demand curve, the greater will be the effect of EU exports on world prices, and the greater will be the volatility of world prices as EU exports vary. Appendix 17A explores these relationships in more detail.

prices. A more robust and better-ordered world market would provide a stronger level of market prices necessitating lower (or perhaps even no) domestic support.

There existed a strong disagreement about what 'fixing-up' the world market actually meant. From a free-trade perspective, the MTNs offered a major opportunity to 'get government out of agriculture' and to eliminate all (distorting) domestic support. This position was echoed in the US opening bid to the UR, and supported by professionals who had long argued the social inequity and economic inefficiency of domestic farm policies. It is interesting to note that the Republican administration in the US had previously presented a draft 1985 Farm Bill to Congress proposing a radical reform and virtual elimination of farm programmes, only for it to be substantially watered down by Congress before becoming the Food Security Act. It is tempting to speculate that the same administration was attempting to achieve in an international forum what it had failed to pass in the domestic setting.

From a more political/pragmatic perspective, the major issue was seen as the appropriate share of the world market to be accorded the major traders and the organisation of trade such that it did not undermine world prices, echoing the revised GATT Article 16 about equitable world market shares. The original position of the EU followed the latter route, being resolved to keep domestic instruments (import levies and export refunds) intact, while being prepared to consider reductions in general levels of support in the longer term only on condition that: (a) general measures of support (under the 'aggregate measure of support' — AMS) included credit for supply controls and (b) that it be allowed to 're-balance' protection by increasing 'bound' tariffs on oilseeds and cereal substitutes to partially balance reductions in cereal protection (Rayner *et al.*, 1993, and Ingersent *et al.*, 1994). The EU "considered that a more urgent short-term priority was to improve international market stability by entering into market management agreements" (Ingersent *et al.*, 1994, p263). In short, a substantial gap existed between those who saw the UR as a means of substantial reform and reduction of domestic farm policies (the US) and those who were determined to preserve their own domestic support systems (the EU).

Resolution of this conflict clearly required a substantial shift in position by at least one of the major players. The course of negotiations has been dealt with in the previous chapter, and is well charted elsewhere (for example Rayner *et al.*, 1993, Ingersent *et al.*, 1994, IATRC, 1994). Only a brief account is necessary here, though an outline chronology is provided in Appendix 17B. Following collapse of negotiations in Montreal in December 1988, rescued in Geneva in April 1989 with the 'mid-term' Geneva Accord agreeing a freeze in support prices and a timetable for the remainder of the negotiations, the antagonists prepared new negotiating positions. The US retreated from its 'zero option' to a proposal for tariffication of all import restrictions, elimination of export subsidies and classification/reduction of internal support measures according to their distorting effect within the AMS. The EU maintained opposition to control of export subsidies, preferring a general phased reduction in an AMS. Aart de Zeeuw, chairman of the negotiating group, attempted to combine these proposals in a draft agreement in June 1990, which was rejected by the EU and talks collapsed again in Brussels in December 1990, the scheduled end of the Round.

The 'independent and separate' review of, and reform proposals for, the CAP under Commissioner Mac Sharry in the first half of 1991 (see in particular, Chapter 5) provided the first major breakthrough for the negotiations, allowing the GATT Director General, Arthur Dunkel, to table his 'Draft Final Act' in December 1991, outlining the major elements of the final agreement. On this basis, the EU and US teams met at The Blair House in Washington to both resolve difficulties over an oilseeds dispute under the settlement procedures of the GATT (see Chapter 16) and agree a mutual position for the UR. The Blair House Accord was subsequently subject to continuing negotiations and brinkmanship leading to the final agreement in principle in Geneva in December 1993 and, at last, signing of the Final Act in Marrakesh on April 15, 1994, seven years and eight months after the start of negotiations.

The 1994 GATT Agriculture Agreement

The 450 page Final Act, and the all important associated 'country schedules' containing the specific commitments of each signatory (in line with details specified in a separate transition document under the title Modalities for the Establishment of Specific Binding Commitments) comprises the 1994 UR Agreement. Percentage reductions in tariffs are specified in the modalities document, and detailed in each country schedule, following acceptance of which the modalities document becomes redundant. This whole package will be the subject of continuing analysis and debate, probably at least until the end of the Agreement in 2001.

However, the major elements appear clear, and there is a strong argument that the *fact* of the agreement is more important and far-reaching than the specific content. Signatories have now accepted the principles that: (a) agriculture is no longer a special case within the GATT, with all waivers and special exemptions for agriculture removed from the trade rules; and (b) domestic farm policies are now subject to international governance through the GATT (now established within the World Trade Organisation (WTO) under the UR), and to binding international commitments itemised in country schedules, especially on border measures. Both facts fundamentally alter the socio-political climate surrounding future agricultural policy developments around the world.

The major elements (see, for example, IATRC, 1994) fall into four areas: market access; export competition; domestic subsidies; sanitary and phytosanitary (SPS) measures, though the last is technically separate from the Agreement on Agriculture. In addition, to assist acceptance of the agreement, a 'Peace Clause' protects certain 'green box' policies from challenge under GATT and exempts other 'blue box' policies from all but countervailing duties, provided support does not exceed 1992 levels. Green box policy instruments are those which are agreed as being non-trade distorting, including R&D and extension services. Blue box policies are those which are accepted as being minimally distorting only for the duration of the UR agreement period of six years, and include the EU area and headage compensation payments and the US deficiency payments (subsequently the transition contract payments under the Federal and Agricultural Improvement and Reform (FAIR) Act, 1996).

The structure of the agreement is identified in Table 17.1 (from IATRC, 1994), which shows the three major areas of agreement (market access, export competition and domestic subsidies) and the three major actions which have been agreed: the definition of new rules; the agreed general rates of reduction in these measures over the period of the agreement — the specifics of which are detailed in the country schedules; and the 'wrinkles' (safeguards, accommodations and guarantees) necessary to achieve the final agreement.

Table 17.1: The main structure of the GATT Agriculture Agreement, 1994.

	Rules	Liberalisation	Safeguards, Accommodations and Guarantees
MARKET ACCESS	• Change non-tariff trade measures to tariffs • Establish tariff quotas. • Bind all tariffs.	• Reduce existing & new tariffs by 36% on average over 6 years. • Reduce tariffs for each item by at least 15%.	• Guaranteed access opportunities to exporters through tariff rate quotas (Min of 5% of domestic markets by end of 6 yrs.) • Special safeguards for importers.
EXPORT COMPETITION	• Defined limits on existing export subsidies. • No new export subsidies.	• Reduce expenditure by 36% over 6 year. • Reduce volume by 21% over 6 years.	• Adherence to food aid rules. • Negotiate later on export credits.
DOMESTIC SUBSIDIES	• 'Green Box' defined for allowable subsidies.	• Aggregate Measure of Support (including all trade-distorting measures) to be reduced by 20% over 6 years.	• Many LDC subsidies exempted. • Payments under 'blue box' production limiting programmes exempted.

Source: IATRC, 1994, p. 7.

Market access

The conversion of virtually all existing non-tariff barriers and unbound tariffs to bound tariffs, and their subsequent reduction by 36% on a simple average basis, with a minimum of 15% reduction for any tariff line, is a major achievement. In practice, the degree of discretion allowed to each country in their conversion procedures has introduced a considerable degree of 'dirty tariffication'. By choosing the appropriate base period prices and definitions of internal prices, countries have been able to set their bound tariffs close to the upper end of the possible scale, exacerbated by the fact that for many commodities (especially in the EU) the base protection rates (internal price less world prices expressed as a percentage of world prices) were historically high.

Thus for the EU, the bound rates are 250% or so for sugar, 237% for beef, 341% for butter and between 150% and 170% for grains. (The Mac Sharry reform led to rather different measures being agreed for grain imports, described in Chapter 5.) The EU has agreed to the average 36% reduction for all these commodities, rather than taking advantage of the minimum required reduction in any single line of 15%, reserving a minimum reduction of 20% for 'sensitive' products such as fruit and vegetables, skim milk powder, olive oil and wine. The effects of the GATT UR Agreement and the European response in the fruit and vegetable sector is particularly complex, and examined in detail by Swinbank and Ritson (1995). The EU has chosen to define minimum access provisions over relatively large aggregates of commodities, thus allowing for the fact that trade preferences within commodity aggregates can offset those for other products within the same aggregate, though actual access commitments are calculated for each tariff line.

Export competition

A major achievement of the Round was the agreement to outlaw any new export subsidies, and to freeze and reduce existing subsidies. "There can no longer be any doubts as to what [maximum] level of export subsidies a country can grant in agricultural trade" (IATRC, 1994, p. 10). This part of the agreement will be the most telling constraint on future development of the CAP. Indeed, the severity of this commitment led to a last minute 'frontloading' provision allowing countries to choose 1991/2 as the starting point for the progressive reduction in export volumes and subsidy expenditure, freeing the EU in particular from the necessity of starting from the 1986-90 export base when cereal exports were significantly lower than in the early 1990s. Subsidised cereal exports have to be reduced by almost 8 million tonnes by 2000, representing a significant share of the total world market. The cut in export subsidies amounts to a substantial saving in EU budget spending of close to ECU 4 billion.

Domestic support

The finally agreed reduction in total AMS of 20% is lower than initially offered by the EU in 1990 of 30%. It is arguable that the initial EU offer would have proved impossible to deliver under the old CAP, hence prompting the 1992 reform package. The Blair House Accord ensured that the new area and headage payments under the reformed CAP are excluded from this provision. In consequence, the actual policy effect of this commitment is not expected to be significant. In fact, the Commission estimates that following the 1992 reforms, the EU support bill as measured by the AMS will be ECU 56 billion, well below the commitment of ECU 61.2 billion. However, it is widely recognised that the Mac Sharry support payments do not really qualify as 'GATT decoupled' (green box), in that they remain coupled to production (crop areas and livestock numbers), and consequently remain on the agenda for further reform or reduction in future rounds or alternative negotiations.

Similarly, the tariffication and import access provisions of the Agreement are not, in practice, expected to lead to significant policy reform pressure within the EU (IATRC, 1994, p. 48). Given that the GATT bound import tariff limits are maxima, actual tariffs can be set below these limits, and thus can continue to be set as variable levies within these bounds. Furthermore, the 'Special Safeguard Provision' allows additional flexibility in conditions of depressed world prices or import surges. Under this provision, countries can impose additional levies of up to one third of normal duties in the event that imports exceed a trigger level, typically set at 105% of base period imports for those products where imports are more than 30% of the domestic market and 100% otherwise. A price trigger also exists under this provision.

European Union policy reform post GATT

The major 'threat' to the reformed CAP arising from the UR is from the volume reduction in subsidised exports. In particular, the EU commitments require modification of the sugar regime, both in reduction in quotas by 4% and of intervention prices of 15% (IATRC, 1994, p. 50). Further modifications of the beef, dairy (particularly cheese) and cereals sectors to meet these volume commitments can also be anticipated in the event that either world prices do not remain at their high 1995/6 levels or that the 1992 reform package does not generate sufficient production reductions.

The most important threat to the coherence of the post-1992 CAP comes from the subsidised export commitments for cereals. Although cheese and sugar may be more in danger of exhibiting UR incompatibility, the cereal situation is considered here to be more fundamental to the future direction of the CAP, and thus cereal contravention of the UR commitments would indicate a more generic reform than in the other sectors. The key question is whether the effects of the recent reform of the cereals regime, which both reduce internal EU market prices and limit areas planted through the set-aside scheme, will be sufficient to comply with the GATT Agreement. There are two component parts to this question: (a) what will be the difference between world prices and domestic EU prices, and hence what level of export subsidy will be required? (b) what will be the level of export volumes? The second question depends in turn on what production and domestic consumption levels will be, taking account of the requirements for import access.

World and EU Cereal Prices: "100 ECU represents the expected world market price on a stabilised world market" (European Commission, 1991, p. 9). If this world price is actually achieved, then the export refund required will be zero and the GATT commitments will be more than fulfilled regardless of export volume from the EU (all exports taking place without subsidy). World prices are, of course, notoriously difficult to forecast. However, as an attempt to put the question in context, Figs.17.1 and 17.2 show the recent history and possible future of world wheat and coarse grain prices respectively, according to the Organisation for Economic Co-operation and Development (OECD).

These figures show the OECD reference prices as the world prices (as used in the Producer Subsidy Equivalent (PSE) calculations) for the EU, compared with EU producer prices, and projections of these prices made by the OECD (1995). These projections were made prior to the escalation of world prices, especially for wheat, in 1995/6, when world prices rose above the upper bound as a result of substantial falls in world stock levels. The upper bound is here defined simply as one standard deviation above the central prediction, the standard deviation in turn relating to the pattern of prices between 1979 and 1992.

Figure 17.1 appears to support the European Commission's view that the new intervention price is close to an expected world price level for wheat, though Fig.17.2 indicates that the same cannot be said for coarse (feed) grains. On the basis of these OECD projections, therefore, the CAP reforms appear consistent with the GATT commitments for wheat, though for coarse grains, a substantial per unit export subsidy will still be required.

Export Volumes: The second key consideration is the volume of exports. Fig.17.3 illustrates the recent history and possible future projections of potential export volumes in the EU, defined here as the OECD projections of production less consumption. On this count, the compatibility of the new CAP with the GATT commitments is reversed, with the wheat sector vulnerable to further required reform while the picture for coarse grains appears to be consistent with the UR agreement. In fact, the OECD projections incorporate compliance with the GATT commitments and show a doubling of EU wheat stocks over the forecast period (1994–2000) from 12.6 to 24.2 mt. On the basis of these projections, therefore, the principal danger to the present CAP from the UR agreement is for the wheat regime, which will come under considerable pressure if world prices fall from historic trend levels.

There is room for debate about these projections, in spite of the care taken by the OECD in adjusting national information and projections to a consistent global market level (OECD, 1995). For instance, the actual world price experience over the 1995/6 period has been very substantially higher than projected here, being at or above the 'upper bound'. On the other hand, the projected ratio of wheat to coarse grain prices by 2000 is rather higher than has been observed over most of the past, suggesting that coarse grain world prices might be rather low or that wheat price projections might be somewhat high. The EU projections show a substantial growth in wheat production relative to consumption, while coarse grain production and consumption remain static, as would be expected given such strength in relative wheat prices. If the price relatives turn out to be closer to those experienced in the past, the potential export volumes might be expected to adjust accordingly, leading to potential problems in GATT compliance for each grain. However, as already noted, the increase in world prices in 1995/6 has been sufficient to eliminate the need for export subsidy completely, leading the EU to impose an export tax on wheat to prevent domestic prices rising to world levels.

Possible CAP reform response: The atypically high world prices for cereals during 1995/6 have resulted in the establishment of a considerable 'buffer' for the EU's GATT commitment on subsidised export volumes. Most cereal exports during 1995/6 were unsubsidised. The UR Agreement on Agriculture (Article 9)

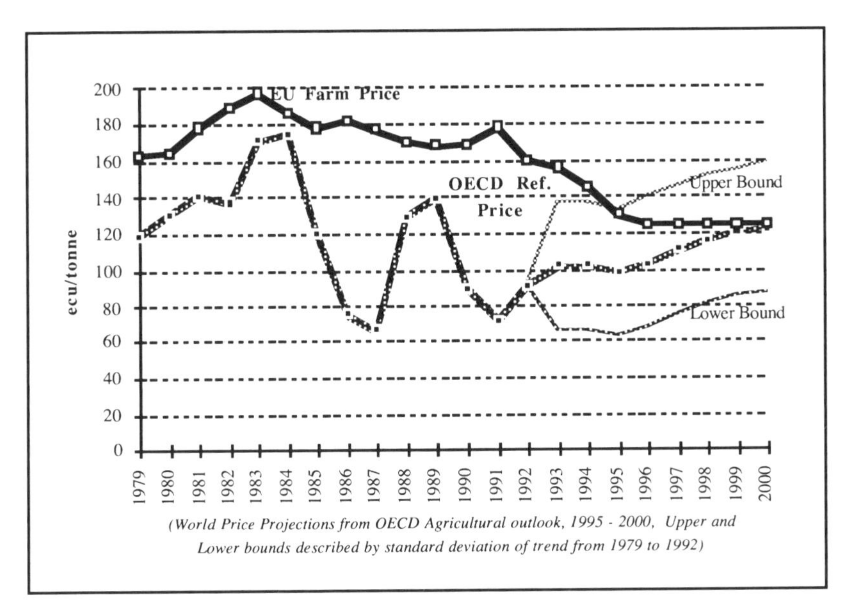

Figure 17.1: World and EU producer prices for wheat.

Source: OECD, 1995 and author's estimates.

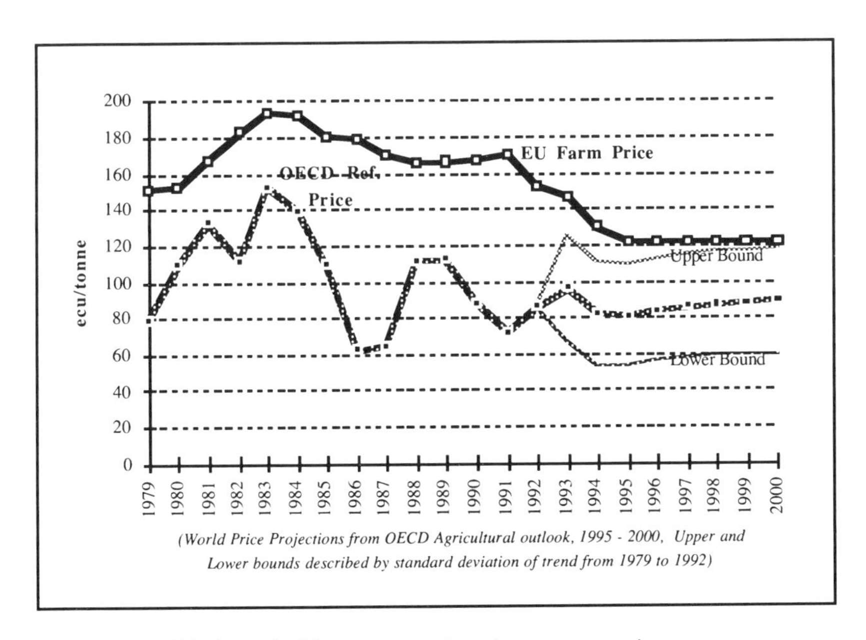

Figure 17.2: World and EU producer prices for coarse grains.

Source: OECD, 1995 and author's estimates.

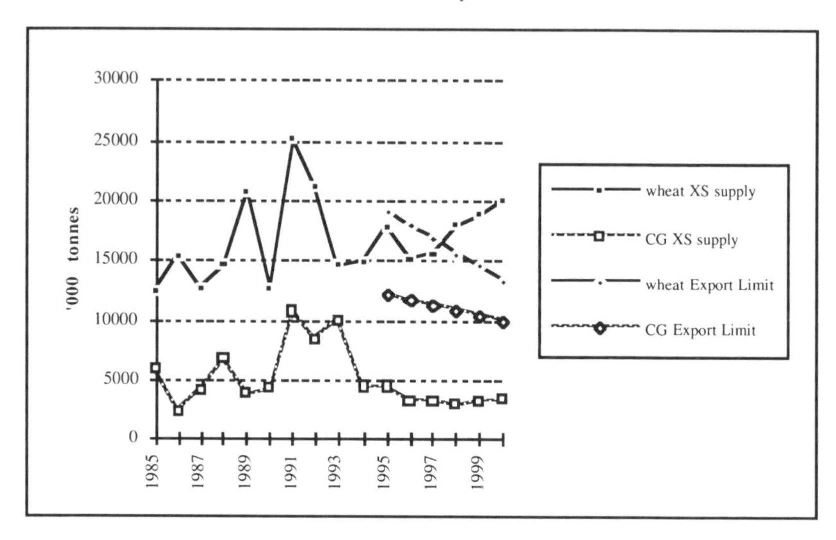

Figure 17.3: EU cereal export projections and GATT limits.

Source: Author's calculations.

provides for the accumulation of unused subsidised export volume allowances from one year to the next, up to but not including the final year of the Agreement (2000). Thus, the 1995/6 (and probably the 1996/7) export seasons provide the EU with the capacity to substantially exceed the nominal annual export limits in the 1997-99 period. However, the EU remains committed to the 21% reduction in subsidised exports by the final year of the agreement, which means that subsidised exports in 2000/01 cannot exceed 13.4 mt. for wheat, regardless of any accumulated allowance from earlier years.

In short, the atypical world market conditions during the mid-1990s have merely delayed, not eliminated, the strong possibility that the CAP remains inconsistent with GATT commitments. Delay may also mean that the eventual inconsistency, and hence the need for urgent policy reform, is correspondingly greater. This danger, and the associated implication that the CAP should be further reformed sooner rather than later, is already being recognised (see, e.g. House of Lords, 1996). Although the international governance of the CAP through the GATT/WTO is not the only, or possibly even the major pressure for further reform of the policy — as further explored in the next chapter — the URA does have substantial implications for the nature of acceptable reform options.

The practical progress towards more liberal trade in farm products under the URA has been limited. However, the spirit and intent of this goal has now been firmly established at the international level, and buttressed through the elevation of the GATT into a fully-fledged World Trade Organisation. For agriculture, the URA includes a specific commitment for a multilateral review of progress, and by implication to pursue further reform, beginning in 1999. Implementation of the URA raises some immediate issues for this review and further negotiations for policy reform. In particular, it is inevitable that there will be renewed pressure on the CAP on two major fronts, regardless of the consistency of the present policy with the current URA: (a) to agree further substantial reductions in tariffs, subsidised exports and other trade distortions; and (b) to complete the 'decoupling' of remaining support payments.

The EU's imposition of export taxes in the face of high world prices is as inconsistent with the aim of liberalised trade as the use of export subsidies. While such a policy may be consistent with a goal of internal market stability (see next section), it denies additional supplies to the world market and hence exacerbates the shortage and high price condition of the world market. The use of such instruments can be expected to be strongly challenged in the next agricultural round, while further agreed reductions in tariffs and export subsidies will place increased pressure on the current CAP.

The issue of 'blue box' measures (especially the compensation payments in the EU) is certain to generate further pressure to make such payments transparently non-distorting, by de-coupling them from production decisions. There can be no doubt that support payments conditional on areas planted and numbers of livestock grazed will encourage farmers to continue to plant crops and keep livestock according to their historically determined entitlements. This reaction must considerably reduce adjustment of production levels to market signals. As a result, the continuation of these conditions of payment make it more likely that the EU will not be able to meet its current UR commitments. It is to be expected, therefore, that these conditions will come under increasing pressure in the future.

There remain two further policy reform issues facing the CAP: the nature and influence of emerging international concerns over farm production and trade; and the context and process of internal EU policy evolution. The next section considers the first of these issues, while the second is the subject of the next chapter.

MTNs post Uruguay

The Uruguay Round is generally assessed as a "major change in attitude of governments towards the international control of national agricultural policy" (IATRC, 1994, p. 86). The agreement has incorporated very substantial rule changes at the expense of substantial action to reduce trade distortions. Although there remains room for argument about the precise meaning of 'trade distortion' and its appropriate measurement in MTNs (Harvey, 1994), the IATRC assessment (1994, p. 88) concludes that "in such future negotiations, countries can concentrate on rates of reduction rather than rules".

However, there appear to be four potential issues which could subvert such apparently free-trade tendencies: food and income security; world market volatility; world market competitive performance; and environmental concerns. As MacLaren (1991) points out, the underlying theoretical support for free trade as an optimal policy is apparently weakened by the existence of differentiated markets, oligopolistic international markets, risk, uncertainty and imperfect information. Furthermore, moves to free-trade inevitably involve changes in the distribution of income, both within and between countries, not all of which will be welcome and which cannot be ignored other than in the clinical and abstract calculus of economic theory. Combined with the substantial domestic political pressures for some continued form of protection, these arguments could prove too great for the WTO and associated MTNs to resist.

Food and income security

The conventional proposition of competitive trade and comparative advantage theory is that there are global efficiency and hence economic welfare gains to be made from free-trade, in exactly similar fashion to domestic efficiency gains identified in Chapters 7 and 8 above. However, this result conveniently ignores the fact that introduction of free(r) trade in a world previously characterised by autarky or severe protection and insulation (see Appendix 17A) will result in welfare losses as well as welfare gains, notwithstanding that the latter are expected to outweigh the former. The losers can be expected to argue for their own protection and, depending on the perceived justice of their case, obtain sympathy and support from the gainers and disinterested parties. Chapters 15 and 16 above identify the potential gainers and losers in the international market as far as the CAP is concerned. In particular, those developing countries heavily dependent on imported food supplies may lose through trade liberalisation (Chapter 15), since world prices are typically expected to increase while reserve stocks and associated food-aid packages decrease. In the longer run, however, increased prices for food products may encourage domestic production in such countries to offset the initial losses, perhaps entirely.

The difficulties faced by such countries have been partially recognised in the UR agreement through dispensations and reductions in their trade liberalisation commitments. However, the international mechanisms available for re-distribution of income and wealth (the World Bank, the IMF, the United Nations and the several international NGOs) are not well-developed. Protection of, or adjustment assistance for, particularly the poorer losers from free(r) trade may be expected to move higher up the international agenda in the future. As they do, issues such as international food reserves, international agricultural assistance programmes, food security insurance accounts and transfers are likely to become more important. This is already illustrated by the organisation of a World Food Summit, scheduled for November 1996, by the Food and Agriculture Organisation (FAO), to establish strategies in pursuit of global food security and access to adequate food for all people.

World market volatility

The issue of production volatility and consequent world market 'instability' can be viewed as the obverse of food security. If domestic production fails, how secure and reliable is the international market in meeting the shortfall? While free markets would be expected to 'buy cheap and sell dear', thus holding stocks over from fat to lean years, is there any guarantee that such markets will store enough and provide 'optimal' stability? A primary objective of many developed country agricultural policies has been to stabilise domestic markets, yet moves towards freer trade will increase the exposure of domestic markets to world price variations. Although a freer world market is expected to be less volatile than one characterised by high insulation rates, it is unlikely to be as stable as the protected domestic markets it replaces. On the other hand, the political volatility and social unsustainability of previous protectionist policies raises the serious question of the actual stability of protected domestic markets, while the effects of world market variations on domestic support costs adds to the argument that perfect insulation and domestic market stability are infeasibly expensive except in the short run.

The adverse consequences of uncertainty for economic efficiency rely on economic agents being risk averse. People or organisations will tend to act more 'conservatively' in the face of uncertainty than otherwise, reducing output and consumption levels and, in effect, using scarce resources to reduce risk while avoiding otherwise profitable investments because of uncertain outcomes. It is thus argued that an 'external' reduction in risk (uncertainty) will lead to a more 'efficient' outcome.

It is widely recognised (e.g. MacLaren, 1991 and references therein) that free markets cannot achieve social optimality if risk and insurance (contingent) markets are absent or incomplete. It is also commonly perceived that risk and contingent markets are likely to be incomplete since it is practically impossible to pre-specify and write contracts for every conceivable future outcome. Risk and insurance markets are also subject to problems of moral hazard, adverse selection and asymmetric or imperfect information. Moral hazard arises through economic actors (consumers, traders, producers or governments) altering their behaviour towards more risky activities as a consequence of stabilisation programmes, thus undermining the bases of such programmes. Adverse selection arises through participation in insurance or stabilisation schemes tending to be only by those facing high risks, thus requiring greater premia (and consequently less incentive to participate) than if the total market were to join. Both problems are well-known in the insurance field, and both are a reflection, in part, of transaction and information costs and of the lack of sufficient information to translate uncertainty into actuarial and hence insurable risk. Even given adequately performing risk and insurance markets, differences in risk aversion and between social versus private costs of capital can lead to strong theoretical arguments in favour of government intervention to solve problems of instability.

Hence, it is often argued that there are strong reasons for government intervention in agriculture to offset the volatility of farm production and associated variability and uncertainty in prices and returns. Although, for example, Newberry and Stiglitz (1981) find that there is likely to be limited static welfare gain from stabilisation, others have argued that the dynamic effects of stabilisation through encouraged investment in new technologies might be substantial (Crawford, 1988, Stiglitz, 1987 — referenced in Munk, 1994, and Chapter 9 of this book).

There is a counter-argument (e.g. McKee and West, 1981). Comparing an uncertain world with a perfectly certain one is equivalent to arguing that a world with cold-fusion nuclear power would be a more efficient world — true but irrelevant. Incomplete markets, in this perspective, are merely a reflection of the fact that the gains to be made from including new contracts or obtaining better information are not offset by the additional real costs of providing these 'goods'. Thus Spriggs and Van Kooten, 1988: "As argued by Arrow and Lind (1970) markets may not exist because of high transactions costs. In such cases there is no market failure, just no market. Secondly, recent research by Myers (1988) suggests that the actual cost of incomplete risk markets may not be very significant" (p. 8).

Gilbert (1992) concludes: "Of course, producers are fundamentally interested in their revenues being smoothed: this is not necessarily achieved through price stabilisation, and perhaps not best achieved in this way. Second, there exist market forces (notably private sector storage) which act to stabilise prices, and international prices may in any case be intermediated by governments in such a way as to stabilise producer prices. There is a clear danger that attempts to stabilise international (terminal) market prices will inhibit these tendencies. Third, it may be argued that commodity futures and options markets allow producers to stabilise their own revenues even given volatile (cash) price distributions. And finally,...calls for price stabilisation may disguise programmes which aim to raise primary prices through regulation of international trade"(p. 8–9).

The potential role of futures and options markets in providing necessary risk spreading and stabilisation facilities is clearly a critical part of the analysis of the case for government intervention to stabilise markets and/or producer returns. Many analysts are convinced that, in the absence of government intervention, risk and insurance markets would be substantially more active for the agricultural sector than at present. Indeed such markets are already increasing in a variety of guises, including the land market where non-standard share and contract farming arrangements are becoming increasingly commonplace. Thus, Gemill (1988) concludes "that the absence of suitable (futures and options) contracts is not a problem: intermediaries, such as merchants, will tailor them to suit the needs of farmers if there is sufficient demand" (p. 458).

However, by their very nature, futures and options markets can provide no more information than is already contained in condensed form in the current spot price. In effect, all futures and options markets provide is a means of spreading risks associated with price variations by substituting risks associated with market performance (in terms of price relationships or 'spreads' and 'basis'), Gilbert (1992 p. 15). Ultimately, the crucial problem of determining a 'planning price', especially for the critical longer term investment or strategic management decisions, is not particularly assisted by such markets. As Spriggs and Van Kooten (1988) point out (p. 6), following Newberry and Stiglitz (1981, p. 238) "The problem is that the generation and dissemination of information is costly and it has public good characteristics. This leads to an under supply of information by the free market" (no matter how complete are contingent and futures markets). "The solution is for government either to improve the quality of information directly, to provide incentives to the private sector to do this or else to reduce the need for information. The introduction of a stabilisation programme is an attempt to reduce the need for information". These authors continue: "To address the information problem, the stabilisation authority would be required to set the support price at the long-run expected level", where the crucial distinction is between random shocks (short run market balancing price changes) and shifts in demand and supply of a fundamental nature (long term price trends) — Tomek (1969).

It is far from clear that the critical information problem can be adequately solved through either direct production of more information by governments or by appropriate incentives to the private sector to generate such information. Certainly there are few practical examples of such action, though the apparent success of "Situation and Outlook" conferences and associated publicly-funded commodity market forecasts in market-oriented countries (US, Canada, Australia) are examples of the perceived benefit of additional information coupled with an established forum in which the value of such information can be openly discussed. Add the very substantial transactions costs associated with risk and insurance markets, and the real possibilities of major economies of scale in covering these if 'managed' by the public sector, and the pragmatic conclusion is that there is a logical case for some insurance/stabilisation provision to an important and highly atomistic industry such as agriculture. Furthermore, stability continues to be a major element of publicly stated objectives for the sector and is unlikely to be ignored on the arguments of pure economic theory.

This conclusion strongly suggests that stability will prove to be a major agenda item in future international agricultural negotiations, on two counts: first, the issue of international action to stabilise world market prices; second, the issue of allowable (i.e. non-distorting or 'green box') domestic farm margin or income stabilisation schemes. It seems possible that some pressure will build up for international stabilisation schemes, either through international buffer stocks or through co-ordinated supply control measures, in spite of the rather sad historical performance of such instruments (see, e.g. Gilbert, 1992). However, theory strongly suggests that effective international action to improve free-market stability should be restricted to: (a) the provision and wide dissemination of better market intelligence; and (b) the possible provision of capital subsidies for stock-holding and futures or forward contracts, to reflect both potential differences between private and social costs of capital and differences in private and social risk-aversion. The issue of legitimate domestic margin or income stabilisation policies will be returned to below (Chapter 18), though it should be clear from the discussion in this chapter that, to be classed as non-distorting, these schemes need to be substantially de-coupled from current production decisions if publicly funded, or else paid for by producers.

World market competitive performance

The competitive performance of the world market depends on two key factors: (a) the absence of distorting policies by national governments; and (b) the competitive performance of the private (free) market. As MacLaren (1991) points out, much commercial policy has been directed towards appropriate 'countervailing' action by governments against policy-induced distortions. It is to be hoped that the URA provides a basis for the more rational resolution of farm trade disputes than the 'beggar-thy-neighbour' trade retaliations of the past, though the dispute settlement procedures to be operated in the future by the WTO are still to be established. In addition, given that free-trade is still an analytical construct rather than a realistic medium-term proposition, arguments about appropriate domestic market defences against continued policy distortions of competing countries are likely to continue to feature in future MTNs.

As far as imperfect competition is concerned, it is difficult to argue with MacLaren's conclusion (1991, p. 273): "the information requirements to implement optimal trade policies in (the face of market imperfections) are just too great, and for this reason, a multilateral free-trade policy may be a useful rule of thumb". However, the issue of international competition policy and appropriate mechanisms and sanctions for the policing of multinational commercial and trading concerns seems likely to grow in importance on the WTO's agenda.

Environmental concerns

A major issue facing the WTO and future MTNs on agricultural policies concerns the appropriate use (if any) of trade policies to ensure and enforce environmental objectives. The 'level playing field' of free-trade is frequently argued to be inappropriate in the face of legitimate differences in national (and even regional) socio-political valuations of non-market environmental benefits and costs of farm production practices. Thus, if EU farmers have to face stringent nitrogen regulations and penalties, it is thought to be 'unfair' to allow competitors' products from less stringently policed regions to 'undermine' the domestic market, and hence legitimate to protect the home market from such 'unfair' competition. The obverse argument is that trade penalties and sanctions are appropriate devices to encourage more environmentally friendly practices elsewhere in the world. As with risk and uncertainty, such arguments (following the public choice approach in Chapter 8) raise the potential problem that environmental objectives will be used as 'stalking horses' for the maintenance and even for the increase in protective trade policies. Chapter 13 has dealt with the domestic issues of farm policy and the environment. The concern in this chapter is the environmental implications for trade policies post-UR.

The preamble to the World Trade Organisation Agreement recognises the need to protect the environment and to promote sustainable development. The Uruguay Round Ministerial Decision on Trade and Environment ensures that linkages between trade policies, environmental policies and sustainable development will be taken up as a priority in the World Trade Organisation. The WTO General Council formally established, in early 1995, a committee on Trade and Environment. Early indications are that this committee recognises and accepts that "the multilateral trading system could make a significant contribution to the promotion of more sustainable agricultural production", where "many delegations emphasised the potential they saw to build on the complementary links that exist between a liberal, well functioning multilateral trading system, better environmental protection and conservation, and the promotion of sustainable development" (WTO, 1995).

The central question facing the WTO concerns the circumstances under which trade restrictions can be considered legitimate and appropriate means of promoting environmental sustainability. Two key aspects to this question are: (a) the extent to which national authorities have the sovereign right to exploit their natural resources in line with their own environmental standards and priorities; and (b) enforcement of responsibilities to ensure that activities within national jurisdiction do not cause environmental damage or loss to the rest of the world.

In this sense, environmental arguments mirror those already addressed in the agricultural trade MTNs – namely the extent to which nations have the sovereign right to support their domestic agricultural sectors and the enforcement of international responsibilities not to distort world markets in so doing. Thus, the UR agreement on 'green box' policies as those which are legitimate expressions of national sovereignty appears to be directly extendible to environmental policies.

However, this characterisation inevitably raises the issue of what environmental assets are to be considered solely within national authority. From a 'deep green' perspective, all natural environments are global not local resources, and require global not local management. The obvious difficulty is that there is no world government capable of implementing such a solution. Some approximation must therefore be sought through international agreement or through existing international organisations. Prevention of illegitimate use of trade restrictions on the basis of environmental concerns conceptually requires three key and prior international agreements: (a) on the definition of 'global commons' – the world's environmental resources requiring international protection; (b) on the appropriate price signals and/or regulations (restrictions on market behaviour) to optimise the preservation of these commons; and (c) on the inalienable rights of nations to value all other environmental resources according to their own preferences, and to be entitled to free trade access to other (world) markets on this basis.

The implication of the last of these three 'priors' is important. The 'level playing field' concept does *not* mean that trading nations (or regions) should have identical environmental conditions or identical social valuations (and hence opportunity costs) of environmental assets, any more than it means that they should have identical costs of land, labour or capital. In effect, different people (countries and regions) are entitled to value their own environments differently, just as they are entitled to different valuations of other resources. It is regional and national differences in these resource endowments, capabilities and social valuations which provide the very basis for economic gains from trade. Without these differences, only variations in consumers' and users' tastes and preferences would provide any scope for gains from trade. There is, therefore, no case for distorting trade because other countries value their environments differently, any more than there is for distortions because other countries have different endowments, and hence values, of land, labour and capital.

Given such prior agreements, trade restrictions could be clearly and unambiguously limited to preservation of global commons (though see below), with any other environmentally based arguments for trade restrictions deemed unacceptable. Agreement on these issues, coupled with an acceptance by all countries to subjugate national and sectional interests to the interests of global sustainability, would provide for the essential elements of a 'world government' capable of designing and implementing appropriate measures for global environmental sustainability. Subsequent agreements on how to encourage participation in sustainable control, how to police participation and how to penalise non-participation seem, by comparison, relatively straightforward. However, international negotiations are quite likely to find the necessary prior agreements 'too hard', and seek to placate domestic political pressures through trade restrictions. Hence, the WTO is destined to deal with environmental trade restrictions in a rather irrational, or more accurately, a rationally ignorant fashion, since 'too hard' implies that the full costs of achieving prior agreement are not offset by the perceived benefits.

Even so, while trade restrictions may be considered legitimate for some environmental concerns, are they necessarily the most appropriate means of achieving environmental objectives? International incentives and penalties must involve use of global (international) systems in which countries participate, and hence which they value. The choice of trade restrictions thus seems inevitable. Clearly, any penalties and incentives have to affect a country's international income or capital movements as the only effective instrument through which international treaties or organisations can realistically affect non-signatory behaviour. Other than the GATT/WTO, the World Bank and the IMF are the two international institutions with the potential economic and political power to police and enforce international environmental agreements, with potentially less damaging and more constructive results.

Some agreements over the appropriate adjustments to 'domestic' (national) market and policy signals in order to promote global sustainability are necessary whatever instruments are chosen for implementation and enforcement. Once these agreements are reached, enforcement might be more appropriately achieved through taxes and subsidies on these international income and expenditure flows operated by the IMF and the World Bank rather than the WTO. Although neither the IMF nor the World Bank were established to manage the world's environmental resource, neither was the GATT, nor is the WTO. In fact, no such organisation yet exists. The point is that it is far from obviously desirable and potentially very dangerous to superimpose this responsibility on the WTO any more than on the World Bank or the IMF. The logic of optimal solutions to problems of global commons seems more in sympathy with the operations of both the later institutions than with the WTO. Indeed, the Bank has already experimented with 'debt-for-nature swaps' (World Bank, 1992).

Conclusions

This chapter has taken a deliberately theoretical and economic approach to the issues raised by the EU's participation in and commitment to international negotiations and 'world governance' of agricultural and related policies. The conclusions from this approach may well seem at odds with a more political interpretation of the Union's likely or desirable responses to (and perhaps even leadership of) future negotiations. In particular the prescriptions for rational future negotiations and agreements advanced in this chapter are based on economic analysis rather than on political acceptability, though the apparent political pressures militating against these 'rational' prescriptions have been noted. The next and concluding chapter seeks to balance the argument by returning to the political economy of the CAP.

References

Anderson, K. and Blackhurst, R. (eds) (1992) *Greening of World Trade Issues,* Harvester Wheatsheaf, Hemel Hempstead.

Arrow, K.J and Lind, R.C. (1970) Uncertainty and the evaluation of public investment decisions, *American Economical Review*, 60 (May) 364–378.

Barrett, S. (1994) Trade Restrictions in International Agreements, *The Globe*, 19, 6–7, UK GER Office, Polaris House, Swindon, June.

Blandford, D. (1990) US Trade Policy and the GATT: Implications for Agriculture, in *Agricultural Policies in the new Decade*, Allen, K. (ed), Resources for the Future and National Planning Association, Washington.

Crawford, V.P. (1988) Long term relationships governed by short term contracts, *American Economic Review*, 78 (3) 485–499.

European Commission (1991) *The Development and Future of the CAP*, Com (91) 258 final/3, Brussels.

Gemmill, G. (1988) The contribution of futures and options markets to a revised agricultural policy, *European Review of Agricultural Economics,* 15(4), 457–475.

Gilbert, C.L. (1992) Commodity Markets, Commodity Futures and International Policy, *Economics Department Paper*, 248, February, Queen Mary and Westfield College, University of London.

Greenaway, D. (1994) The Uruguay Round: Agenda, Expectations and Outcomes, Chapter 2 in *Agriculture in the Uruguay Round*, Ingersent, K.A., Rayner, A.J and Hine, R.C. (eds), Macmillan, London.

Harvey, D.R. (1994) Policy Reform after the Uruguay Round, Chapter 11 in *Agriculture in the Uruguay Round,* Ingersent, K., Rayner, A.J. and Hine, R.C. (eds), Macmillan, 223–269.

House of Lords (1996) *Enlargement and Common Agricultural Policy Reform*, Select Committee on the European Communities, 12th Report, Session, 1995-96, HMSO, London.

IATRC (1994) *The Uruguay Round Agreement on Agriculture: an Evaluation*, Commissioned Paper No. 9, The International Trade Research Consortium, (Chairman, Prof. A. McCalla, University of California, Davis).

Ingersent, K.A., Rayner, A.J. and Hine, R.C. (1994) Agriculture in the Uruguay Round: an Assessment, Chapter 12 in *Agriculture in the Uruguay Round*, Ingersent, K.A., Rayner, A.J and Hine, R.C. (eds), Macmillan, London.

Josling, T.J. (1977) Government Price Policies and the Structure of International Trade, *Journal of Agricultural Economics*, 28(3), 261–276.

Johnson, D.G. (1973) *World Agriculture in Disarray*, Macmillan, London (second edition, 1991).

Lloyd, P.J. (1992) The Problem of Optimal Environmental Choice, Chapter 3, in Anderson and Blackhurst (eds) *Greening of World Trade Issues,* Harvester Wheatsheaf, Hemel Hempstead, 49–72.

McCalla, A.F. and Josling, T.E. (1985) *Agricultural Policies and World Markets*, Macmillan.

MacLaren, D. (1991) Agricultural Trade Policy Analysis and International Trade Theory: a Review of Recent Developments, *Journal of Agricultural Economics*, 42(3), 250–297.

McKee, M. and West, E.G. (1981) The theory of second best: a solution in search of a problem? *Economic Inquiry*, 1 (3), 436–448.

Munk, K.J. (1994) Explaining Agricultural Policy, Chapter C, European Commission, 1994: EC Agricultural Policy for the 21st Century, *European Economy, Reports and Studies*, 4.

Myers, R.J. (1988) The value of ideal contingency markets in agriculture, *American Journal of Agricultural Economics*, 7 (May).

Newbery, D. and Stiglitz, J. (1981) *The Theory of Commodity Price Stabilisation: a study in the economics of risk*, Oxford, Clarendon Press.

OECD (1995) *The Agricultural Outlook, 1995–2000*, Paris.

Rayner, A.J., Ingersent, K.A. and Hine, R.C. (1993) Agriculture in the Uruguay Round: an Assessment, *Economic Journal*, 103, 1513–1527.

Ronningen, V. and Dixit, P. (1989) *Economic Implications of Agricultural policy Reforms in Industrial Market Economies*, USDA Staff Report, AGE 89–363, Washington.

Spriggs, J and Van Kooten, G.C. (1988) Rationale for Government Intervention in Canadian Agriculture: A Review of Stabilisation Programs, *Canadian Journal of Agricultural Economics*, 30(1), 1–21.

Stiglitz, J. (1987) Some theoretical aspects of agricultural policies, *The World Bank Research Observer*.

Swinbank, A. and Ritson, C. (1995) The Impact of the GATT Agreement on EU Fruit and Vegetable Policy, *Food Policy*, August.

Tracy, M. (1994) The Spirit of Stresa, *European Review of Agricultural Economics,* 21(3/4), 357–374.

Tyers, R. and Anderson, K. (1986) Distortions in World Food Markets: A Quantitative Assessment, background paper for the *World Bank's World Development Report*.

Tyers, R. and Anderson, K. (1992) *Disarray in World Food Markets: a Quantitative Assessment*, Cambridge University Press, Cambridge.

Tomek, W.G. (1969) Stability for primary products: means to what ends? *USAID Prices Research Project Occasional Paper* 28, Cornell University, Ithaca, New York.

Warley, T.K. (1989) Agriculture in the GATT: Past and future, in *Agriculture and Government in an Interdependent World*, Proceedings of the XXth Conference of the International Association of Agricultural Economists, Dartmouth.

World Bank (1992) World development report 1992, Washington D.C., pp 169.

World Trade Organisation (1995) Trade and Environment Bulletin No. 5, Geneva.

Zeitz, J. (1988) Strategies for dealing with agriculture in the Uruguay Round, Journal of Agricultural Economics, 39(3), 382–386.

Appendix 17A Some further analytics of world price determination

To explore the determinants of the elasticity of the Rest of the World excess demand curve, we need to employ some elementary algebra, beginning from the definition of excess demand in the rest of the world (Xd) as the difference between demand in the rest of the world (Dr) and Supply in the rest of the world (Sr) at each and every price:

$Xd = Dr - Sr$, by definition.
So: $\partial Xd/\partial P = \partial Dr/\partial P - \partial Sr/\partial P$; as the response of each side of the equation to a change (∂) in the (world) price;
Then, multiplying the LHS by P/Xd to turn it into an elasticity, which in turn means multiplying each term on the RHS by the same term, to preserve the equality, gives:
$\partial Xd/\partial P * P/Xd = \partial Dr/\partial P * P/Xd - \partial Sr/\partial P * P/Xd$ = Elasticity of RoW Xd, (Exd);
Now multiplying the first RHS term by Dr/Dr (ie 1) and the second term by Sr/Sr, and re-arranging terms gives an expression in terms of elasticities on the RHS as well. So:
$Exd = [\partial Dr/\partial P * P/Xd * Dr/Dr] - [\partial Sr/\partial P * P/Xd * Sr/Sr]$
$Exd = [\partial Dr/\partial P * P/Dr * Dr/Xd] - [\partial Sr/\partial P * P/Sr * Sr/Xd]$;
$= [ED(row)] * Dr/Xd - [ES(row)] * Sr/Xd$

What these expressions mean is that the elasticity of the excess demand curve for the rest of the world (Exd) is the weighted sum of the elasticities of demand (ED(row)) and supply (ES(row)) with respect to the world price in the rest of the world, the weights being the ratios of demand and supply (i.e. consumption and production) in the rest of the world to EU exports (which equals Xd in order for world markets to clear). Typical values of domestic elasticities lie between 0 and 1 for farm products, while EU exports are typically rather small compared with production and consumption levels in the rest of the world, which makes the elasticity of excess demand facing the Union rather large (the Xd curve in the diagram is relatively 'flat' and the effect of EU exports on world prices is small).

BUT:
The elasticities above are with respect to the world price, not to the ruling domestic prices in the rest of the world. Many countries have protected and insulated their domestic agricultural industries from the effects of world prices. If domestic prices do not change as world prices change, i.e. the domestic policies *insulate* the domestic markets, then the elasticities of supply and demand in the above expression are zero, since perfect insulation prevents domestic supply and demand responding to world price changes. In this case, the elasticity of excess demand will also be zero, i.e. the excess demand curve will be vertical. The 'flexibility' of world prices to changes in EU exports will be infinite in this case, with any increase in EEC exports depressing world prices to zero, or any decrease driving world prices up without limit. The consequences of this extreme situation are explored by Josling (1977).

However, few countries can afford totally to insulate their domestic markets from the world price (e.g. the bigger the gap between domestic and world prices, the more costly in one way or another are domestic policies). Hence there is some logic for movements in world prices to be translated, at least partially, into movements in domestic prices. There is also some empirical evidence of this, through correlations of world and domestic price changes. Protection, i.e. a 'wedge' between the domestic price and the world price, with domestic and world prices tending to move together, will make the elasticities more inelastic, *cet. par.* but will not reduce them to zero. The more insulative and protective are domestic support policies around the world with respect to the world price, the more inelastic (the less flexible) the Xd facing the home country, with obvious consequences for domestic policies at home.

We can see this as follows:

go back to the expression for the Exd:

Exd = ED(row) * Dr/Xd - ES(row) * Sr/Xd

= ∂Dr/∂P * P/Dr *Dr/Xd - ∂Sr/∂P * P/Sr * Sr/Xd;

The 'true' domestic elasticities (about which we usually feel fairly comfortable and confident) are with respect to domestic prices (Pd)

Exd = ∂Dr/∂Pd * Pd/Dr, where Pd is the domestic price in the RoW, rather than the world price.

Similarly: ES(row) = ∂Sr/∂Pd * Pd/Sr

Now, to retrieve the expression for the Exd, we need to 'transform' these domestic elasticities so that they reflect responses to the world price. We need the elasticities of domestic prices with respect to world prices —E(Pd/P):

E(Pd/P) = [(∂Pd/∂P) * (P/Pd)]

Now multiply each term in the RHS by this elasticity:

∂Dr/∂P * P/Dr * [(∂Pd/∂P) * (P/Pd)] *Dr/Xd - ∂Sr/∂P * [(∂Pd/∂P) * (P/Pd)] * P/Sr * Sr/Xd

{The terms in Pd cancel out, leaving the expression as before, so we haven't changed the expression, merely expanded it.}

Now we have the complete expression:

Exd = ED(row) *E(Pd/P) * Dr/Xd - ES(row) * E(Pd/P) * Sr/Xd

where E(Pd/P) = [(∂Pd/∂P) * (P/Pd)]

and the first term in the RHS of E(Pd/P) is: (∂Pd/∂P) - the *insulation* of the policies in the rest of the world, since this measures the rate of change of domestic prices in response to world price changes. (Note, in principle, the EU's old CAP set domestic prices which did not alter whatever the level of world price, through *variable* import levies and export refunds, so this term might be thought to be zero, making the EXd of the EC, which the rest of the world faces, zero. However, the consequences of world price changes will show up in the EC budget, which gets bigger as world prices fall, and hence exerts some downward pressure in turn on domestic (EC) support prices, so (∂Pd/∂P) is not zero for the EC, and has been estimated as about 0.5.)

The second term (P/Pd) measures the extent to which domestic prices are different from world prices (usually above in developed economies), i.e. the 'protection' offered by the EC policy — actually the inverse of protection as normally measured — so that Pd > P and P/Pd < 1 (in the EC, for cereals under the old CAP, about 0.5). The consequence of insulative and protective policies is thus to reduce the elasticity of excess demand compared with free-trade, and in so doing make the world market price more volatile in the face of either demand or supply shocks.

Appendix 17B Chronology of the Uruguay Round

1982	GATT Ministerial meeting to draw up agenda.
1984	Japan and EC endorse agenda.
1985/6	India and Brazil arguments with US about priorities for round — Agriculture versus other (new) issues — especially intellectual property, capital and investment.
Sept. 1986	Group of 38 declaration — Punta del Este, Uruguay, opening the round.
Dec. 88/ April 89	Mid Term Review, Montreal, followed by "Geneva freeze and progressive reduction" of support was agreed.
April 1990	Supposed end of Proposal stage of negotiations.
June 1990	Aart de Zeeuw's "basis for negotiations" paper, on the grounds of the existing proposals from the US and Cairns group, but rejected by the EC.
July 1990	Supposed date for acceptance of outline agreement.
Oct.15, 1990	Deadline for final offers (EC unable to meet).
Nov. 1990	EC final offer finally tabled (reduce support by 30%), but well short of US and Cairns Group targets (US suggesting 75% reductions).
Dec. 1990	Final Ministerial meeting, Brussels, to end round, failed. Cairns Group withdrew from all negotiations because of lack of progress on agriculture.
Feb. 1991	Negotiations re-started, all contracting parties undertook to reach "specific and binding commitments to reduce farm income support in each of three areas: internal assistance, border protection and export subsidies."
May 1991	Negotiations still bogged down as battle between EC and US.
June 1991	US Congress extends "fast track" negotiating mandate for two years (under which Congress may only vote on the final agreement as a whole, rather than on a 'line-by-line' basis, which would have effectively ensured that any final agreement could never be ratified by the US).
July 1991	Mac Sharry proposals for EC CAP reform lend weight to argument that progress is finally being made.
Dec.1991	Arthur Dunkel (GATT Director General) releases "compromise paper" as a basis for resumed negotiations. Main provisions: (a) Export competition: export subsidy spending to be reduced by 36% and subsidised quantities by 24% from 86/90 average levels; (b) Internal Support, measured basis 1986/88, to be reduced by 20%; and (c) Market Access: import

	restrictions to be converted to tariffs and reduced by 36% on average across the board, with individual commodities reduced by a minimum of 15%; to provide for minimum access to protected markets of at least 86/88 average levels of imports or 3% of domestic consumption by 1993 to 5% by 1999, whichever is greater.
May 1992	Mac Sharry reform package agreed by EC Council of Ministers (with unvoiced disagreement by French farmers). Concept of 'blue box' appears – containing policies which are regarded as transitionary and intended to be phased out over time.
July 1992	Further breakdown in negotiations, pending US elections and EC Maastricht debates?
Nov. 1992	US threatens trade retaliation over ongoing oilseed dispute under general GATT provisions, retaliation to begin on December 5th. Bilateral negotiations between EC and US continue.
Nov. 1992	Blair House Accord (Presidential Official Guest House in Washington, DC across road from the White House) between EC and US signed, as modification to the Dunkel Draft following nearly a year of negotiations between EC and US: providing for, over 6 years: 21% reduction by volume and 36% cut in spending on subsidised exports (from 1986/90 base) and 20% cut in domestic subsidies as measured by a total Aggregate Measure of Support (AMS), with export reductions to be made in equal instalments over the six years, against a base period of 1986/88, and with agreed exemptions of certain policies from GATT challenges. The Blair House Accord also included resolution to a number of long-standing Atlantic trade disputes, including EC production subsidies on oilseeds; restrictions on imports of corn gluten feed and access to Spain for US feed grains.
	The quantity and expenditure cuts agreed are to apply on a product by product basis, with no aggregation of product groups (to total grain, all milk etc.). Cuts in the first year must be the full 1/6th of the total cut, though some limited flexibility is allowed thereafter within the overall cut. The AMS to exclude 'green box' policies as specified in the Dunkel Draft Act (conservation measures, crop insurance and disaster payments, extension programmes, and income payments not tied to products or to current production). Blair House specified that direct payments under production limiting programmes would also be excluded from the required cuts *for the period covered by the Uruguay Agreement* (leaving the question open as to what might happen thereafter). Criteria for excluding direct payments: a) payments based on fixed areas and yields; b) payments made on ≤ 85% of base level production; c) livestock payments on a fixed number of livestock.

Neither US deficiency payments nor EC direct payments subject to reduction during the Uruguay Agreement (again leaving the question open as to what might happen thereafter).

Sept. 20/21 1993 Brussels Meeting of EU's foreign, farm and trade ministers: French express dissatisfaction with Blair House and request re-opening negotiations, especially in subsidised export reduction, with a view to altering at least the equal instalments provision in favour of greater cuts towards the end of the period, plus exempting disposal of current stocks and allowing the cuts to be made in aggregate rather than product by product. Council did not agree to re-open negotiations, but to send the Trade Commissioner (Sir Leon Brittan) to Washington to seek "interpretations, amplifications and additions" to Blair House with the US Trade Representative (Micky Kantor). US reaction — Blair House a "hard-struck bargain" (Clinton) which cannot be re-opened, and anyway was not seen by the US as a particularly good settlement. Mr. Balladur (French PM) continues to threaten to veto the accord and hence de-rail the Round, if his concerns are not met. However, the 'Luxembourg compromise', which allows member states to invoke special national interest reasons to disagree with a Community decision and hence block it, only delivers a veto if, by the fact of invoking it, a delegation can swing a sufficient number of (other) delegations to its side to create a blocking minority. In view of the seriousness of blocking the GATT, this was felt to be extremely unlikely. UK (effectively) threatens to boycott all EU business in the event of veto, thus 'suspending' EU business requiring unanimous decisions, including developments of Maastricht. Disagreements finally resolved through agreement to allow 'front-loading' of export volume reductions.

Dec. 15, 1993 GATT Agreement reached in principle in Geneva, following final reconciliation of EU and US disagreements between Trade Representatives Brittan and Kantor one week earlier. Main features of the Blair House Accord left intact.

April 15, 1994 Formal Agreement signed in Marrakesh. Agreement to come into force July 1, 1995 and to last until June 30, 2001. A new World Trade Organisation to be established with effect from January 1, 1995. "Technically speaking, the GATT (1947) Agreement, which current contracting parties (members) have accepted, has been converted into GATT (1994) through the Uruguay Round outcome. Countries can still be signatories to GATT (1947) but not GATT (1994). However, most countries will accede to the WTO, which includes GATT (1994) along with the General Agreement on Trade in Services (GATS) and the Agreement on Intellectual Property Rights (TRIPS). Thus one can still talk of GATT as being a set of trade rules for goods (including agriculture), if no longer as a separate institution" (IATRC, 1994).

PART V

The CAP and the Future

Chapter 18

The CAP in the 21st Century

David Harvey

Introduction

The last chapter indicated that the future of all domestic agricultural policies, including the CAP, is now subject to international governance, if not government. The era of highly protective and insulative domestic policies is over, though the shape of future policies is still embryonic. This is a remarkable shift. It brings in its wake increasing concerns at the international and inter-regional level with market stability, food security, competition policy and the relationships between environmental and trade policies. Yet domestic issues will remain strong determinants of the evolution of farm policies. In the case of the CAP, these seem certain to be dominated by environmental concerns coupled with rural development issues and the future expansion of the EU to include the CEE countries. Gone are the days of glacial policy development, melted only by budgetary crises, in spite of commentators' continual concerns over efficiency. That, too, is remarkable.

So what will the CAP look like in ten years time? The answer is, of course, highly speculative. While it is tempting simply to parade prejudices and subjective/normative suggestions, such a carnival is of little import other than entertainment. What is needed for genuine policy analysis is a framework within which readers may substitute their own judgements and experience and thus reach different conclusions, or at least judge the merit of the speculations presented. Yet, as argued in Chapter 8 , no such agreed and commonly accepted framework yet exists. Thus the agenda for this chapter is two-fold: an outline of such a generally systematic framework; and the use of the framework to predict the future shape of the CAP, including in the penultimate section a suggested outline for a new CAP for the next millennium.

It is recognised that this is a tall order, and that it is unlikely that all readers will agree with the framework, still less the implications or inferences drawn from it. However, such a structure may allow others to dissect the logic from the prejudice, the positive from the normative, the outrageous from the possible, and thus eventually the future from the past. It will also be apparent that this structure in itself is also an agenda for the future. Analysis will have to co-evolve with the policy if either are to survive into the next century. In the event that either or both do not, the question becomes one of what the replacements will look like, so in both cases the structural approach appears the only alternative.

An evolutionary framework

"The reasonable man adapts himself to the world; the unreasonable one persists in trying to adapt the world to himself. Therefore all progress depends on the unreasonable man" — G.B. Shaw

In place of the rather unsatisfactory nature of the present analytical frameworks for understanding the development of policies, this chapter outlines an alternative, based on evolutionary economics. "Public policies may reflect not changes in objective conditions but shifts in values, or understandings" (Nelson and Winter, 1982). Allanson *et al.* (1994) develop this approach in the context of rural sustainability, which substantially influenced the development of the following arguments.

The historical underpinnings of an evolutionary approach to economic behaviour have been dealt with elsewhere (e.g. Clark and Juma, 1988). Notwithstanding the serious dangers associated with socio-biology, there is considerable attraction in the concept of social (human) systems evolving rather than simply working, and thus considerable force to the objections of the Austrian school to the presumption of neo-classical economics that the world is mechanistic or clockwork. From an evolutionary perspective, not only does the clock behave in an extremely 'fuzzy' fashion, so that the time it tells is only 'average', subject to considerable variation or 'noise', but, even more importantly, the process of telling the time actually triggers changes in the clock's mechanism the next time round. It is this latter point which is absent in the Austrian objections to the neo-classical school, since both enshrine the concept of unalterable 'laws' of economic behaviour.

The evolutionary perspective incorporates diversity (noise) as the critical driving force of economic change and development. It is the 'experiments' (either conscious or subconscious) which allow the existing socio-economic order to be tested against the contextual environment. Thus, Nelson and Winter (1974) propose an evolutionary model of economic growth (NW) which relies on firm heterogeneity, rather than as in the vast majority of neo-classical analysis, either assuming all firms are the same or conducting the analysis on the basis of an average or representative firm. In their words: "the model comprises a number of very simple firms" (operating at full capacity but otherwise satisficing), "interacting in an equally simple selection environment. Technically advanced firms reinvest their profits and expand, thereby driving up the wage rate facing other firms. Firms with low rates of return look for better techniques...rejecting technical regress in favour of the *status quo* (so) progress is achieved on average. Imitation helps to keep the technical race fairly close, but at any given time there is considerable cross-sectional dispersion in factor ratios, efficiency and rates of return. How do the quantitative results look? In a word...plausible" (1974, p.896)

Nelson and Winter conclude that even a highly simplified "model within an evolutionary theory is quite capable of generating aggregate time series with characteristics corresponding to those of economic growth in the United States. One does not have to extrapolate the performance of evolutionary theory very far beyond the present primitive level in order to conclude that neo-classical models are unlikely to be decisively superior" (1974, p.899). There is no equilibrium in this model, the results cannot be described as optimum (there are always better but unfound and unused techniques), there is no production function — the apparatus of neo-classical economics is not necessary to generate realistic real-world observations. Nelson and Winter (1982) and Dosi *et al.* (1988) provide substantial amplification of these ideas and concepts.

"From such a perspective the concept of a 'social optimum' disappears. Occupying a central place in the policy analysis are now the notions that society ought to be engaging in experimentation and that information and feedback from that experimentation will be the central concern in guiding the evolution of the economic system" (Nelson and Soete, 1988, p.633).

An evolutionary approach to policy development

Such a perspective rings several important bells for the policy analyst. Non-optimal policies are continually observed; it is difficult to project likely policy change from formal models; public choice analysts differ substantially about the explanations of past policy decisions (see, e.g. de Gorter and Tsur, 1991). Yet (MacLaren, 1992) there appears to be a "conservative social welfare function" or inertia; policy change depends on the context and circumstances facing the sector and policy makers, in a way which conventional models find difficult to incorporate; there is an apparent crisis policy management process and somewhat discontinuous policy change.

Development of a fully-fledged evolutionary model of the policy process is beyond the bounds of this chapter, indeed it has yet to be done. However, the broad elements of such a model can be outlined. Suppose that policy actors (ministers, other politicians and political parties, bureaucrats, pressure groups, treasury ministers and officials etc.) are treated like firms in the NW model, making (proposing) policy elements (instruments and settings). The 'yield' of these proposals depends on the extent to which they are 'fitted' to the *political climate* (voters and constituency preferences and opinion, often weighted through interest groups) and also to the *socio-economic terrain* (the effects such policy settings are expected to have on the performance of the sector and its relationships with the rest of the economy). Together, the climate and terrain determine the structure and nature of the objective set which the system can be seen as if trying to satisfy. In effect, the policy organism is pictured as producing a constellation of potential policy options, more or less well fitted to the current socio-political environment (climate and terrain). The policy selection process results in the development of those expected to be best fitted to the environment, leaving the rest in embryonic form. As the environment changes, in a coevolutionary fashion since these political and economic systems can also be viewed in evolutionary terms, so the selection of dominant (active) policy species will change.

Such a model clearly runs the risk of becoming at least as complex as the system being described, and potentially taking longer to run than the real thing. In any event, for the purposes of this chapter, a 'reduced form' is clearly needed. Consider the development of the dominant species of policy as a farm policy organism (as a sub-set of the government/political process as a whole). This organism is a collection of currently active policy instruments and an associated decision-making process embodied in an institutional complex, satisfying (not optimising) the current policy objectives, which include the costs of the policy set. The objectives, in turn, are a reflection of the political climate and socio-economic terrain in which the policy's survival is determined. Policy costs could be reflected as a set of constraints — the food supply for the organism — rather than (negative) objectives. However, the policy process seldom identifies costs as an explicit and rigid constraint, preferring to treat these in a similar fashion to the achievement of objectives.

The extent to which particular choices satisfy the several objectives depends on the external context and conditions in which the policies operate (the terrain). The definition of the objectives is dependent on the political climate, conditioned by socio-economic performance. The 'satisficing' levels of objective-achievement (echoing Fearne, 1989) are dependent on the political structure of the decision-making group — the institutions. The internal 'self-organised' structure of the organism, in turn, is subject to evolutionary change, possibly at discrete intervals (elections and/or crises), in response to the macro-performance of the complete policy-organism. The decision-making process is both noisy and fuzzy, since choice(s) of instruments and levels of settings reflect the uncertainty and lack of information of the policy makers, while the outcome of a collection of individuals and groups trying to reach agreement (and also needing to sell such agreements to their constituencies) is somewhat unpredictable.

This model incorporates a substantial element of inertia providing the terrain and climate (together making the socio-political environment) does not change substantially; it is likely to generate both mixed and demonstrably inefficient policy sets; it allows for the influence of terrain and climate change and for these (in conjunction with the policy choices) to feedback to changes in the structure and institutions of the decision-making process. It thus has the potential to meet the major deficiencies of the public choice literature, as identified for instance by MacLaren (1992).

In pictorial terms, such a framework appears similar to that proposed by Moyer and Josling (1990). However, the distinctive features of this outline are: (a) the fuzzy nature of the policy choices; (b) the independence of the choice set (here the pool of variation) from interest-group proposals (though inclusive of these proposals); (c) the explicit driving force of the evolutionary system as one of achieving a satisficing performance rather than as a bargain between competing groups; and (d) the specification in a form capable of simulation modelling enabling qualified prediction and experimentation. Such a simulation system has not yet been built. However, it is possible to pursue the conceptual model in an applied setting — the CAP cereals policy — in qualitative terms.

The evolution of the CAP — the model in practice

Although this chapter is supposed to be about the future, the proposal of a new framework for the prediction of the future requires confrontation with observable data in order to 'verify' the structure. A recent account and discussion of the prevailing debate and issues surrounding the CAP, against which this section might be compared, is contained in Kjeldahl and Tracy (1994), in which Nedergaard explores a more conventional public choice analysis of the present policy.

In the interests of brevity and simplicity, the story will begin in 1973, the date of the first enlargement of the EC and the CAP, and be confined to cereal policy as the cornerstone of the CAP. Additionally, only the most salient changes in terrain and political climate will be considered, while structural changes of the policy organism will be largely ignored, since these have barely yet emerged. The 'starting' point is an established CAP cereals regime characterised by historically high internal prices relative to prevailing world prices, defended through intervention buying, import levies and export subsidies. This organism had already generated signs of over-production and unsatisfactory cost within the original six member states, and the Mansholt Plan (essentially to 'downsize' the EC farm sector) was in the potential policy choice set, as were the UK's deficiency payment and minimum import price systems. Indeed, for some commodities, aspects of these instruments were already being used.

The broad quantifiable background (the socio-economic terrain) within which the policy organism has evolved since then is outlined in Table 18.1. The crude and debatable nature of most of these data is freely acknowledged. However, they are taken here as being reasonable approximations of *perceptions* of the organism's surroundings, which are more important in determining the evolution than more 'accurate' representations of the real environment.

The policy has not prevented either the continued decline in agriculture relative to the rest of the economy (1, showing agricultural gross value added (GVA) as a proportion of EU GDP) or the decline in farm labour (2). Nor has it been able to close the gap between incomes earned in farming and those earned elsewhere in the economy (3, measured here simply as the ratio of rows 1 and 2 — as the average GVA per head in agriculture as a proportion of average GDP per head in the whole economy). Against a rising level of unemployment (4, and associated social and economic re-structuring problems), the policy resulted in a rising total cost (taxpayer plus consumer costs (5)), a large rise in taxpayer costs (6), a growing proportion of farm GVA being accounted for by these support costs (7), and a growth of the cost as a fraction of total GDP (8), though many of the latter 'trends' appear to have stabilised somewhat during the last few years.

During the mid-1970s, the world commodity price boom substantially reduced the protection rates of the policy (PSEs turned negative for cereals in the Community during 1973–1975). Coupled with the expansion of the Community to include the UK, then a substantial importer, the threats of surpluses and increased budgetary cost receded while the popular concerns over the ability of the world to feed itself and associated conviction that world prices would stay firm, reduced the pressures which had encouraged the development of the

Table 18.1: The socio-economic terrain of the CAP.

Agriculture's place in the EU	1973	1980	1986	1992
1. Ag GVA as % total GDP	3.80	2.54	2.16	2.02
2. % pop. in agric.	10	9	8	5.8
3. Ag. relative income	0.38	0.28	0.27	0.35
4. EU unemployment rate (%)	2.6	6.1	11.9	11.2
5. Total support cost (bn. ecu)	3.2	30.8	65.4	64.0
6. Taxpayer support cost (bn. ecu)	4.0	16.0	22.9	35.8
7. Support cost as % GVA (agric.)	8.82	48.28	85.83	66.05
8. Support cost as % GDP	0.33	1.23	1.86	1.33

Mansholt Plan. The echo of the situation in world markets in the mid-1990s is almost uncanny.

The policy organism evolved to take advantage of the higher world prices and lack of pressure on over-supply, and pursued the evolutionary line towards higher protection rates (the line of least resistance). By the end of the 1970s, it became apparent that the 'food supply' and benign environment allowing such an organism to thrive were at an end. The 'ice age' of growing surpluses and escalating budgetary costs produced a range of viable mutations in the organism to cope with the colder climate. As an aside, the evolution of UK agriculture during this period also shows characteristics difficult to identify from a strictly neo-classical account of the sector. Entry to the CAP in 1973, amidst the well-established perception that accession would raise farm prices, coincided with the rapid escalation of world prices and thus of market prices in the UK (not fully within the CAP until 1979). The rise in farm profitability was (it can be argued) misinterpreted to be a consequence of the CAP rather than world market prices, and hence regarded as more secure. Coupled with rapid inflation and negative real interest rates, the 'fittest' response of the farm sector was to increase borrowings and invest heavily in capital equipment and productive capacity, in contrast to the typical industry response to less secure profit improvements – an increase in land prices and little else. The 'ice age' in the farm policy environment coincided with tight money markets and strongly positive real interest rates, leaving the farm sector substantially over-borrowed and over-equipped, thus, in the evolutionary perspective, ill-adapted. It took much of the 1980s for the UK farming sector to correct for this evolutionary blind-alley.

By the early 1980s, the policy organism can be seen as searching for adaptations to cope with the new environment. 'Prudent price' strategies, co-responsibility levies, and guaranteed thresholds appeared, but none were sufficiently well-adapted to the new environment to prosper in the longer term. Nevertheless, the policy set was not sufficiently inconsistent with the political landscape of the member states to demand a thorough overhaul (Harvey, 1982). By 1984, the environment surrounding the dairy sector in particular was especially severe, particularly in the budgetary/surplus dimension. Action on the price axis having proved insufficiently well-adapted, the organism developed supply control in the form of production quotas, albeit against considerable opposition, especially from the UK. Given the latter's pheno and genotype, such resistance was to be expected. However, within the closed European Community and for the well-suited dairy sector, this adaptation proved very well fitted to the new climate, though the co-evolution of the dairy sector and the policy implementation structure to cope with the new policy organism took some time to occur.

However, by 1986, the incipient problems of surpluses and escalating budgetary spending had emerged in the cereals sector, where the adaptation of supply control was both less well suited to the structure of the sector than for dairy, and also potentially more far reaching given the place of cereals in the farm sector organism's 'food chain'. Furthermore, the prevailing environment now included a growing ecological dimension to the political climate, though highly variable in manifestation depending on the variety of terrains within the Community, and included socio-economic considerations about rural communities as well as concerns over the natural ecosystem. This ecological climate influenced the budgetary terrain, eroding the primitive concern over the size of the budget to reveal a (possibly more fundamental and impermeable) concern over 'value for money'. The increasing persistence of the ecological climate had also exposed considerable internal difficulties for the policy organism which seemed likely to promote internal (though variegated) reforms in time.

Against this background, it is not surprising that the policy organism sought relief in international negotiation, 'seeing' an opportunity to change the nature of the terrain (world prices) and thus ease the threats of the micro (domestic) environment. In other words, the argument is that the EC was willing to sign up to the Punta del Este declaration partly (and perhaps primarily) as a means of resolving internal farm policy difficulties rather than as a response to either external pressure or a conception of the wider benefits of a new multilateral trade agreement. Once entered, however, the macro-climate of international negotiation became a major part of the organism's environment. An immediate (though perhaps not well appreciated) consequence of this environmental change was the addition of (or at least added weight to) a new set of policy options — especially those of the US: set-aside and acreage-restricted deficiency payments, and also loan rates.

The organism's behaviour during the early phases of the GATT negotiations clearly supports the proposition that it was seeking to change the environment to suit itself rather than willing to adapt itself to a new environment. How long and to what extent it might have been able to continue this trajectory is now unclear, because in 1989 the organism's immediate environment suffered the cataclysmic shock of the collapse of the Berlin wall. This 'earthquake' fundamentally altered both the terrain and the political climate surrounding the CAP. Immediate absorption of East Germany into the organism was a near inevitable consequence, and added to the strong climatic pressure to prepare for the absorption of siblings elsewhere in central and eastern Europe. West Germany could be content with a farm policy designed to support traditional small and largely uneconomic farms with artificially high prices. A unified Germany, however, could not be burdened with such a policy. It now needed a policy which would encourage and sustain the potentially large commercial contribution of the eastern landers' farms. The outcome was the agreement to the 1992 reform package, incorporating as far as cereals policy is concerned two major changes in policy direction: (a) the replacement of isolated and internally supported market prices with price reductions and (area-related) compensation payments; and (b) the requirement (for larger producers) for set-aside of areas planted in order to receive compensation payments. This package constitutes a radical change in policy direction, notwithstanding a substantial weakening of the initial proposals in the final agreement (see Table 18.2).

Throughout the discussion on this reform package, the Commission maintained the fiction that it was independent of the GATT negotiations. The 1990 initial EU offer to GATT (a reduction in AMS of 30%), however, was clearly incompatible with the old CAP and some reform was obviously going to be necessary if the Uruguay Round was to be successfully concluded. Nevertheless, it can be argued that continual resistance to international pressure (especially the US) within the UR negotiations, and persistence in distancing the internal reform debate from these pressures, may have strengthened the Commission's hand in obtaining final agreement within the EU (IATRC, 1994, p47). Few in Europe were prepared to pass the ostensible control over farm policy to the international negotiators. The policy organism had been bred and conditioned to be isolated from and independent of international pressures.

The influence of the GATT negotiations in encouraging semi-decoupled compensation and set-aside is clear and typically well-acknowledged, including the assertion by the European Commission that the new lower intervention price is set at a world free-trade level (see Chapter 17 above). While a strong supply control policy (presumably through set-aside or similar form of area control) might have seemed a viable response to GATT pressure for CAP reform in the early stages of the negotiations, the lack of progress on a 'managed world market' agenda must have provided the Commission with strong signals that the GATT environment would not permit such an option to survive. Once this 'fact' had been assimilated, only two courses were then viable: an effective blocking of an agricultural agreement under the GATT (on the precarious assumption that other countries would eventually allow the rest of the agreement to go through without agriculture); or acceptance that internal market support prices would have to be reduced. The latter option required some form of compensation payment scheme to make it acceptable to the farm lobbies, and it is argued here was substantially assisted (if not actually pre-conditioned) by the unification of Germany.

Given the compensatory nature of the new payments, it seems almost inevitable that these payments should be linked to areas of cereals. The requirement that producers should plant their land in order to obtain payments can be explained as a 'natural' evolution from the previous market-based support system and an unwillingness of the political decision makers to live with a complete de-coupling of payments immediately. However, an initial proposal for reducing dairy support included a lump-sum payment through a 'CAP bond' for a 5% cut in quota, indicating that this option was at least considered seriously in some quarters.

Table 18.2: Progress of Mac Sharry Reform proposals, EC.

COMMODITY	MEASURE:	DRAFT PROPOSAL JANUARY, 91	FINAL PROPOSAL JULY, 91	AGREEMENT MAY, 92
Cereals	Target Price Intervention Price Threshold Price	100ecu/t 90ecu.t (from 155ecu/t) ns	100ecu/t 90ecu/t 110ecu/t	110ecu/t 100ecu/t 155ecu/t
	Co-responsibility	abolished	abolished	abolished
	Set-Aside	≤30ha: 0; 31–80ha: 25% >80ha: 35% (rotational)	≤ 20ha: 0; > 20ha: 15% (rotational)	≤ 20ha: 0; > 20ha: 15% (non-rotational allowed at higher rate; + regional base)
	Compensation Payments: :	≤30ha: full; 31–80ha: -25% >80ha: -35%	full	full
	Set-aside:	none	≤50ha: full; >50ha: none	full
Oilseeds & Protein Crops		as for cereals	as for cereals	as for cereals
Milk	Quota:	cut by 4.5 to 5% (with 'extensive' modulation)	cut by 5% (inc. 91/2 price agreement cut of 2%)	cuts to be determined later
	Prices: Target: Butter SMP	reduced by 10% reduced by 15% reduced by 5%	reduced by 10% reduced by 15% reduced by 5%	none reduced by 5% none
	Compensation Payments	≤ 15 cows (≤1LU/ha) 45 ecu/cow	Quota: 100ecu/kg over 10 years as a bond Price: 75ecu/cow, ≤40cows s.t. stocking rates	none
	Co-responsibility	abolished	abolished	retained

Table 18.2 continued: Progress of Mac Sharry Reform proposals, EC .

Beef	Intervention Price:	reduced by 15% with safety net	reduced by 15%	reduced by 15% with safety net (Iq restricted)
	Compensation: male beef premium:	raised by 80ecu/hd. limited to 1LU/ha, ≤90 LUs	raised by 140ecu/hd. ltd. to 1LU/ha, ≤90 LUs	raised by 140ecu/hd. ≤2LUs/ha.; ≤90 LUs?
	suckler cow premium:	no change in rate; limited to 1LU/ha, ≤90 LUs	raised by 35 ecu/hd. ltd. to 1LU/ha, ≤90 LUs	raised by 80ecu/hd. ≤ 2LUs/ha; no headage limit
	special premia:	none	none	i) early season slaughter ii) extensive (≤1.4LU/ha) 60ecu, 30ecu/hd respectively
Sheep	Ewe Premium:	≤350 hd. (750 in LFAs)	≤350 hd. (750 in LFAs)	≤500hd. (1000hd. in LFAs) 50% premia payable over these limits.

Notes: The Draft proposal, January, 1991, was not officially released but was reported, *inter alia,* in Agra Europe, January 18th, 1991.
The Final Proposal: European Commission: Development & Future of the CAP COM (91) 258 Final, 22.7.91, a follow up to the Reflections Paper (COM(91) 100, 1.2.91, which contained no specific proposals for levels of support, rather concentrated on the framework for reform.
The Agreement was reported in Agra Europe, 22.5.92, followed by various regulations in the EC Official Journal (e.g. cereals — OJ No. L 181/ p12–39, 1.7.92). Only full post-transitional changes are recorded here.

The inclusion of set-aside in the package is more difficult to reconcile with most logical analyses of the policy options. However, within the evolutionary framework, this part of the reform can be seen as: (a) a mimicry of an apparently acceptable policy option used by the other major negotiator — the US, and thus defensible within the GATT negotiations; (b) a potential negotiating weapon, as evidence of the EU's willingness to make a 'down-payment' on the objective of stabilising world cereal prices at a competitive level; or (c) a 'throwback' to the genotype of supply control, countering the illogical but pervasive view that price reductions alone would not be sufficient to remove the surplus production problem.

Central and eastern European liberalisation also appears to have been a strong influence, though this is not supported by reports of the policy-making decisions (Franklin and Ockenden, 1995). However, it does seem clear that the only basis on which the CEE agriculture sectors can be admitted to an EU free-trade area without compromising the CAP is if the latter is reformed so that internal market prices are close to their free-trade world competitive levels. In addition, the substantial reduction in the internal EU price (so long as the EU remains on a net-export basis) is a necessary improvement as far as internal ecological considerations are concerned, reducing the incentive for intensive (high input) production techniques and allowing if not encouraging the 'development' of land use in more environmentally friendly ways. (See Chapter 14.)

The recent history and latest reform of the policy thus appears broadly consistent with the evolutionary 'model'. While it is also consistent with the major thrust of neo-classical analysis — that support prices should be reduced to world competitive levels — the timing and direction of the reforms have not been well-predicted by any neo-classical analysis. Public choice analysis of the policy has also left unanswered questions of when and if support to the farm sector would be changed, and if so how and by how much. Reference back to the Munk analysis outlined in Chapter 8 above, for instance, shows that conventional public choice analysis (though highly aggregated in this case), can provide relatively strong indications of trends in the socio-political environment. However, reflection of the evolution (or even joint dependence) of the political processes determining the shape, rate and timing of policy change is seldom captured adequately by such theories and models.

Implications for the future

The CAP an endangered species? Some symptoms

"No European Union topic provokes more anguish than reform of the Common Agricultural Policy. All previous changes to the system, which consumes half the EU's 70bn ecu (£55bn) annual budget, have met strong resistance and have only become reality after long and agonised negotiations. Resistance to further changes remains strong, particularly as the most recent reforms...are not all in place."[1] Commenting on four recent reports to the European Commission,[2] unanimous in their conclusions that further reform is necessary anyway, and is an absolute requirement for the integration of the CEE countries (Czech Republic, Poland, Slovakia, Hungary, Bulgaria and Romania), this newspaper report quotes Commission officials as saying that the conclusions are "politically naive" and that "we need political decisions first".

Following the protracted and enervating negotiations over the Mac Sharry reforms and the GATT Agreement, policy makers were entitled to a certain lethargy in the face of calls for further reform. Similarly, it is understandable that the outgoing Commission in 1994 should hold a somewhat complacent view of the compatibility between the 1992 reforms and GATT commitments. It is not difficult to understand a view of the medium term future of the CAP which holds that there is neither sufficient political pressure nor enough social gain to make pro-active effort for further reform worthwhile.

However, the 1992 CAP reforms are far from perfectly fitted to the new policy environment, suggesting that even the reformed CAP is endangered, if not destined for extinction. A number of unresolved issues appear to be likely to cause serious discomfort for the present policy organism, among which the most obvious are as follows:

- the compatibility of the 1992 reforms with GATT commitments, especially on subsidised cereal exports;
- the political sustainability of government cheques (the compensation payments) both as line-items in government budgets and as payments for ill-defined and increasingly questionable non-market benefits from commercial agriculture;
- increased regulation and control over farming (especially the set-aside controls), seen as inconsistent with the development of a competitive agriculture by commercial farmers and as costly and subject to considerable fraud and policing costs by the bureaucracy;
- the continued lack of integration of the policy with either the growing ecological concerns or with continued concerns over rural development and the threat of rural 'desertification' with removal of farm support;
- incompatibility between the reformed CAP and prospects of CEE enlargement;
- unexplored but potentially damaging incompatibilities between the reformed CAP (especially quotas, including those established for cereal and livestock compensation payments, and set-aside) with the concepts and spirit of the Single European Market and European unification;
- related questions about the necessity for 'financial solidarity' under which the European Commission is responsible for 100% of the budgetary costs of the market support policy (and hence for the full cost of the compensation payments).

From an evolutionary perspective: (a) the policy environment has now changed fundamentally; and (b) the new policy framework and process (organism) is one undergoing inherent change and adaptation, itself generating an inherent tendency for further reform. Whilst the previous policy of isolation from world markets and single-minded focus on agricultural issues was stable and broadly acceptable so long as it remained undisturbed, recent events have destroyed this balance while a new homeostasis is far from being established. In such circumstances, further and perhaps rapid change is inevitable.

An evolutionary perspective strongly suggests that the environment within which the future policy organism will develop now entails a substantially different terrain and climate than pre-1992. There are three major dimensions of this change — two environmental shocks and a mutation — echoing some of the 'superficial' dissonance between the current policy and its environment outlined above, and affecting the future evolution of the policy in important and conceptually distinct directions.

Two environmental shocks and a mutation

The processes of GATT/WTO (including the formal review of the agriculture part of the agreement in 1999) and prospects for enlargement to include CEE countries represent new and substantial external regulation of the current policy trajectory, well-described as a substantial shock to the political climate surrounding the CAP. The presumption must be that the CAP can only comply with this international government through a reduction of internal EU prices to demonstrably competitive world levels. In addition, the design and implementation of compensation payments is substantially restricted by this international regulation. The passing of the present compensation arrangements as within the 'blue box', and thus non-trade-distorting, is widely understood to be a convenient fiction for the purposes of the current agreement only. Future agreements will have the clarification of non-distorting measures at the top of their agenda. The definition of such measures has already been agreed in principle — that they should neither be related to product *nor to production.* Furthermore, the entitlement of CEE producers to compensation payments is both logically questionable and subject to severe budgetary limit, further reinforcing the conclusion that these payments must become fully de-coupled for the policy to survive.

The partial de-coupling of farm support from market prices and the replacement with compensation payments has opened up a 'Pandora's Box' of debatable issues concerning the reasons for and legitimacy of farm support, none of which are satisfactorily (that is politically sustainable) resolved under the present policy, though none were seriously open to debate (and thus influence on policy direction) under the previous incarnation of the CAP. This is a major example of policy change generating internal mutation within the policy organism.

Commercial farm prosperity is now increasingly widely understood to be unsustainable through market price support, but only through an internationally competitive industry. Farm incomes, or even farm revenues, are recognised as being insufficient to secure rural economic health or environmental sustainability, and perhaps not even necessary. It follows that reliance on line-items of the budget for support or compensation entails a concomitant responsibility among recipients to justify that support through delivery of socially desirable products and practices which would not otherwise be forthcoming through the market mechanism. Meanwhile, compensation implies a distinct and finite sum, reflected in the concerns the present commercial farming sector in the EU has about the future of annual payments as well as about their distribution.[3]

This growing dissatisfaction of the commercial sector of European agriculture appears quite different from the stance the industry was able to take under the previous policy environment, and much more likely to generate further policy evolution. An obvious direction is towards a bond scheme, following Tangermann (1991), at least made available on a voluntary basis in return for release from set-aside and other production restrictions. The size of this bond payment is indicated, as an upper limit, by the estimate of producer surplus gain associated with market protection policies under the current CAP (see Table 8.1, above). This total annual gain could be converted to a lump-sum (the value of the bond) through conventional compounding arithmetic, given assumptions about the appropriate time period and discount rate. In practical terms, some account would then need to be taken of the appropriate extent of compensation, which need not be 100% given political judgements about the necessity for full compensation in the light of equity (interpersonal comparison) considerations.

Such a further reform would protect farmers from the continual erosion of their compensation payments (justified in exactly the same way as redundancy payments and pension enhancements in other declining or downsized industries), and would also save the bureaucracies considerable and ongoing implementation and policing costs, which would be incurred under the bond option on a once-and-for-all basis. Once accepted, bond compensation for removal of market price support raises the question of whether long adjustment periods to the new regime are now needed for the commercial sector. Since most of the adjustment problems concern the de-valuation of the asset base and consequent re-adjustment of the fixed cost structure of the farm business, a bond payment might well provide a sufficient adjustment cushion to allow very rapid transition to the new regime. At least some farmers would benefit from and be prepared to take advantage of such an option.

There are, of course, competitors to these arguments. Political decision-makers may be unwilling to relinquish their control over annual payments, in turn promising continued political support in return for the dependence. However, as soon as these competitors are discussed, their double-edged natures as far as commercial farmers are concerned are brutally exposed, encouraging them to at least think of the once-and-for all alternative. The distinct prospect for internal, farmer-driven reform is a novel and exciting prospect for the CAP, but is entirely consistent with the framework advanced here.

However, a more legitimate reason for the persistence of annual payments results from the second major shock in the policy environment — budgetary value for money, opening up new territories for the CAP organism. It may be recognised that farm support payments may not be sufficient to ensure the sustainability of either rural economies or natural environments. Nevertheless, this is not the same thing as arguing that withdrawal of such support (even with compensation through bonds) might not harm the ability of the farm sector to contribute to both sustainabilities. Indeed, if there are real social benefits from a more economically secure 'farming' population than the free market would provide (as there may be, though this is substantially under-researched), then some annual payments might well be justified. The two possible grounds would be: (a) the contribution of agricultural 'surplus' to the economic development of rural areas; and (b) the necessity of paying for ecological and landscape aspects of the natural environment (christened Conservation, Amenity, Recreation and Environmental — CARE — goods, McInerney, 1986) over and above the payments a competitive market might provide or to which society might be reasonably entitled as of right.

Two new territories for farm policy and required adaptations

The new policy environment has highlighted rural development as a major rationale for 'farm' policy. This has potential implications both for the geographical and individual distribution and for the method of providing compensation. Since compensation is a monetary equivalent of those resources which are 'surplus' to competitive requirements in the farm sector, there is some argument that it would be in society's interest to encourage retention of such a surplus within some rural areas, logically involving some annual and conditional payment stream. However, other aspects of the policy's environment mean that the form of such payments will be likely to be substantially different from conventional farm policy, and much more likely to emerge as integrated rural development (or in many cases, preservation) programmes. As presaged in the LFA payments, such programmes will take location rather than occupation as the primary characteristic determining eligibility, but will increasingly require demonstration of their contribution to the whole rural economy rather than simply to agriculture for their continuation.

The environmental protection rationale for farm policy is more problematic (Harvey, 1991a, and Harvey and White, 1995.) In essence, however, social values (positive or negative) over and above those signalled to land users through free market prices have to be reflected back to these users, either through (annual) taxes and subsidies or through regulation. Thus, wider issues than food security, agricultural and rural prosperity are at stake in future policy direction, so that social values of products, production processes and land use are central to the legitimacy of both policy and market processes, and farm policy becomes a rural social and environmental policy package.

The central importance of social values, which include for these purposes concerns over animal welfare, health and safety, raises a number of fundamental issues for the evolution of the CAP. The requirements for both locally differentiated policy settings and for local determination of rural social and environmental values undermine both the case for a rigid 'common' policy with common settings and implementation, and for common financing of measures solely from the European budget. These characteristics point to potentially rapid and far-reaching metamorphosis of the European Union's farm and rural policy organism in two major directions.

Two evolutionary 'predictions'

The spatial dimensions and differentiation of the new policy environment require spatially differentiated policy responses, strongly echoing some arguments for a re-nationalisation of the CAP (Kjeldahl and Tracy, 1994), but going further to involve substantial regionalisation and localisation of the ecological and rural development aspects of the policy, and thus involving a cross-fertilisation of the CAP with the increasingly important European Regional and Social Funds. There appears to be a powerful set of constituencies in favour of a more equitable distribution of economic activity between favoured and less favoured areas than would necessarily be achieved through the unhindered operation of market forces. There has also been in the past a presumption in favour of a larger agricultural sector than would be the consequence of an unhindered market-place, largely as a reflection of the socio-political concern over security of food supplies. Since this security is now of merely historical interest, it is now the geographical distribution of incomes and economic activity which is of major importance as far as farm policy is concerned. In this case, conventional economic analysis suggests that policy concerns should now be about the provision of an adequate infrastructure of communication and transport links, and of a pattern of communal and social services, sufficient to support sustainable local rural economies. It will not, in the future, be about farming for food production.

This development, however, raises the considerable problems associated with appropriate definitions and enforcements of 'level playing fields' within the EU, with potential ramifications to the international arena. As noted in the previous chapter, the 'level playing field' concept does *not* mean that trading nations (or regions) should have identical environmental conditions or identical social valuations (and hence opportunity costs) of environmental assets, any more than it means that they should have identical costs of land, labour or capital. Thus, the extension of European competition policy (see, e.g. Woolcock, 1994) to embrace agricultural and land use policy also becomes a major part of the environment with which the evolving CAP must come to terms. In turn, this puts the development of the Single European Market, and also of the EU itself, discussed at the 1996 Intergovernmental Conference, in a central position in the evolution of the policy.

However, it is also plausible that there is a concern over an 'optimal' structure of agriculture — in terms of farm sizes and types — in particular regions, both as this contributes to a socially acceptable and desirable landscape as well as the (arguable) contribution to the pattern of rural employment, activity and social structure. Encouragement of such an ill-defined optimal structure, loosely characterised as the preservation of the 'traditional family farm', may also be an effective force in favour of more or less traditional forms of farm support (even if barely justified on rational or logical grounds). Nevertheless, the co-existence of such concerns within the constellation of other political environment characteristics points to specific locally targeted policy instruments rather than to the universal support characteristic of the current CAP's ancestor.

The more fundamental implication of these political environment changes concerns the internal structure of the policy organism — its decision-making and implementation institutions (organs). In this new environment, a key role is played by social valuations of, especially, ecological and countryside aspects of agriculture and land use. These social valuations are critical to the future legitimacy of differentiated 'intervention' in agricultural markets and are also crucial to the implementation of appropriate policy instruments. Yet they are fundamentally local in character, depending on the characteristics and environmental potential of the local land base and ecology, as well as the largely local population interested and thus willing to pay for the conservation of this base.

While it might be administratively convenient if there were a clear correspondence between CAP compensation payments and payments to encourage CARE good provision, there is no logical connection between the two payments. There is no reason to suppose that the total compensation payment required to assist adjustment to lower market prices or compensate for removal of market support will correspond to the total payment society is willing to make for the provision of CARE goods. Furthermore, even in the unlikely event that these two sums do correspond, there is no reason to suppose that those who need (deserve) payment for the provision of CARE goods will do so in exact correspondence with their compensation entitlements. The distribution of compensation payments will be quite different from that of necessary CARE good payments. The clear implication is that concepts of 'cross-compliance' — where receipt of compensation payments should be conditional on the provision of an appropriate package of CARE goods — has limited logical support. This is not to say, however, that cross-compliance will not feature at least as transition policy, and that the inertia of the policy organism will ensure the survival of such an artefact until its internal contradictions become unsupportable. If history is any guide, this could take some considerable time. On the other hand, the collapse of the iron curtain suggests that some large policy changes can happen surprisingly suddenly, and become the cause rather than the consequence of other socio-economic changes.

A possible policy framework for the future of the CAP

The implication of the arguments in this book, especially Chapters 7, 8, 13 and 17, is that present product-related support systems should be completely decoupled from production. As indicated above, decoupling immediately raises the problem of what the support payments are for, answers to which reflect the inherent heterogeneity of the European farm sector. The implication is that — echoing the argument in Chapter 4 — no single policy instrument can realistically be expected to satisfy the triumvirate of sustainable reasons for providing support to farmers: compensation for removal of product-related support; care of the environment; sustenance of the rural economy. The challenge, therefore, is to design a new package which both retains the spirit of decoupling and provides for flexibility in application to suit spatially variegated conditions amongst a widely differentiated population of farmers' businesses, families and circumstances. Only a policy system exhibiting these characteristics can be expected to survive. What might such a policy look like?

The key factors applying to the design and implementation of policies as far as rural land use (and thus farm policies) are concerned can be summarised as:

a) farm (and rural) incomes are properly seen as consequences of, rather than precursors to, the delivery of socially desirable products and asset maintenance;
b) social optimality depends on getting both the product price and CARE good signals to land users 'right' (either through prices or regulations), which means reducing product support prices to free trade world levels and providing appropriate incentives, penalties and regulations reflecting social valuations for non-market CARE goods;
c) continued 'coupling' of farm support policies to farm production decisions (as is still evident in the conditionality of compensation payments on planting crops and grazing livestock, and in the set-aside requirements) cannot be consistent with sustainable agriculture or land use;
d) trade policies are inappropriate or second-best mechanisms for achievement of environmental and sustainable policy objectives;

e) spatial and business heterogeneity and multiple, regionally diverse objectives require a multiplicity of policy instruments.

Based on these implications, it is possible to outline a framework for an effective and efficient policy mix to achieve the principal objectives of an efficient and competitive industry managing its resources in a socially optimal fashion.

This framework — Table 18.3 — assumes that maintenance of farm incomes can no longer either be regarded as an appropriate objective or can be achieved through commodity support programmes. However, it incorporates the notion of *limited* and production/product neutral compensation for removal of previous support, both on theoretical and practical political grounds, and also retains some limited defence against 'predatory' policies of foreign suppliers or their policy-makers through import tariffs to defend a 'free trade' market price.

The essential elements of this framework represent a relatively minor evolution from the present EU policy for cereals (though not as yet for other products, especially milk and sugar). The key differences are: (a) that compensation payments should be explicitly limited both in total and per farm; (b) that compensation should be independent of all product/production, including areas and livestock numbers; and (c) that more flexibility is included in the ways in which this compensation can be made to a widely heterogeneous farming sector and population within Europe. Within these parameters, producers should be given the option of choosing the form in which they would prefer to receive their compensation entitlements, within nationally and regionally defined constraints reflecting the social justification for continued support.

The framework implies a considerable element of 'renationalisation' of the CAP, in the sense that socio-political decisions on the appropriate levels and mechanisms for the delivery of farm sector support should be devolved from Brussels to the member states and regionally differentiated within member states. However, as Wilkinson (1994), points out, "the type of renationalisation we are discussing should better be called 'targeting' " and "a fine line has to be drawn between *renationalisation* and *subsidiarity*" (p. 31).

An essential element of farm policy is the provision for stable markets, or more importantly, returns. The arguments and analysis above (Chapter 17) leads to the following conclusions.

a) The public good characteristics of information, coupled with the substantial transactions costs of risk and insurance markets, the potential economies of scale in providing for these, and the difficulties of adverse selection and moral hazard for the private markets, all point to a substantial case for some form of government intervention for stabilisation purposes. The transactions cost argument (including the specification of the appropriate contract) is extended by the propensity of governments both at home and abroad to intervene in agriculture, thus changing potential market outcomes in an unpredictable way.
b) There remains a strong political argument that stability cannot be ignored and that 'governments need to be seen to be doing something' at least until public opinion catches up with economic theory. This argument can be extended in the EU case by the observation that resistance to necessary (and cost-

reducing) policy reform from the farming sector is likely to be substantially greater if the proposed policy reform has nothing to say about provision of stability or stop-loss insurance.

c) There are examples, especially from North America, of practical means of providing at least some form of 'stop-loss' or 'fail-safe' long term planning stability or insurance — the distinctions become somewhat blurred at the practical policy level.

Table 18.3: An agricultural policy framework for the EU for the next century.

Eliminate all internal support prices, intervention activities, export subsidies and import levies and substitute:

1 Common import tariffs such that the internal price is, on average, within the boundaries of best estimates of a free trade world price (unless this proves above currently bound tariff levels under GATT, in which case the latter should apply).

---

2 Provide limited compensation to existing producers for this policy change on the following conditions:

i compensation payments to be limited in aggregate to an agreed fraction of base production levels;
ii total (EU funded) compensation payment be agreed at EU level within absolute upper limit already committed in GATT;
iii distribution of EU total between member states to be determined on basis of base level production shares;
iv distribution of compensation to farmers to be determined by member states;
v whatever the form of compensation payments, all should be independent of either current production or current products (including planted areas and stocking levels);
vi producers to elect to take compensation payments from within following options, subject to nationally and regionally determined constraints reflecting the justification for support:
- fixed annual payment of explicitly limited duration;
- 'CAP bond';
- 'down payment' for farm-level income stabilisation account — see Table 18.4 below.

---

3 Within these provisions, member states to be free to decide:

i the farm-level distribution of compensation;
ii the extent to which national 'top-loading' shall be applied, subject to EU agreed upper limits

---

Building on these arguments, and taking account of the substantial dangers of political capture of stabilisation programmes as stalking horses for (increased) support, as well as the continued problems of moral hazard and adverse selection, the following outline proposal is made for a European 'stabilisation' programme (Table 18.4), in an essentially similar fashion to those suggested by Harrington and Doering (1993), for the US.

The essential features of these plans are:

a) that they are both voluntary, and require contributions from participating farmers;
b) that they are limited to 'stop-loss' or 'fail-safe' levels of revenue protection;
c) that they should be actuarially sound, with the premia covering the payouts;
d) that provision is made for farmers to withdraw from the programmes;
e) that they contain sufficient flexibility to be tailored at the national level to suit particular farm types or regions, including the possibility for the Revenue Insurance Plan (RIP) that different types of farm with different circumstances might enjoy different rates of farmer contribution to the plan. Although there may seem to be some advantage in 'rolling up' such programmes over the farm level mix of commodities to make the programmes farm based rather than commodity based, this would reduce flexibility as far as farmers are concerned if they wished to adjust their production mix.

Moral hazard and adverse selection problems are minimised in these plans through the incorporation of observable farm determined performance (yield levels) in the definition of the contract. The plans do not conflict with the purposes and functions of legitimate and generally efficient futures and options markets, which would still be expected to operate for short-term market price-risk spreading. Fixing the 'stop-loss' price as a fraction of longer-term moving average market prices (which are otherwise unsupported) minimises the danger of such programmes being used as a stalking horse by farm lobbies, while explicit farm business level limits ensure that 'rent capture' by farmers themselves is limited, whilst also bolstering the argument that they can be considered production neutral. The danger that these stop-loss price floors might distort production and marketing decisions is greater the closer the floor is set to current moving average (or other trend) price levels.

A key question for such a proposal is how it might be introduced in the present EU system, where historic market prices contain a large element of support, and where farmers are totally unused to such a proposition. As far as market prices are concerned, it is suggested that the appropriate 'market' prices to use for the present (until world market liberalisation is much more complete) would be the OECD annually computed reference price (the world price) plus the agreed 'world price defence tariff' outlined in Table 18.3 above. If necessary, this EU wide price could be adjusted to national or regional levels through the application of annually revised estimates of EU spatial price relationships. To encourage farmers to take advantage of these plans, it is suggested that farmers be offered the opportunity to substitute limited annual compensation payments for a specified 'premium holiday' for these plans, which could be phased out over time rather than suddenly closed off. This may well provide an important and valuable political rationale for compensation, especially for the larger commercial farmers, and would warrant a discount on the arithmetic entitlement to compensation for these larger farmers. Determination of an appropriate discount might be achieved through a 'tender' system, whereby farmers are invited to 'bid' current entitlements to (indefinite and insecure) compensation payments for specific premium holiday provisions. Similarly, the option of substituting a lump-sum payment (bond) for the present annual compensation payments could also be provided through a bid/tender system.

Table 18.4: Framework for production and revenue insurance programmes.

Voluntary tri-partite farm level yield insurance and revenue insurance plans as follows: ('tbd' in the following signifies 'to be determined' through negotiation. Moving averages could be considered shorthand for any reasonable and reliable trend estimate)
1 Production Insurance Plan – PIP: Registered farmers pay actuarially-sound annual premia insuring a critical production level set at not more than 75% (tbd) of a 10 year (tbd) moving average of farm annual 'yields' (appropriately defined for livestock). Payments to be made on shortfalls below this level at registered market prices, subject to an upper limit per farm business (tbd). Farm 'PIP accounts' to be closed only on condition that notional balances are 'accounts owing'.
2 Revenue Insurance Plan – RIP: Registered farmers pay a fraction (tbd) of actuarially-sound annual premia insuring a critical indexed gross revenue level, set at not more than 80% (tbd) of a 10 year (tbd) moving average of market prices indexed by cash costs of production. Payments to be made on the basis of moving average farm 'yields', subject to an upper limit per farm business (tbd). Balance of annual premia necessary to ensure actuarial integrity of programmes to be contributed by EU and member states according to an agreed formula (tbd). Farm 'accounts' to be closed only on condition that notional balances are 'accounts owing'.

Turning to the question of appropriate policy instruments for CARE good provision, Hodge (1988) has suggested that Conservation, Amenity and Recreation Trusts (CARTS) may prove a useful mechanism for solving the twin problems of how much people are willing to pay for various elements of the natural environment and of providing the instruments through which such environments can be encouraged and paid for, simultaneously providing for the legitimising of the payments. The central idea is that there already exist a number of voluntary institutions concerned with the preservation and enhancement of the natural environment. These institutions depend on there being a public willingness-to-pay for CARE goods through membership subscriptions and donations, and have evolved to implement a variety of schemes (varying from direct ownership and management of land through negotiation of land use practices) to provide these goods for their members (and, of course to 'free-riders' who choose not to join).

As Hodge notes, the literature suggests that the free-rider problem will typically lead to an under-provision of public goods through the voluntary club route. Some public support is, therefore, justified. A more general application is suggested by Hodge to involve a public subsidy to such CARTS in proportion to their membership income, taken here as an indication (though biased downwards because of the free-rider problem) of the public's willingness-to-pay. Such a mechanism would provide for the continual demonstration of the legitimacy of the 'policy', while also allowing individuals (through their CART membership rights) to actively participate in the determination of the types and varieties of CARE goods provided. While this specific policy development may or may not prove a viable direction for a CAP organism to take, it is consistent with the evolutionary pressures facing the organism, and points to markedly different institutional structure than has been evident in the past, serving as a useful illustration of possible future developments.

Conclusions

In conclusion, the arguments of this chapter strongly suggest that we are entering a new era of policy development and evolution within the European Union. This is markedly at odds with both the historical development of the policy pre-1992 and with the opinions of many commentators both within and outside the European policy process. It is also significantly at odds with previous public choice accounts of potential policy development, for example, Nedergaard (1994), who 'forecasts' growing bureaucratisation of the policy, or Moyer and Josling, who 'forecast' continued crisis management (and short-term response) and continued monopolistic farm pressure groups, leading to a strong presumption in favour of the status quo. The arguments of this chapter suggest that a fundamental change is now inevitable.

It has to be admitted that the evolutionary story is, at this stage, no more than a parable. However, this feature alone certainly does not distinguish the 'theory' from its competitors. Nevertheless, the parable is metaphorically rich, incorporating much of the 'conventional' wisdom about policy developments and is capable, at least in principle, of formalisation. Notice that arguments in favour of an evolutionary approach do not (at the policy level) necessarily entail the denial of neo-classical theories and approaches — these must stand or fall on their own merits and may provide at least workable models of economic mechanisms as relationships between *Homus economicus* and the political environment.[4]

Against this background, the penultimate section of this chapter has outlined a possible alternative framework for the future development of farm policy in the EU which might serve to satisfy both the logical imperative of sustainable evolution of policy and the potentially conflicting pressures on policy makers for further policy reform. Reform (or, better, evolution) is inevitable. It is hoped that this book provides the background against which this future evolution can be both better understood and, more importantly, better guided.

References

Allanson, P., Murdoch, J., Lowe, P and Garrod, G. (1994) *The Rural Economy: an Evolutionary Perspective*, CRE Working Paper 1, Centre for Rural Economy, Department of Agricultural Economics and Food Marketing, The University of Newcastle upon Tyne.

Boulding, K.E. (1991) Economics and Social Systems, Chapter 38 in *Routledge Companion to Contemporary Economic Thought*, Greenaway, D., Bleaney, M. and Stewart, I.M.T. (eds), 742–758.

Clark, N. and Juma, C. (1988) Evolutionary Theories in economic thought, Chapter 9 in: Dosi, G., Freeman, C., Nelson, R., Silverberg, G. and Soete, L. (eds) *Technical Change and Economic Theory*, London, Pinter, 197–218.

de Gorter, H. and Tsur, Y. (1991) Explaining Price Policy Bias in agriculture: the calculus of support-maximising politicians, *American Journal of Agricultural Economics*, 73(4), 1244–1254.

Dosi, G., Freeman, C., Nelson, R., Silverberg, G. and Soete, L. (eds) (1988) *Technical Change and Economic Theory*, London, Pinter.

Fearne, A.P. (1989) A Satisficing model of CAP decision making, *Journal of Agricultural Economics*, 40(1), 71–81.

Franklin, M. and Ockenden, J. (1995) *Future Directions for EU Agricultural Policy- report of a series of Roundtable Fora*, Royal Institute of International Affairs, London.

Harrington, D.H. and Doering, O.C. (1993) Agricultural Policy Reform: a Proposal, *Choices*, American Agricultural Economics Association, 1st quarter, 14–17, 40–41.

Harvey, D.R. (1982) National Interests and the CAP, *Food Policy*, 7(3), 174–190.

Harvey, D.R. (1989) Alternatives to present price policies for the CAP, *European Review of Agricultural Economics*, 16, 83–111.

Harvey, D.R. (1991a) Agriculture and the Environment: The Way Ahead?, Chapter 15 in Hanley, N. (ed), *Farming and the Countryside: An Economic Analysis of External Costs and Benefits*, CAB International, Wallingford, 275–321.

Harvey, D.R. (1991b) The Producer Entitlement Guarantee (PEG) Option, Chapter 17, *The Common Agricultural Policy and the World Economy,* Essays in honour of Professor John Ashton, Ritson, C. and Harvey, D.R. (eds), CAB International, Wallingford, 311–336.

Harvey, D.R. (1994) Policy Reform after the Uruguay Round, Chapter 11 in: *Agriculture in the Uruguay Round*, Ingersent, K., Rayner, A.J. and Hine, R.C. (eds), Macmillan, Basingstoke, 223–269.

Harvey, D.R. (1995) European Union Cereals Policy: an Evolutionary Interpretation, *Australian Journal of Agricultural Economics*, 35(3), December, 193–217.

Harvey, D.R. and White, B. (1995) Regional Economics Approach: Quantitative Models in Integrated Scenario Studies, Chapter 2.2, *Scenario Studies for the Rural Environment,* Schoute, J.F.Th., Finke, P.A., Veeneklaas, F.R., and Wolfert, H.P. (eds), Kluwer Academic Publishers, Dordrecht, 55–74.

Hodge, I.D. (1988) Property Institutions and Environmental Improvement, *Journal of Agricultural Economics*, 39(3), 369–375.

IATRC (1994) *The Uruguay Round Agreement on Agriculture: an Evaluation*, Commissioned Paper No. 9, The International Trade Research Consortium, (Chairman, Prof. A. McCalla, University of California, Davis).

Kjeldahl, R. and Tracy, M. (eds) (1994) *Renationalisation of the Common Agricultural Policy*, Institute of Agricultural Economics, Copenhagen jointly with Agricultural Policy Studies, La Hutte, Belgium.

MacClaren, D. (1992) The political economy of agricultural policy reform in the European Community and Australia. *Journal of Agricultural Economics* 43 (3), 424 – 439.

McInerney, J.P. (1986) Agricultural Policy at the Crossroads, Gilg, A.W. (ed), *Countryside Planning Yearbook*, Volume 7, 44–75, Geo Books, London.

Moyer, H.W. and Josling, T.E. (1990) *Agricultural Policy Reform: Politics and Process in the EC & US*, New York and London, Harvester Wheatsheaf.

National Farmers Union (1994) *Real Choices: a discussion document*, Long Term Strategy Group, NFU, London, March.

Nedergaard, K. (1994) The Political Economy of CAP Reform, in: Kjeldahl, R. and Tracy, M. (eds) *Renationalisation of the Common Agricultural Policy*, Institute of Agricultural Economics, Copenhagen jointly with Agricultural Policy Studies, La Hutte, Belgium, 85–103.

Nelson, R.R. and Soete, L.L.G. (1988) Policy Conclusions, in: Dosi, G., Freeman, C., Nelson, R., Silverberg, G. and Soete, L. (eds) *Technical Change and Economic Theory*, London, Pinter.

Nelson, R.R. and Winter, S.G. (1974) Neoclassical versus Evolutionary Theories of Economic Growth: Critique and Prospectus, *Economic Journal*, 84, 886–905.

Nelson, R.R. and Winter, S.G. (1982) *An Evolutionary Theory of Economic Change*, Cambridge, Mass., Belknap Press of Harvard University Press.

Tangermann, S. (1991) A bond scheme for supporting farm incomes, in Chapter 10, *The Changing Role of the CAP*, Marsh, J., Green, B., Kearney, B., Mahe, L., Tangermann, S. and Tarditi, S., Belhaven, London, 95–101.

Wilkinson, A. (1994) Renationalisation: an Evolving Debate, Chapter 3 in: Kjeldahl, R. and Tracy, M. (eds) *Renationalisation of the Common Agricultural Policy*, Institute of Agricultural Economics, Copenhagen jointly with Agricultural Policy Studies, Belgium, 23–32.

Woolcock, S. (1994) The Single European Market: Centralisation or Competition among National Rules, Royal Institute of International Affairs (European Programme), London.

---

[1]Caroline Southy, Financial Times, 16.1.95, p2. Note also, the UK Minister of Agriculture — Rt. Hon. William Waldegrave, speaking at the Oxford Farming Conference (January 6, 1995): "there is now likely to be a pause while the so-called 'peace clause' operates for a decade before further steps follow" though he went on that "nonetheless, the process is in train".

[2]By, respectively, A. Buckwell, Wye College, London. UK; S. Tangermann, Gottingen, Germany; S. Tarditi, Siena, Italy; L. Mahe, Rennes, France.

[3]Witness, for example, the English NFU's publication (1994) of a fundamental

re-consideration of the role of farm support in commercial agriculture and serious consideration of the options available, beginning a continuing internal debate which is qualitatively different from any such debate in the recent past. Similar new debates are now beginning elsewhere in Europe.

[4]The arguments here are capable of much wider integration than simply within the narrow philosophy of economics. It is not beyond the realms of possibility that it could provide a framework for the eventual development of that chimera — a unified social science. However, before any reader gets carried away with this pipe-dream, a careful reading of Issac Asimov's Foundation Saga in five volumes (Grafton Books, London) is in order. On the other hand, Boulding (1991), for one, appears to share a similar dream.

Index